BIOLOGICAL OCEANOGRAPHY

AN INTRODUCTION

Related titles

ELDER & WILLIAMS
Fluid Physics for Oceanographers and Physicists, 2nd edition

OPEN UNIVERSITY
The Ocean Basins: Their Structure and Evolution
Seawater: Its Composition, Properties and Behaviour, 2nd edition
Ocean Circulation
Waves, Tides and Shallow-Water Processes
Ocean Chemistry and Deep-Sea Sediments
Case Studies in Oceanography and Marine Affairs

PARSONS, TAKAHASHI & HARGRAVE
Biological Oceanographic Processes, 3rd edition

PICKARD & EMERY
Descriptive Physical Oceanography: An Introduction, 5th edition

POND & PICKARD
Introductory Dynamical Oceanography, 2nd edition

TAIT & DIPPER
Elements of Marine Ecology, 4th edition

TOMCZAK & GODFREY
Regional Oceanography: An Introduction

BIOLOGICAL OCEANOGRAPHY

AN INTRODUCTION

Second Edition

CAROL M. LALLI

and

TIMOTHY R. PARSONS

University of British Columbia, Vancouver, Canada

ELSEVIER
BUTTERWORTH
HEINEMANN

AMSTERDAM · BOSTON · HEIDELBERG · LONDON · NEW YORK · OXFORD
PARIS · SAN DIEGO · SAN FRANCISCO · SINGAPORE · SYDNEY · TOKYO

Elsevier Butterworth-Heinemann
Linacre House, Jordan Hill, Oxford OX2 8DP
200 Wheeler Road, Burlington MA 01803

First published 1993
Reprinted with corrections 1994, 1995
Second edition 1997
Reprinted 1999, 2000, 2001, 2002, 2004

British Library Cataloguing in Publication Data
Lalli, Carol M.
Biological oceanography: an introduction. - 2nd ed.
1. Marine biology   2. Marine ecology
I. Title   II. Parsons, Timothy R. (Timothy Richard), 1932-
574.9'2

Library of Congress Cataloging in Publication Data
Lalli, Carol M.
Biological oceanography: an introduction / Carol M. Lalli and
Timothy R. Parsons. - 2nd ed.
p. cm
Includes bibliographical references and index.
1. Marine biology   2. Marine ecology   3. Oceanography.
I. Parsons, Timothy Richard, 1932-   . II. Title.
QH91.L35   96-42139
574.92-dc20   CIP

ISBN 0 7506 3384 0

Cover illustration: Corals and associated fish of a reef off Tahiti.
(Photo courtesy of J. B. Lewis)

For information on all Butterworth-Heinemann publications
visit our website at www.bh.com

Printed in China

CONTENTS

CHAPTER 4   ZOOPLANKTON

CHAPTER 5   ENERGY FLOW AND MINERAL CYCLING

CHAPTER 6   NEKTON AND FISHERIES OCEANOGRAPHY

CHAPTER 9 — HUMAN IMPACTS ON MARINE BIOTA

APPENDIX 1 — GEOLOGIC TIME SCALE

APPENDIX 2 — CONVERSIONS

SUGGESTED FURTHER READING

ABOUT THIS VOLUME

This volume is complementary to the Open University Series on oceanography. It is designed so that it can be read on its own, like any other textbook, or studied as part of S330 *Oceanography*, a third level course for Open University students. The science of oceanography as a whole is multidisciplinary. However, different aspects fall naturally within the scope of one or other of the major 'traditional' disciplines.Thus, you will get the most of this volume if you have some previous experience of studying biology. The other volumes in the Open University Series lie more within the fields of physics, geology or chemistry (and their associated sub-branches).

Chapter 1 begins by describing unique properties that affect life in the sea, and by making comparisons with life on land. The major categories used to define marine environments and marine organisms are introduced, and basic ecological terms and concepts that are central to studies of biological oceanography are reviewed. The last section of Chapter 1 outlines the historical development of this scientific discipline.

Chapter 2 considers some physical and chemical features of the oceans including light, temperature, salinity, density, and pressure, all of which greatly influence the conditions under which marine organisms live. Major water current patterns are described because they transport many marine organisms, as well as dissolved gases and other chemical substances, and thus they affect the distributions of species and the size of populations in particular areas.

Chapter 3 introduces the various types of phytoplankton, and describes the way these floating plants manufacture energy-rich organic compounds by the process of photosynthesis. Plants require energy from solar radiation, and the effects of diminishing light levels with increasing depth are considered in detail. Essential nutrients like nitrate and phosphate are present in relatively low amounts in the lighted surface waters of the ocean, and the consequences of varying nutrient concentrations on plant growth are explored. Simple mathematical expressions are used to describe the relationships between plant growth and light intensity or nutrient concentration. Finally, vertical water movements that cause geographic differences in biological productivity are examined.

Chapter 4 describes the major types of zooplankton and their general life history patterns and feeding mechanisms. Vertical distribution of these animals is considered in relation to environmental differences with depth. Many animals migrate vertically in the water column, either daily or seasonally, and the consequences of these migrations are considered. Broad geographic patterns of distribution are also described, as well as smaller scale patterns established by a variety of physical and biological influences. Lastly, there is a discussion of long-term (decadal) changes in the abundance and species composition of zooplankton communities.

Chapter 5 explores the flow of energy through marine food chains and food webs, and explains the importance of using these ecological concepts to predict yields of fish based on measurements of plant growth in particular geographic areas. Different experimental approaches are also described that can be used to study biological oceanographic issues concerned with the dynamics of marine food chains. Finally, the important chemical changes that occur during the recycling of minerals are considered, with particular emphasis on nitrogen and carbon cycles.

Chapter 6 begins by describing the various types of nekton including the larger crustaceans, squid, marine reptiles, marine mammals, seabirds, and fish. The ecological importance of these larger animals is emphasized, as well as the consequences of their exploitation. The second half of this chapter looks at fisheries management problems and at the causes of fluctuation in the abundance of fish stocks.

Chapter 7 describes the major types of plants and animals that live on the seafloor, concluding with a section that explains how these organisms are sampled, and how their rate of growth (or production) can be determined.

Chapter 8 reviews the environmental conditions and general ecology of different bottom communities ranging from temperate intertidal communities, to tropical coral reefs and mangrove swamps, to deep-sea assemblages.

Chapter 9 considers the many ways in which human populations cause changes in marine ecosystems. Attention is given to problems of overfishing and exploitation of marine resources, and to the impacts of various types of pollution. There is also consideration of the changes caused by the accidental or deliberate introduction of marine species into new environments. Specific impacts on estuaries, mangrove swamps, and coral reefs are discussed.

You will find questions designed to help you to develop arguments and/or test your own understanding as your read, with answers provided at the back of this volume. Important technical terms are printed in **bold** type where they are first introduced or defined.

The oceans occupy about 71% of the Earth's surface. The deepest parts of the seafloor are almost 11 000 m from the sea surface, and the average depth of the oceans is about 3800 m. The total volume of the marine environment (about 1370×10^6 km^3) provides approximately 300 times more space for life than that provided by land and freshwater combined. The name given to our planet, 'Earth', is a synonym for dry land, but it is a misnomer in that it does not describe the dominant feature of the planet — which is a vast expanse of blue water.

The age of Earth is thought to be about 4600 million years. The ocean and atmosphere formed as the planet cooled, some time between 4400 and 3500 million years ago, the latter date marking the appearance of the first forms of life (see the Geologic Time Scale, Appendix 1). The earliest organisms are believed to have originated in the ancient oceans, many millions of years before any forms of life appeared on dry land. All known phyla (both extinct and extant) originated in the sea, although some later migrated into freshwater or terrestrial environments. Today there are more phyla of animals in the oceans than in freshwater or on land, but the majority of all described animal species are non-marine. The difference in the number of species is believed to be due largely to the greater variety of habitats on land.

1.1 SPECIAL PROPERTIES AFFECTING LIFE IN THE SEA

Why should life have arisen in the sea, and not on land?

Marine and terrestrial environments provide very different physical conditions for life. Seawater has a much higher density than air, and consequently there is a major difference in the way gravity affects organisms living in seawater and those living in air. Whereas terrestrial plants and animals generally require large proportions of skeletal material (e.g. tree trunks, bones) to hold themselves erect or to move against the force of gravity, marine species are buoyed up by water and do not store large amounts of energy in skeletal material. The majority of marine plants are microscopic, floating species; many marine animals are invertebrates without massive skeletons; and fish have small bones. Floating and swimming require little energy expenditure compared with walking or flying through air. Overcoming the effects of gravity has been energetically expensive for terrestrial animals, and perhaps it should not be surprising that the first forms of life and all phyla evolved where the buoyancy of the environment permitted greater energy conservation.

Two other features of the ocean are especially conducive to life. Water is a fundamental constituent of all living organisms, and it is close to being a universal solvent with the ability to dissolve more substances than any other liquid. Whereas water can be in short supply on land and thus limiting to life, this is obviously not the case in the marine environment. Secondly, the temperature of the oceans does not vary as drastically as it does in air.

On the other hand, certain properties of the sea are less favourable for life than conditions on land. Plant growth in the sea is limited by light because only about 50% of the total solar radiation actually penetrates the sea surface, and much of this disappears rapidly with depth. Marine plants can grow only within the sunlit surface region, which extends down to a few metres in turbid water or, at the most, to several hundred metres depth in clear water. The vast majority of the marine environment is in perpetual darkness, yet most animal life in the sea depends either directly or indirectly on plant production near the sea surface. Marine plant growth is also limited by the availability of essential nutrients, such as nitrates and phosphates, that are present in very small quantities in seawater compared with concentrations in soil. On land, nutrients required by plants are generated nearby from the decaying remains of earlier generations of plants. In the sea, much decaying matter sinks to depths below the surface zone of plant production, and nutrients released from this material can only be returned to the sunlit area by physical movements of water.

The greatest environmental fluctuations occur at or near the sea surface, where interactions with the atmosphere result in an exchange of gases, produce variations in temperature and salinity, and create water turbulence from winds. Deeper in the water column, conditions become more constant. Vertical gradients in environmental parameters are predominant features of the oceans, and these establish depth zones with different types of living conditions. Not only does light diminish with depth, but temperature decreases to a constant value of 2–4°C, and food becomes increasingly scarce. On the other hand, hydrostatic pressure increases with depth, and nutrients become more concentrated. Because of the depth-related changes in environmental conditions, many marine animals tend to be restricted to distinctive vertical zones. On a horizontal scale, geographic barriers within the water column are set by physical and chemical differences in seawater.

Much of this text deals with descriptions of marine communities and the interactions between physical, chemical and biological properties that determine the nature of these associations. Some attention is also given to the exploitation of marine biological resources. Despite the fact that the oceans occupy almost three-quarters of the Earth's surface, only 2% of the present total human food consumption comes from marine species. However, this is an important nutritional source because it represents about 20% of the high-quality animal protein consumed in the human diet. Although a greater total amount of organic matter is produced annually in the ocean than on land, the economic utilization of the marine production is much less effective. One branch of biological oceanography, fisheries oceanography, is a rapidly developing field that addresses the issue of fish production in the sea.

1.2 CLASSIFICATIONS OF MARINE ENVIRONMENTS AND MARINE ORGANISMS

The world's oceans can be subdivided into a number of marine environments (Figure 1.1). The most basic division separates the pelagic and benthic realms. The **pelagic environment** (pelagic meaning 'open sea') is that of the water column, from the surface to the greatest depths. The **benthic environment** (benthic meaning 'bottom') encompasses the seafloor

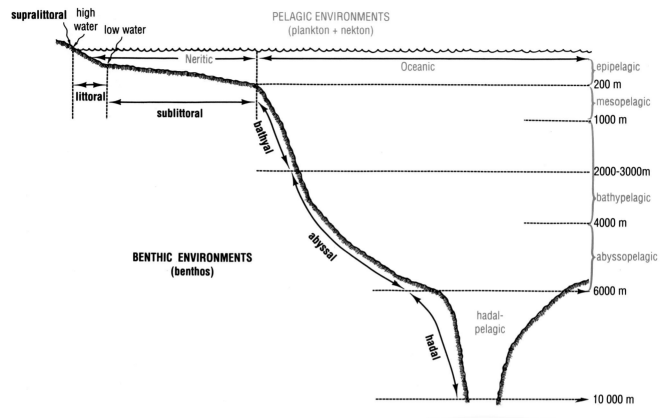

Figure 1.1   The basic ecological divisions of the ocean. The neritic (or inshore) pelagic zone is separated from the oceanic (or offshore) pelagic zone by the edge of the continental shelf, which is generally at about 200 m depth. Benthic habitats are in bold type; pelagic divisions are in blue. (Not to scale.)

and includes such areas as shores, littoral or intertidal areas, coral reefs, and the deep seabed.

Another basic division separates the vast open ocean, the **oceanic** environment, from the inshore **neritic** zone. This division is based on depth and distance from land, and the separation is conventionally made at the 200 m depth limit which generally marks the edge of the continental shelf (Figure 1.1). In some areas like the west coast of South America where the shelf is very narrow, the neritic zone will extend only a very slight distance from shore. In other areas (e.g. off the north-east coast of the United States), the neritic zone may extend several hundred kilometres from land. Overall, continental shelves underlie about 8% of the total ocean, an area equal to about that of Europe and South America combined.

Further divisions of the pelagic and benthic environments can be made which divide them into distinctive ecological zones based on depth and/or bottom topography. These will be considered in later chapters.

Marine organisms can be classified according to which of the marine environments they inhabit. Thus there are **oceanic species** and **neritic species** depending upon whether the organisms are found in offshore or coastal waters, respectively. Similarly, plants or animals that live in association with the seafloor are collectively called **benthos**. The benthos includes attached seaweeds, sessile animals like sponges and barnacles, and those animals that crawl on or burrow into the substrate. Additional subdivisions of the benthos are given in Chapter 7.

The pelagic environment supports two basic types of marine organisms. One type comprises the **plankton**, or those organisms whose powers of locomotion are such that they are incapable of making their way against a current and thus are passively transported by currents in the sea. The word plankton comes from the Greek *planktos*, meaning that which is passively drifting or wandering. Depending upon whether a planktonic organism is a plant or animal, a distinction is made between **phytoplankton** and **zooplankton**. Although many planktonic species are of microscopic dimensions, the term is not synonymous with small size as some of the zooplankton include jellyfish of several metres in diameter. Nor are all plankton completely passive; most, including many of the phytoplankton, are capable of swimming. The remaining inhabitants of the pelagic environment form the **nekton**. These are free-swimming animals that, in contrast to plankton, are strong enough to swim against currents and are therefore independent of water movements. The category of nekton includes fish, squid, and marine mammals.

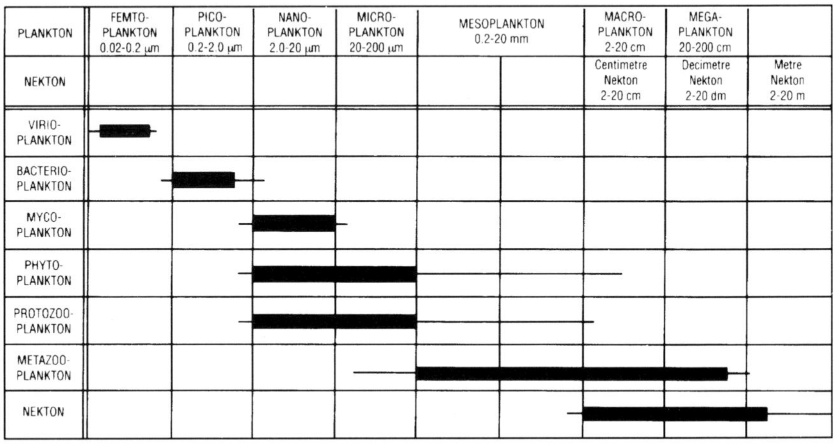

PLANKTON	FEMTO-PLANKTON 0.02-0.2 μm	PICO-PLANKTON 0.2-2.0 μm	NANO-PLANKTON 2.0-20 μm	MICRO-PLANKTON 20-200 μm	MESOPLANKTON 0.2-20 mm	MACRO-PLANKTON 2-20 cm	MEGA-PLANKTON 20-200 cm	
NEKTON						Centimetre Nekton 2-20 cm	Decimetre Nekton 2-20 dm	Metre Nekton 2-20 m
VIRIO-PLANKTON	▬							
BACTERIO-PLANKTON		▬						
MYCO-PLANKTON			▬					
PHYTO-PLANKTON			▬▬▬					
PROTOZOO-PLANKTON			▬▬▬					
METAZOO-PLANKTON				▬▬▬▬▬				
NEKTON						▬▬▬▬		

Figure 1.2 A grade scale for the size classification of pelagic organisms.

Finally, it is often convenient to characterize pelagic organisms according to size. Figure 1.2 presents one scheme for dividing plankton and nekton into size categories. The categories encompass the smallest and most primitive members of the pelagic environment, the viruses and bacteria, which are placed in the femtoplankton (<0.2 μm) and picoplankton (0.2–2.0 μm) categories, respectively. The largest nekton form the other extreme of the size range, with species of whales that can be 20 m or more in length. Phytoplankton and zooplankton (the latter including both unicellular protozoans and multicellular metazoans) comprise the intermediate size categories.

QUESTION 1.1 In Figure 1.2, why is there overlap in size between the largest plankton (macro- and megaplankton) and the smallest nekton?

1.3 BASIC ECOLOGICAL TERMS AND CONCEPTS

Basic ecological concepts are central to many studies of biological oceanography, and certain ecological terms will be used throughout this text. Marine organisms can be considered either individually or, more commonly in ecological studies, on different collective levels. A **species** is defined as a distinctive group of interbreeding individuals that is reproductively isolated from other such groups. A **population** refers to a group of individuals of one species living in a particular place, and **population density** refers to the number of individuals per unit area (or per unit volume of water). The various populations of micro-organisms, plants, and animals that inhabit the same physical area make up an ecological **community**.

The **habitat** of an organism is the place where it lives, but the term also may refer to the place occupied by an entire community. The **environment** consists of both nonliving **abiotic** (physical and chemical) components like temperature and nutrient concentrations, and **biotic** components that include the other organisms and species with which an organism interacts (e.g. predators, parasites, competitors, and mates).

The highest level of ecological integration is the **ecosystem**, which encompasses one or more communities in a large geographic area and includes the abiotic environment in which the organisms live. Examples of ecosystems could include estuaries (see Section 8.5 for the component communities of estuaries), or the total pelagic water column (with different communities at different depths). **Species diversity** is often used to describe the simplicity (or complexity) of communities and ecosystems; it can be defined in several ways but, unless otherwise stated, the term is used throughout this book to mean total number of species.

SI units (International System of Units) are now widely used in sciences since the system was adopted by the Conférence générale des poids et mesures (CGPM) in 1960. However, much of the ecological literature relevant to biological oceanography has used other types of units, and we have continued to report results as they were originally published. Conversions for more commonly used units can be found in Appendix 2.

1.3.1 *r*- AND *K*-SELECTION

Each plant or animal species, whether pelagic or benthic, or marine or terrestrial, has its own unique suite of biological and ecological features that define its life history. Two very different types of life histories are recognized as representing the extremes along a broad continuum of patterns. One type is known as *r***-selected**, and this life history pattern is demonstrated by **opportunistic** species. The contrasting type of life history is called *K***-selected**, and organisms with this complement of characteristics are referred to as **equilibrium** species. The terms *r* and *K* refer to different portions of growth curves (Figure 1.3); *r* is the intrinsic rate at which a population can increase, and *K* is the maximum population density that can be supported by the environment. Populations that are kept at low densities by abiotic or biotic environmental factors are influenced largely by the parameter *r*, whereas populations that are at or near the carrying capacity of the environment will be influenced by the parameter *K*.

Generally, *r*-selected species are of relatively small size, and they reach sexual maturity early. They usually produce many young several times per year. These opportunistic species typically live in variable or unpredictable environments, and they are able to respond quickly to favourable conditions or new habitats by rapid rates of colonization and reproduction. However, *r*-selected species typically have little ability to compete with other species, and they have high mortality. Consequently, populations of opportunistic species tend to be short-lived. *r*-selection leads to high biological

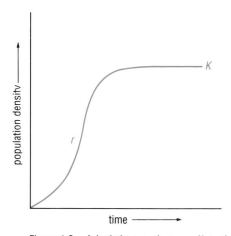

Figure 1.3 A logistic growth curve. Note that the population initially grows rapidly, then it slows and eventually ceases to grow as it reaches the carrying capacity of the environment. *K* is the maximal population size at carrying capacity, and *r* is the rate of population growth.

productivity, as the organisms devote a large proportion of their available energy to rapid growth and reproduction.

At the other extreme, fairly constant and/or predictable habitats favour K-selected species that are larger in size and slow-growing, but long-lived. They take longer to reach reproductive maturity and produce fewer young, but death rates are fairly low. They are particularly adapted to live in areas that are not subject to frequent disturbance, as they require sufficient time to complete their life cycles. Equilibrium species tend to build up their population sizes to the maximum that the environment will sustain (i.e. to the carrying capacity of the environment), and they sustain this population size for long periods by utilizing their resources very efficiently.

Table 1.1 summarizes other differences in these very different life histories. As mentioned above, the concept of r- and K-selection can be applied to pelagic as well as to benthic marine organisms, and it is also used in terrestrial ecology. It should be noted that the different patterns are applied in a relative sense to organisms. For example, although phytoplankton as a whole have short life cycles compared to whales, within the group there are both r-selected and K-selected phytoplankton species. It should further be stressed that the majority of organisms possess a mixture of r- and K-features. However, the importance of examining contrasting life styles lies in recognition of the ways in which different species deal with competition, predation, and environmental change. Throughout this book, you will be encouraged to learn about the many different life history patterns of marine plants and animals, and to compare them on the basis of r- and K- strategies.

Table 1.1 A comparison of the life history patterns exhibited by r- and K-selected marine species.

	r-selected opportunistic species	K-selected equilibrium species
Climate	variable/unpredictable	constant/predictable
Adult size	small	large
Growth rate	rapid	slow
Time of sexual maturity	early	late
Reproduction periods	many	few
Number of young	many	few
Dispersal ability	high	low
Population size	variable; usually below carrying capacity of environment	relatively constant; at or near carrying capacity
Competitive ability	low	high
Mortality rate	high; independent of population density	lower; density dependent
Life span	short (<1 yr)	long (>1 yr)
Pelagic/benthic ratio	high	low

1.4 THE HISTORICAL DEVELOPMENT OF BIOLOGICAL OCEANOGRAPHY

Human interest in the biology of the oceans can be traced back to observations made in the fourth century B.C. by Aristotle, who described and catalogued 180 species of marine animals. The great sea-going expeditions of the fifteenth and sixteenth centuries increased geographical knowledge of the oceans and added incidental observations on biology, but modern studies on the biology of the oceans did not really start until the middle of the nineteenth century.

The British naturalist, **Edward Forbes** (1815–54) (Figure 1.4), is often credited with being a founding father of oceanography, as he was one of the first persons to conduct systematically designed studies of the marine biota. He pioneered in the use of a dredge for obtaining samples of benthic marine animals, and he recognized that different species occupy different depth zones. His book, *The Natural History of the European Seas*, was published five years after his death, at the same time as Darwin's *The Origin of Species*. Unfortunately, Forbes is often remembered for the 1843 publication of his **azoic hypothesis**, which claimed that marine organisms could not exist at depths exceeding about 300 fathoms (550 m). Forbes was unaware that there were already records of life from deeper areas of the sea. In 1818, John Ross had obtained bottom samples containing worms and a starfish from about 1920 m in Baffin Bay, west of Greenland. His nephew, James Ross, led an expedition to the Antarctic in 1839–43 and collected benthic animals from as deep as 730 m. Such was Forbes's influence, however, that proponents of the azoic hypothesis clung to their beliefs despite increasing contrary evidence. The idea that the cold, dark reaches of the ocean could not possibly harbour any sort of life was finally refuted in 1860, when a submarine cable was brought up for repair from more than 1830 m and was found to have encrusting animals growing on it. Now there was a growing impetus to organize deep-sea expeditions to study this vast unknown environment and its inhabitants.

Charles Wyville Thomson became Edward Forbes's successor as professor of natural philosophy at the University of Edinburgh. In 1873, he published one of the first texts of oceanography, *The Depths of the Sea*, based on a review of early expeditions. Thomson also became organizer and leader of the first oceanographic expedition to circumnavigate the world. This was the *Challenger* **Expedition** of 1872–76, which travelled 110 900 km visiting all the major oceans except for the Arctic. The expedition was organized by The Royal Society specifically to survey the oceans with respect to their physical features, chemistry, and biology. The ship was a British naval sailing vessel with auxiliary steam, specially outfitted for scientific work (Figures 1.5 and 1.6). Besides Thomson, two other naturalists completed the journey. **Henry N. Moseley** was described as an indefatigable scientist and an enthusiastic amateur artist who contributed many drawings to the final official report. **John Murray**, a Canadian-born Scotsman, carried out his duties as a naturalist and later was instrumental in publishing the results of the voyage. The contrasts between the scientists' viewpoint of the voyage, and that of the ship's crew can be seen below.

Figure 1.4   Edward Forbes, a founding father of modern oceanography.

Figure 1.5 H.M.S. *Challenger*.

'Strange and beautiful things were brought to us from time to time, which seemed to give us a glimpse of the edge of some unfamiliar world.' C. Wyville Thomson, *The Challenger Expedition* (1876).

From a naval officer's diary: 'Dredging was our *bête noire*. The romance of deep-sea dredging or trawling in the *Challenger*, when repeated several hundred times, was regarded from two points of view; the one was the naval officer's who had to stand for 10 or 12 hours at a stretch, carrying on the work ... the other was the naturalist's ... to whom some new worm, coral, or echinoderm is a joy forever, who

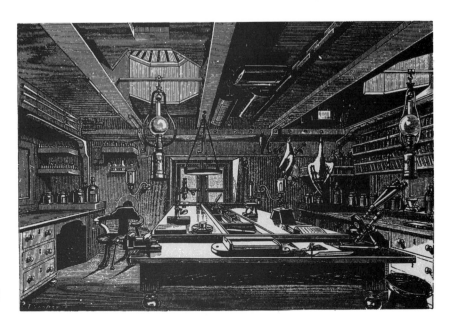

Figure 1.6 The zoological laboratory on board H.M.S. *Challenger*.

retires to a comfortable cabin to describe with enthusiasm this new animal, which we, without much enthusiasm, and with much weariness of spirit, to the rumbling tune of the donkey engine only, had dragged up for him from the bottom of the sea.'

Although a vast amount of information about the ocean had already been amassed by the time of the *Challenger* Expedition, much had been collected incidentally or in bits and pieces by individual scientists. The *Challenger* voyage attempted to integrate biology, chemistry, geology, and physical phenomena, and it established systematic data collection using standardized methods. For these reasons, the *Challenger* Expedition is considered to mark the beginning of modern oceanography. Over 76 scientists analysed the collections made during the voyage, and it took 19 years before all 50 volumes of final reports were published under the direction and financial patronage of John Murray. The expedition produced a basic map of the seafloor, and proved without doubt the existence of life at great depths. The biological samples yielded 715 new genera and 4417 new species of marine

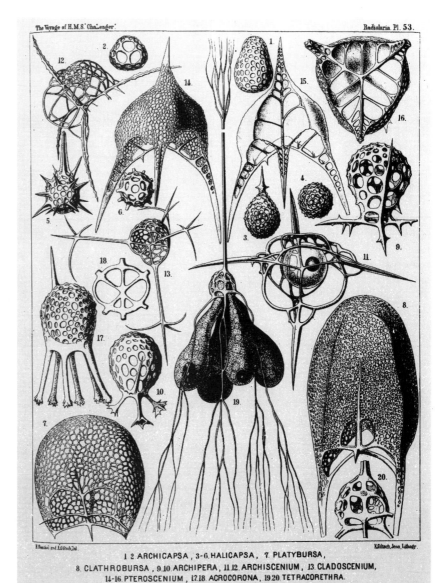

Figure 1.7 Some of the 3508 new species of radiolarians collected by the *Challenger* Expedition, and described and illustrated by the German zoologist Ernst Haeckel.

organisms; 3508 of them were new species of Radiolaria (a group of protozoans) (Figure 1.7), all described by the great German biologist **Ernst Haeckel**.

> The monograph on echinoderms was researched and complied by **Alexander Agassiz** of the United States, who said 'I felt when I got through that I never wanted to see another sea urchin and hoped they would gradually become extinct...'

The *Challenger* Expedition also discovered the true nature of *Bathybius haeckelii*, described by **Thomas Huxley** in 1868. *Bathybius* was a thin layer of mucus-like jelly covering the surface of preserved mud samples, and Huxley believed that it represented a primordial living slime which carpeted the deep seafloor. This 'organism', however, proved to be a precipitate of calcium sulphate, the result of mixing alcohol with seawater to preserve bottom samples.

> 'Never did an expedition cost so little and produce such momentous results for human knowledge.' — Ray Lancaster

Many other expeditions followed on the path blazed by the *Challenger*, and a few of those that made major contributions to biological oceanography are listed in Table 1.2. John Murray went on to organize the *Michael Sars* expedition of 1910, and in 1912 he co-authored a classic text in general oceanography, *The Depths of the Ocean*, with **Johan Hjort**, a Norwegian scientist.

Table 1.2 Major biological oceanographic expeditions.

Vessel	Country	Dates	Major objectives/advances
Challenger	U.K.	1872–76	Global biological collections; existence of life in deepest waters
Blake	U.S.A.	1877–86	Dredging; Caribbean and Gulf of Mexico collections
Princesse Alice I and II; *Hirondelle* I and II	Monaco	1886–1922	Deep-sea collections
Albatross	U.S.A.	1887–1925	Deep sea; Pacific and Indian Ocean collections
National	Germany	1889	Plankton collections
Valdivia	Germany	1898–99	Vertical distribution of pelagic organisms; deep-sea biology
Michael Sars	Norway	1904–13	Mid- and deep-water collections; North Atlantic
Dana I and II	Denmark	1921–36	Deep-water global collections; fishery research
Discovery I and II	U.K.	1925–39	Antarctic ecology
Meteor	Germany	1925–38	Atlantic biology
Galathea	Denmark	1950–52	Deep-sea dredging to 10 000 m; global collections
Vitiaz	U.S.S.R.	1957–60	Biology of trenches
Trieste (bathyscaphe)	Swiss/U.S.	1960	Deepest manned dive (10 916 m, Mariana Trench)
Alvin (submersible)	U.S.A.	1977	Discovery of deep-sea hot springs

Work was being conducted on marine plankton even before the time of the *Challenger*. The first person reported to have studied marine plankton was a surgeon and amateur naturalist, **J. Vaughan Thompson**, who towed a simple fine-meshed net to collect plankton off the coast of Ireland in 1828. His studies resulted in the first description of the planktonic stages of crabs. **Charles Darwin** also used a similar net to collect marine plankton during his stint as an unpaid, and seasick, naturalist on the voyage of the *Beagle*, from 1831 to 1836. In 1847, **Joseph Hooker** recognized that the diatoms collected in plankton nets were plants, and he suggested that they played the same ecological role in the sea as green plants do on land. However, it was not until 1887 that the term 'plankton' (see Section 1.2) was actually defined by **Victor Hensen**, a professor at Kiel University who also led the first oceanographic expedition entirely devoted to quantitative collections of plankton (the German 'Plankton Expedition' on board the *National*, see Table 1.2) The word 'plankton' was more critically defined in 1890 by Ernst Haeckel, and today the word encompasses all drifting organisms including plants (phytoplankton), animals (zooplankton), and bacteria (bacterioplankton).

Monographs on many different groups of zooplankton were available by the end of the nineteenth century, and taxonomic guides to phytoplankton were beginning to appear. Increasing attention was also given to those organisms that were too small to be collected by nets, as indicated below.

> '... **H. H. Gran** [a Norwegian scientist] now commenced using his big steam centrifuge for centrifuging the water samples from different depths, (and) he continued to avail himself of its help until the end of the cruise. By means of it he was able to collect in a little drop below the microscope all the most minute organisms, and in spite of the movements of the little ship and the vibration from the propeller, he was able with his microscope to study the many hitherto unknown forms in their living state, to draw them, and to count the number of the different species.' — John Murray's account of a cruise on *Michael Sars*, as related in *The Depths of the Ocean* (1912).

The late 1800s and early 1900s also marked the establishment of several marine and oceanographic laboratories, many of which were founded by biologists. In Europe, the German zoologist, **Anton Dohrn**, established the Stazione Zoologica de Napoli in 1872; the station was unique at that time in that its facilities were available to visiting scientists of other countries. The Marine Biological Association of the United Kingdom started a laboratory in Plymouth, England, in 1888. In 1906, **Prince Albert I** of Monaco established an oceanographic museum and aquarium to house extensive collections made by his research ships (see Table 1.2). In America, **Louis Agassiz** (father of Alexander Agassiz) established the first marine biological laboratory on the east coast in 1873; this was later (1888) moved to Woods Hole where it became the Marine Biological Laboratory. During this period, **Spencer Baird** started the first of a series of laboratories devoted to fisheries studies in Woods Hole and, in 1930, the Woods Hole Oceanographic Institution was officially established. On the west coast of America, **William Ritter** (a student of Alexander Agassiz) founded a research organization in 1905 that eventually (in 1924) became the Scripps Institution of Oceanography in La Jolla, California. Now, almost all countries bordering the ocean have oceanographic or fisheries stations.

The history of biological oceanography is as much intertwined with the chemistry of seawater as it is with the nature of marine animals and plants.

An understanding of the ecological role of phytoplankton required measurements of nutrients in seawater, and these were first carried out by the German chemists **Brandt** (1899) and **Raben** (1905). Later, the English chemist **H. W. Harvey** extended the earlier work of measuring nitrates and phosphates to include other nutrients such as iron and manganese.

The beginning of an ecological understanding of the sea came from the first textbooks that attempted to integrate biological data with the physical and chemical properties of the sea. One of the earliest and most comprehensive texts in this respect was *The Oceans* (1942) by Sverdrup, Johnson, and Fleming. Later ecological works included the 1957 publication of Volume 1, *Ecology*, of the *Treatise on Marine Ecology and Paleoecology* edited by J. W. Hedgpeth; G. Riley's *Theory of Food Chain Relations in the Oceans* (1963); and the delightfully written and illustrated text by **Alister Hardy**, titled *The Open Sea: Its Natural History* (1965).

> Alister Hardy was a wonderful storyteller, equally at home describing his hot-air ballooning experiences in England, or relating shipboard experiences like the following from a voyage to Antarctic waters on board *Discovery I*. 'Expecting a host of surface life we slung a bo'sun's chair (a board supported by ropes on each side like a swing) close to the water . . . right in front of the bows themselves. Here Kemp and I took turns with a hand-net and bucket. For sheer pleasure it was ideal: swinging in mid-air and gently rising and falling with the swell over the deep blue surface which occasionally rose to bathe and cool one's legs; one advanced like a gliding and soaring bird with nothing in front of one but the virgin ocean, as yet quite undisturbed by the bows behind . . . I rode in triumph, fishing out treasure after treasure as they came floating towards me on the very gentle undulating swell. An experience never to be forgotten.' From *Great Waters* (1967).

One unfortunate development in the history of biological oceanography was that much of the study of the top marine predators, the fish, fell under a separate discipline, that of **fishery sciences**. This came about because the most abundant marine fish form the basis of commercial fisheries. The branch of fishery sciences was founded in about 1890, led by **Alexander Agassiz** in the United States, **Frank Buckland** in England, and **W. C. McIntosh** in Scotland, all of whom looked to ocean science as a means of improving fish catches.

In 1902, the **International Council for the Exploration of the Sea** (ICES) was established under the auspices of King Oscar II of Sweden. This organization attempted to integrate physical studies of the oceans with biological investigations of fish, but this was not completely successful. Scientists trained in physics or chemistry and those trained in biology used different methods and approaches, and they proceeded to work independently of each other. ICES proved unable to force effective legislation concerning control of endangered stocks, or overfishing, and the organization was not as innovative in developing new fishing techniques and discovering new fisheries stocks as the fishermen themselves. After the war, ICES sponsored co-operative expeditions in the North Sea and North Atlantic, paid for by the national institutions in each participating country.

Fisheries management strategies tended to concentrate on economic models based on fish abundance and catch, while ignoring the rest of the biology of the sea. Classic texts on fisheries dealt primarily with the effects of harvesting on fish population size (e.g. *On the Dynamics of Exploited Fish Populations* by Beverton and Holt, 1957). Increasing human populations and increased demand for food resources have driven commercial fisheries to expand their fleets while developing new ways to locate fish schools and to harvest the stocks more efficiently. Diminishing stocks of exploited species have alerted fisheries scientists to the appreciation that the abundance of fish in the oceans is not only related to the numbers of fish removed, but is also greatly influenced by ocean climate. A relatively new field known as **fisheries oceanography** has developed which attempts to relate oceanographic data to fluctuations in fish stocks.

Biological oceanography began as a descriptive science, and basic observations on the biology of marine organisms and their environments continue to be an important aspect of biological oceanography. However, the development of new techniques and apparatus has changed the scope and scale of oceanographic research. Sonar, originally developed during World War II to detect enemy submarines, was later employed to study the topography of the seafloor, find fish schools, and, most recently, to locate and follow concentrations of larger zooplankton. Submarines and scuba diving became more sophisticated, and now both are used to obtain *in situ* observations of marine life. Underwater sound recording with hydrophones is now employed to study communication in marine mammals and the echo-location of prey by mammals and some fish. Computers have greatly decreased the time needed to analyse data routinely, and they are useful tools in simulating oceanographic events. The development of satellites and remote sensing has made it possible to map the ocean temperature and to trace ocean currents. The scale of research in biological oceanography now extends from laboratory studies on the effects of environmental change on single phytoplankton cells, to employing satellite imagery to obtain global patterns of plant production at the sea surface.

There is now an awareness that meteorological events in the atmosphere and climatic changes in the ocean and on land are connected over vast distances, and that humans can also produce impacts on the sea which can be measured on a global scale. The latter include the air- and water-borne dispersal of pesticides and other chemicals, and the overexploitation of fish and marine mammal stocks. As human impacts on the ocean increase, an expanding base of knowledge about the ecology of the seas becomes essential to address questions of exploitation and pollution with more certainty.

QUESTION 1.2 Why is the history of biological oceanography so recent compared with the development of terrestrial biology?

1.5 SUMMARY OF CHAPTER 1

1 The marine environment provides about 300 times more inhabitable space for living organisms than that provided by land and freshwater combined. All known phyla of plants and animals originated in the sea, and there are presently more phyla represented in the oceans than on land.

2 In comparison to life in air, the fluid nature of the ocean provides a buoyant environment in which the effects of gravity on living organisms are reduced. Because of this, marine organisms do not have to invest energy in

building large proportions of skeletal material, and they expend comparatively little energy in maintaining buoyancy and in locomotion.

3   Plant growth in the ocean is limited to the near-surface regions because light does not penetrate very far in seawater, and it is further limited by the low concentrations of essential nutrients (e.g. nitrate and phosphate) that are present at these depths. Because almost all life in the sea depends directly or indirectly on plants, the total plant production at the surface determines the amount of animals that can be produced.

4   Vertical gradients in environmental parameters (e.g. light, temperature, pressure) establish depth ranges with distinctive environmental characteristics.

5   Despite the vast extent of the marine environment, only 2% of the human diet comes from marine resources. However, this represents 20% of the high-quality animal protein consumed by humans.

6   The benthic environment encompasses the seafloor, and those species of plants and animals that live on or within the seabed form the benthos. The pelagic environment is that of the water column, from the sea surface to the waters immediately above the seafloor; inshore waters form the neritic zone, and offshore waters form the oceanic region. Plankton and nekton inhabit the pelagic environment; the distinction between the two groups of organisms is based on relative swimming ability, with nektonic species being stronger swimmers that are able to move independently of current direction.

7   Pelagic organisms can be classified into size categories ranging from femtoplankton (viruses) through intermediate sizes to the largest nekton (whales).

8   On an ecological scale, organisms can be considered individually or in assemblages that include populations of a single species, or communities made up of the populations of many interacting species. The highest unit of ecological integration is the ecosystem, which encompasses one or more communities as well as surrounding environment.

9   The life history patterns of all species form a continuum that ranges between the extremes described by *r*-selection and *K*-selection. Opportunistic species are adapted to live in variable or transitory environments by having short life cycles, production of many young, and high dispersal ability; however, these *r*-selected species have high mortality rates and their populations are often of short duration. *K*-selected species live in stable environments and usually have population densities near the carrying capacity of the environment; these equilibrium species typically have longer life spans, produce relatively few young, and have comparatively low death rates.

10   Edward Forbes (1815–54) is regarded as the founding father of biological oceanography, and the *Challenger* Expedition of 1872–76 marks the beginning of systematic oceanographic studies that integrate physical phenomena, water chemistry, and biology.

11   New techniques developed in the mid- to late-1900s expanded the scope and scale of oceanographic research. These include sonar, submarines, scuba diving, underwater sound recording, and remote sensing from satellites, all of which are now used to investigate life in the sea.

Now try the following questions to consolidate your understanding of this Chapter.

QUESTION 1.3 When did the first vertebrates appear in the ocean? Refer to the Geologic Time Scale in Appendix 1. *550 ma*

QUESTION 1.4 What are some characteristic features of the environment at 3000 m depth in the water column? *No light, cold. High pressure.*

QUESTION 1.5 What is the greatest depth reached by a manned diving vessel in the oceans? Refer to Table 1.2. *10916 m.*

In order to understand the ecology of the seas, it is necessary to understand the abiotic physical and chemical constraints of the marine environment to which the resident organisms are adapted. Some of these ecological constraints derive from the nature of seawater itself, with special properties related to the fluid nature of water and to the chemicals dissolved in this fluid. Other abiotic environmental features important to life in the sea result from the interplay between the Earth's atmosphere and the sea surface.

2.1 SOLAR RADIATION

Sunlight is as essential to life in the sea as it is to life on land. Some fraction of the solar radiation penetrating into the sea is absorbed by plants during photosynthesis, and this energy is used in the conversion of inorganic matter

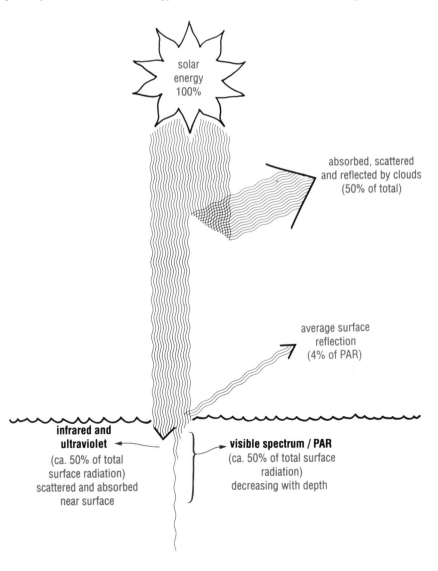

Figure 2.1 A schematic illustration of the passage of solar radiation through the atmosphere and sea surface, and the proportion of photosynthetically active radiation (PAR) available at depth in the sea.

to organic compounds. Some wavelengths of light are absorbed by water molecules and are converted to heat, which establishes the temperature regime of the oceans. In addition, light in the sea controls thé maximum depth distribution of plants and of some animals. Vision in animals is dependent on light, and certain physiological rhythms such as migrations and breeding periods may be set by periodic light changes.

2.1.1 RADIATION AT THE SEA SURFACE

Biological oceanographers have tended to use a variety of units to measure solar radiation at the sea surface, and to measure light intensity at depth in the sea. For that reason, conversions between different units are given in

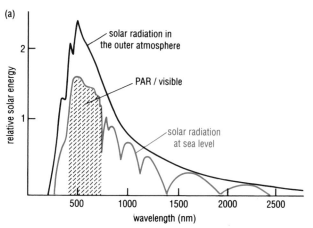

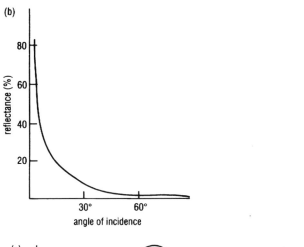

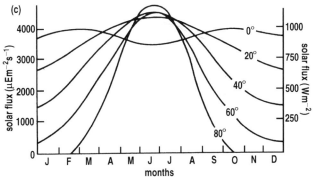

Figure 2.2(a) Solar radiation spectra before and after passage through the atmosphere, showing the zone of PAR (photosynthetically active radiation) or visible light.

(b) Percentage of light reflected from a calm sea surface as a function of sun angle.

(c) Changes in overhead solar radiation received at the sea surface with latitude and season in the Northern Hemisphere (contour lines represent latitude).

Appendix 2, and these should be referred to when necessary. Two light units used for biological studies in the sea are the einstein (E), which measures photons (one einstein is a mole of photons, or 6.02×10^{23} photons), and the watt (W), which measures the energy of radiation. The energy of radiation depends on the wavelength of the light, but for photosynthetic radiation (400 to 700 nm), one W m^{-2} is approximately equal to 4.16 μE m^{-2} s^{-1}.

Solar radiation coming from the Sun to the outside of the Earth's atmosphere is fairly constant (Figures 2.1 and 2.2a). About half of this energy is absorbed and scattered in the various layers of the atmosphere, so that the amount reaching the sea surface is about 50% of that received at the top of the atmosphere. Some of this is reflected back into the atmosphere from the sea surface (Figures 2.1 and 2.2b). The amount reflected depends on the Sun angle and becomes very large below a Sun angle of 5° to the horizon. During any day, the actual amount of radiation reaching the sea surface at any point is thus a function of the Sun angle, the length of the day, and weather conditions. The Sun angle is determined by the time of year, time of day, and by the latitude. At the Equator, radiation from an overhead Sun is fairly constant throughout the year but, at 50° N, the seasonal variation in incident radiation ranges from about 1000 μE m^{-2} s^{-1} in January to over 4000 μE m^{-2} s^{-1} in June (Figure 2.2c).

QUESTION 2.1 What is the approximate maximum solar radiation received at the surface of the Arabian Sea off Bombay (latitude *ca.* 20 N) in (a) September and (b) January? Refer to Figure 2.2c.

A summary of temporal variations in radiation at the sea surface is given in relative units in Figure 2.3. The **diel** variation is the change in solar radiation over 24 hours (i.e. the difference between day and night). **Diurnal** variations are those that occur during hours of daylight due, for example, to cloud cover. Seasonal variations are most marked at high latitudes. This is particularly so within the Arctic Circle where there can be 24 hours of sunlight on the ocean surface during the summer. Differences in the input of surface solar radiation account for much of the difference in photosynthesis by phytoplankton at discrete localities in the ocean.

2.1.2 RADIATION IN THE SEA

In comparison with other liquids, water is relatively transparent to solar radiation, but much less so than air. Of the sunlight penetrating the sea surface, about 50% is composed of wavelengths longer than about 780 nm. This **infrared** radiation is quickly absorbed and converted to heat in the upper few metres (Figure 2.1). **Ultraviolet** radiation (< 380 nm) forms only a small fraction of the total radiation, and it also is usually rapidly scattered and absorbed, except in the very clearest ocean waters (Figure 2.4). The remaining 50% of the radiation comprises the **visible spectrum**, with wavelengths of between approximately 400 and 700 nm that penetrate deeper in the sea. These are of particular importance for animals with vision, and because they are also approximately the same wavelengths used by plants in photosynthesis. These wavelengths are often referred to as **photosynthetically active radiation (PAR)**. The maximum intensity of PAR with the Sun directly overhead is about 2000 μE m^{-2} s^{-1}. Obviously this value will vary with Sun angle, and it decreases to zero as the Sun approaches the horizon.

$a = 3700\ \mu E\ m^{-2} s^{-1}$

$b = 2900\ - \cdot\ -$

Figure 2.3 Temporal variations in surface solar radiation. (Relative scales).

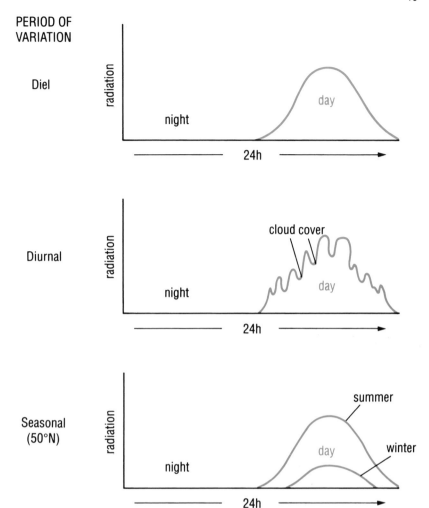

PERIOD OF VARIATION

Diel

Diurnal

Seasonal (50°N)

Latitudinal (equator and Arctic summer)

Figure 2.4 The penetration of light of different wavelengths into clear oceanic water. The lines indicate the depths of penetration for 10% and 1% of the surface light levels.

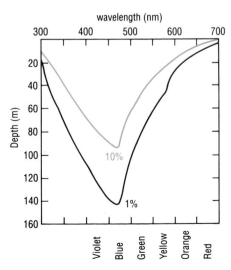

As light passes through water, it is both scattered and absorbed, with different wavelengths of the visible spectrum penetrating to different depths (Figure 2.4). Red light (*ca.* 650 nm) is quickly absorbed, with only about 1% still remaining at 10 m in very clear seawater. Blue light (*ca.* 450 nm) penetrates deepest, with about 1% remaining at 150 m in clear water.

There is an exponential decrease of light intensity with depth. An extinction coefficient, k, is calculated to express this attenuation of light. The extinction coefficient of seawater can be calculated from measurements taken with a radiation meter lowered into the sea, and using the following equation in which I_0 is the surface radiation and I_D the radiation at depth:

$$k = \frac{\log_e(I_0) - \log_e(I_D)}{\text{depth (m)}}$$

(2.1)

QUESTION 2.2 The radiation at 10 m is 50% of the surface radiation as measured with a radiation meter. What is the extinction coefficient?

As you might have inferred from Figure 2.4, the extinction coefficient, k, is different for various wavelengths of light. It is about 0.035 m^{-1} for blue light, but about 0.140 m^{-1} for red light. However, if many particles are present in the water, the blue light is scattered more than the red, and this will affect the colour spectrum of undersea light, resulting in a shift of the most deeply-penetrating wavelength toward a green colour (Figure 2.4). The extinction coefficient is also affected by the amount of coloured, dissolved, organic material in seawater, and by the amount of chlorophyll contained in living phytoplankton and in plant debris. In the clearest ocean water of the tropics, light which is visually detectable by a deep-sea fish may penetrate to more than 1000 m (Figure 2.5). In turbid coastal waters, scattering and absorption of light are increased by the presence of much silt and numerous phytoplankton, and the same amount of light may not reach 20 m.

QUESTION 2.3 Novice scuba divers are often disappointed at seeing a coral reef for the first time because of the monotony of colour compared with colour photographs and films of reefs. Why is this so?

Three vertical ecological zones in the water column are defined by the relative penetration of light in the sea (Figure 2.5). The shallowest zone is called the **euphotic zone**, and it is defined as that region in which light is sufficient to support the growth and reproduction of plants. Here, there is sufficient light for plant production by photosynthesis to exceed the loss of material that takes place through plant respiration (see Section 3.2). The

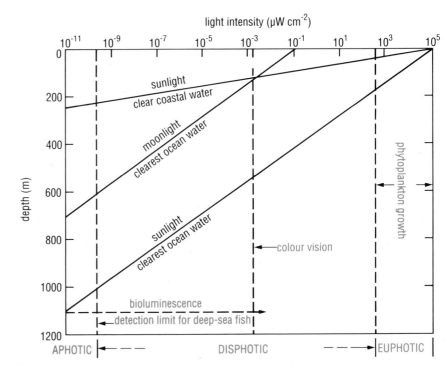

Figure 2.5 The vertical ecological zones established by light penetration in the sea. Note that the light intensity scale is logarithmic with depth. The positions of the vertical broken lines delimiting the three ecological light zones are approximate only (see text).

amount of light required for photosynthetic production to just balance respiratory losses in plants is known as the **compensation light intensity**. The depth at which photosynthetic production is balanced by plant respiration is called the **compensation depth**, and it defines the lower boundary of the euphotic zone. Thus the euphotic zone extends from the surface to a depth of just a few metres in turbid inshore regions, to a maximum depth of about 150 m in very clear, tropical oceanic water. In any region, the compensation depth (D_C), and thus the depth of the euphotic zone, can be calculated from:

$$D_C = \frac{\log_e(I_0) - \log_e(I_C)}{k} \qquad (2.2)$$

Surface radiation (I_0) is measured directly, and k is calculated from equation 2.1 assuming a wavelength of 550 nm. The value used for the compensation light intensity (I_C) varies with different species of phytoplankton as well as with the previous history of light adaptation of any particular species. For example, heavily shaded phytoplankton can adapt to lower compensation light intensities. In general, however, values for I_C will range between 1 and 10 μE m^{-2} s^{-1}.

Below the euphotic zone is the dimly lighted **disphotic zone**, a region where fish and some invertebrates can see, but where light is too low for positive net photosynthesis (i.e. loss of plant material through respiration exceeds plant production by photosynthesis over 24 hours). However, living phytoplankton which have sunk from the euphotic zone may be present here.

The deepest and largest region in the open ocean is the dark **aphotic zone**; this extends from below the disphotic zone to the seafloor. Here, sunlight cannot be detected by any biological system. This vast region does not support plant life, and is spatially removed from the initial link in the marine food chain.

QUESTION 2.4 What do you think is the biological role of moonlight (see Figure 2.5) in the sea?

2.2 TEMPERATURE

Water temperature is one of the most important physical properties of the marine environment as it exerts an influence on many physical, chemical, geochemical, and biological events. Temperature controls the rates at which chemical reactions and biological processes (such as metabolism and growth) take place. Temperature and salinity variations combine to determine the density of seawater, which in turn greatly influences vertical water movements with consequent changes in chemical and biological events within the water column. Water temperature partly determines the concentration of dissolved gases in seawater; these include oxygen and carbon dioxide, which are profoundly linked with biological processes. Temperature is also one of the most important abiotic factors influencing the distribution of marine species.

2.2.1 SEA SURFACE TEMPERATURES

There is a continuous exchange of heat and water between the ocean and atmosphere. The seas are heated primarily by the infrared wavelengths of

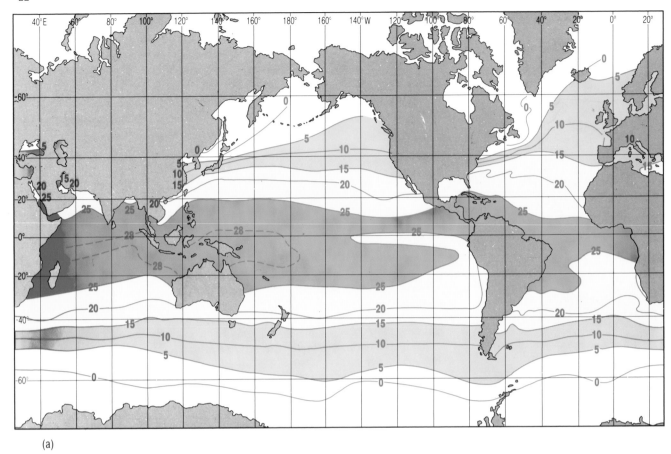

Figure 2.6 The global distribution of sea surface temperatures (°C) in (a) February and (b) August.

solar radiation. The radiant energy of these wavelengths is quickly transformed into heat by absorption. The heating effect of sunlight is confined to the immediate surface of the ocean, with 98% of the infrared spectrum being absorbed within about the first metre of the water column.

Sea surface temperatures vary with latitude (Figure 2.6). Surface temperatures can exceed 30°C in the tropical open ocean, and approach 40°C in shallow tropical lagoons. At the other extreme, water surface temperatures in polar regions may be as low as −1.9°C the freezing point of typical seawater. The moderate regime of surface seawater temperature is in sharp contrast to air temperatures affecting terrestrial ecosystems; these range from as high as 58°C (in northern Africa during summer) to −89°C in the Antarctic during winter (Figure 2.7). The temperature regime of the oceans is buffered by certain physical properties of water. Water has a very high specific heat, meaning that it can absorb or lose large quantities of heat with little change in temperature. Furthermore, the oceans are cooled primarily by evaporation and, because the latent heat of evaporation for water is the highest of all substances, great quantities of heat can be transferred and stored in water vapour with relatively little change in water temperature.

It is sometimes convenient to designate biogeographic zones based on sea surface temperatures. The following zones lie within the boundaries set by the annual average surface temperatures given in the right column:

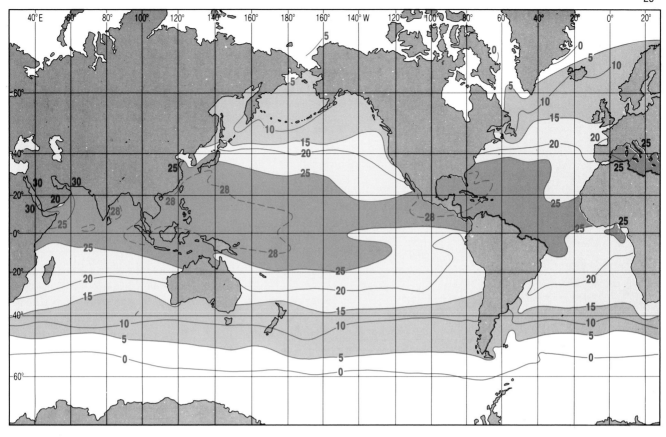

(b)

Tropical	25°C
Subtropical	15°C
Temperate	5°C (northern limit)
	2°C (southern limit)
Polar	< 0–2° or 5°C

Temperate zones in both hemispheres are characterized by a mixture of subpolar and subtropical water, and by having the maximal annual temperature range. Although attempts have been made to ascribe latitudinal limits to these temperature zones, this has little or no ecological significance in the pelagic environment where currents displace water of different temperatures away from their origins, and where water temperature changes very gradually through mixing. In pelagic communities, faunal boundaries follow certain **isotherms** (lines of equal temperature), or are more precisely described by combinations of temperature and salinity which define distinctive bodies of water (see Section 2.4).

The mean daily variation in surface temperature in the open ocean is very small, generally less than 0.3°C, and it is usually imperceptible at 10 m depth. Even in shallow water, the daily surface temperature change is less than 2°C. Temperature changes over 24-hour periods are therefore of little importance to plankton and fish, unlike residents of intertidal and terrestrial ecosystems which can be subjected to very considerable differences between day and nighttime temperatures.

Annual surface temperature fluctuations (compare Figures 2.6a and b) are very small in Antarctic waters and are less than 2–5°C in Arctic and tropical seas. In temperate and subtropical areas, they are large enough to influence biological events significantly. In the open ocean at latitudes of 30–40°, where clear skies permit the maximal heat gain in summer and maximal heat loss in winter, the annual variation is about 6–7°C. However, the western areas of the North Pacific and North Atlantic have annual variations of up to 18°C because of the prevailing westerly winds that bring very cold continental air masses over these regions in winter and warm continental air in summer. In shallow marginal seas, and coastal areas generally, the fluctuation in water temperature closely parallels air temperature, and annual variations may exceed 10°C.

In addition to daily and seasonal variations in surface temperature, there are longer term climatic changes that affect marine ecology. Some of these events are deduced only from changes in seafloor sediments that suggest dramatic temperature changes in the overlying water during the geological past. Other climatic changes can be observed in the present time and include major perturbations such as the El Niño events in the Pacific Ocean. These are cyclical changes in sea surface temperature occurring every two to ten years that have widespread impacts on marine ecology as well as on global weather. An El Niño can have a catastrophic impact on commercial fisheries in affected areas, and the details of this event are considered in Section 6.7.2.

2.2.2 VERTICAL TEMPERATURE DISTRIBUTION

Turbulent mixing produced by winds and waves transfers heat downward from the surface. In low and mid-latitudes, this creates a surface **mixed layer** of water of almost uniform temperature which may be a few metres deep to several hundred metres deep (Figure 2.8). Below this mixed layer, at depths of 200–300 m in the open ocean, the temperature begins to decrease rapidly down to about 1000 m. The water layer within which the temperature gradient is steepest is known as the **permanent thermocline**. The temperature difference through this layer may be as large as 20°C. The permanent thermocline coincides with a change in water density between the warmer low-density surface waters and the underlying cold dense bottom waters. The region of rapid density change is known as the **pycnocline**, and it acts as a barrier to vertical water circulation; thus it also affects the vertical distribution of certain chemicals which play a role in the biology of the seas. The sharp gradients in temperature and density also may act as a restriction to vertical movements of animals.

Temperature decreases gradually below the permanent thermocline. The thermal stratification of the oceans is shown schematically in Figure 2.9. In most oceanic areas, the water temperature at 2000–3000 m never rises above 4°C regardless of latitude. At greater depths, the temperature declines to between about 0°C and 3°C. The temperature of deep water at the Equator is within a few degrees of that of deep water in polar regions. The only exceptions to cold deep conditions are found in certain localized areas of the deep sea, where bottom water temperature may be elevated by geothermal activity (see Section 8.9)

In temperate climates, **seasonal thermoclines** (Figure 2.8) are established in the surface layer during the summer. These result from increased solar radiation that elevates surface temperature at a time when winds are lessened. Thus there is little turbulent mixing to promote downward movement of heat, and a thermal stratification is set up in the near-surface waters. This phenomenon persists until autumn, when the surface water is cooled

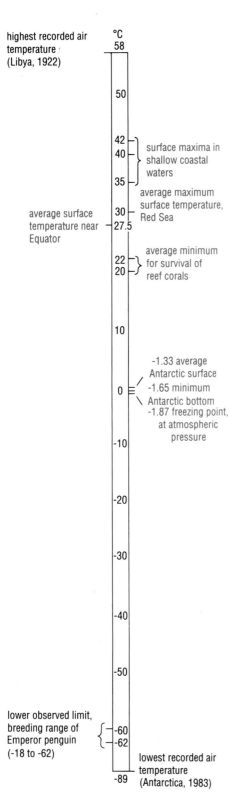

Figure 2.7 Temperature ranges in the sea (blue) and on land (black).

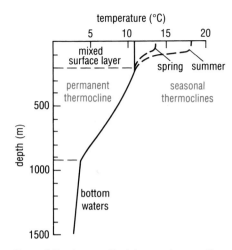

Figure 2.8 A generalized temperature profile for temperate latitudes. The solid line shows the winter condition with a mixed surface layer of homogeneous temperature overlying the permanent thermocline. The dashed lines show the formation of seasonal thermoclines that develop in the surface water in spring and summer due to elevated solar radiation and warming coinciding with lessened wind.

and increasing winds cause sufficient turbulence to mix the upper layers and break down the thermocline. Because permanent and seasonal thermoclines greatly affect biological productivity on a global and temporal scale, respectively, more will be said about thermal stratification in later sections.

In what way does seawater temperature affect faunal distributions?

The physiological ability to cope with environmental temperatures plays a large role in determining the distributional limits of marine organisms. The majority of marine animals (i.e. invertebrates and fish) are **poikilothermic** species with a varying body temperature that approximately follows the ambient water temperature, but marine mammals are **homoiothermic** and maintain a constant body temperature. Animals that can exist in environments with a wide temperature range are known as **eurythermic**. Such species tend to have wide distributional ranges or they live in regions of considerable temperature fluctuation, such as temperate intertidal zones. Those species that are restricted to narrow temperature limits are called **stenothermic**. They include such groups as reef-building corals, which require a minimum temperature of 20°C, as well as those species that are restricted to cold waters. The geographic range of cold-stenothermic species may be very wide; for example, some species that are found at shallow depths in the Arctic are also present at depths of 2000–3000 m in Equatorial areas where similar cold temperatures prevail.

2.3 SALINITY

Salinity refers to the salt content of seawater. For our purposes, **salinity** can be simply defined as the total weight (in grammes) of inorganic salts dissolved in 1 kg of seawater. However, salinity is not measured by weight because it is difficult and tedious to dry all the salts in seawater. Salinity is more easily and routinely determined with a salinometer that measures electrical conductivity, which increases with increasing salt content. The major elements are present in the form of ions, with sodium and chloride predominating. The ten major constituents listed in Table 2.1 make up about 99.99% of all the dissolved substances in the ocean.

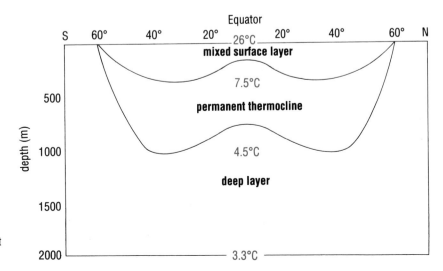

Figure 2.9 A generalized and schematic cross-section, showing the main thermal layers of the oceans and their average temperatures at the Equator.

Why does this list not include such common and biologically important elements as oxygen, nitrogen, and iron?

Other elements and compounds in the oceans are present at lower concentrations than those listed in Table 2.1. Some of these, like oxygen and carbon dioxide, exist as dissolved gases and will be considered later. Those elements that are linked with biological processes (e.g. nitrogen in the form of nitrate) exhibit highly variable concentrations, unlike the ions listed in the table. Although there also are dissolved organic compounds in seawater, all are in concentrations too low to affect salinity.

Table 2.1 The major constituents of seawater with a salinity of 35.

Ion	Concentration $(g\ kg^{-1})$	% by weight of all salts in the sea
Chloride (Cl^-)	18.98	55.04
Sodium (Na^+)	10.56	30.61
Sulphate (SO_4^{2-})	2.65	7.68
Magnesium (Mg^{2+})	1.27	3.69
Calcium (Ca^{2+})	0.40	1.16
Potassium (K^+)	0.38	1.10
Bicarbonate (HCO_3^-)	0.14	0.41
Bromide (Br^-)	0.07	0.19
Borate (mainly H_3BO_3)	0.03	0.07
Strontium (Sr^{2+})	0.01	0.04

Because the concentrations of most of the major constituents are not significantly affected by biological and chemical reactions, they are said to show **conservative behaviour**. This property results in the **constancy of composition of seawater**. That is, the total salinity may vary, but the relative proportion of each major ion to the total remains stable, as do the ratios of the concentrations of each major ion to the others. These ionic ratios depart from normal only in localized regions such as estuaries, which receive an inflow of freshwater containing different relative proportions of major ions.

2.3.1 RANGE AND DISTRIBUTION OF SALINITY

Variability in salinity is linked with global climate. Salinity in surface waters is increased by the removal of water through evaporation, and it is decreased primarily through the addition of freshwater via precipitation, either in the form of rain or snow, or from river inflow. At higher latitudes, salinity also is decreased by ice and snow melt.

The average salinity of the oceans is about 35, and variability in the global distribution of surface salinity in the open ocean is shown in Figure 2.10. Salinity values closely follow the curve for evaporation minus precipitation shown in Figure 2.11. Note that the highest salinity values are found at about 20–30° latitude in both hemispheres, in areas having high evaporation and low precipitation. Low salinities are found in polar areas, which have high precipitation as well as melting ice, and in areas influenced by polar water.

Certain marine areas have salinities outside the range of those in the open ocean. These generally occur in inshore and shallow areas that are exposed to coastal runoff or river inflow, or that have limited mixing with the open

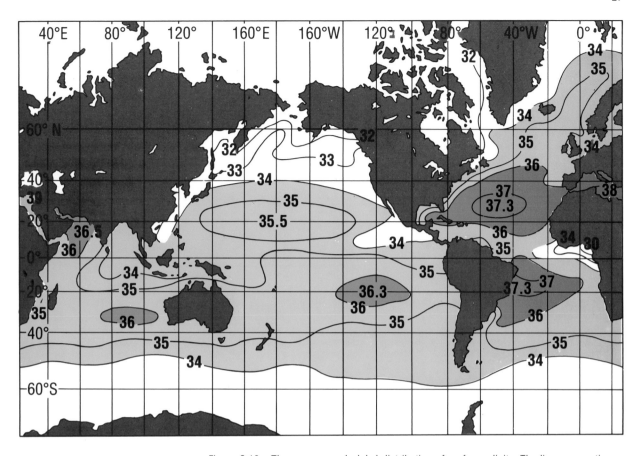

Figure 2.10 The mean annual global distribution of surface salinity. The lines connecting points of equal salinity are **isohalines**.

sea. The salinity ranges given below roughly characterize particular types of marine environments:

Open ocean	32–38 (average, 35)
Shallow coastal areas	27–30
Estuaries	0–30
Semi-enclosed seas	< 25
(e.g. Baltic Sea)	
Hypersaline environments	> 40
(e.g. Red Sea; tropical coastal lagoons)	

Estuaries (0–30) and Semi-enclosed seas (< 25) are **brackish** water.

The range of salinity in surface waters is much greater than that in deeper layers because fluctuations result primarily from sea surface–atmosphere interactions. Figure 2.12 displays the distribution of salinity with depth in the Atlantic Ocean. An area where salinity changes rapidly with depth is called a **halocline**. Such zones exist in low and mid-latitudes and lie from the bottom of the mixed layer to about 1000 m. Below this depth, salinity is 34.5–35.0 at all latitudes.

Diurnal variations in salinity are usually very small, apart from intertidal areas or shallow lagoons where evaporation and precipitation effects may be intense. Seasonal variations in salinity are also very small, except in inshore shallow waters. The average annual variation in surface salinity of the open ocean is about 0.3.

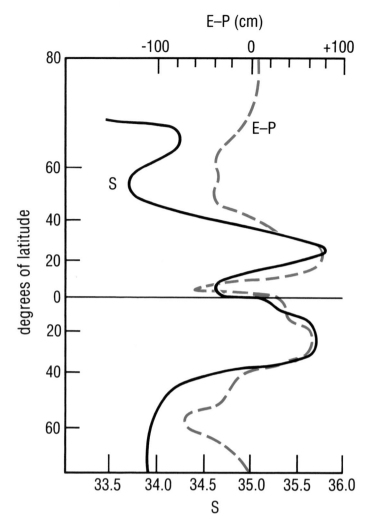

Figure 2.11   The distribution of average surface salinity (S, black line) plotted against the difference between average annual evaporation and precipitation (E - P, blue line) at different latitudes.

2.3.2   BIOLOGICAL IMPORTANCE OF SALINITY

In most marine invertebrates and primitive fish (sharks, rays), the salt content of the blood and body fluids is about the same as in seawater of average salinity. In bony fish (teleosts), the salt concentration of the blood is only about 30–50% of the ambient salinity. This has several physiological consequences. Because there is a tendency for water to move across semipermeable membranes from a zone of low salt concentration to one of high concentration (a process called **osmosis**), marine teleost fish tend to lose water and thus increase their internal salt concentration. These animals have evolved various physiological mechanisms of **osmoregulation** that counteract this problem. Most marine fish, for example, excrete very small quantities of urine and secrete salts across the gills. This type of **active transport**, in which the kidneys work against the normal osmotic trend, requires an expenditure of energy. Sea turtles, seabirds, and marine mammals also exhibit various means of maintaining osmotic balance with their environment.

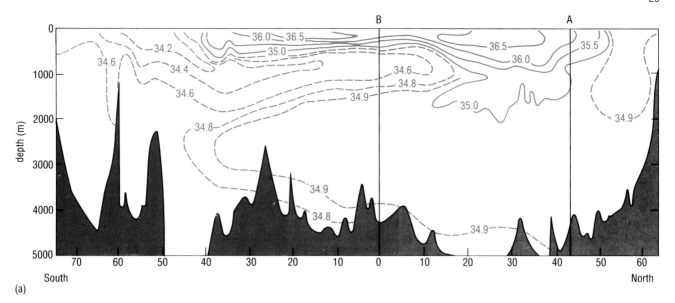

(a)

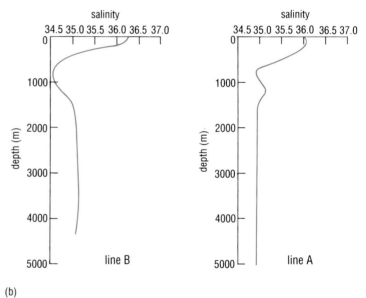

(b)

Figure 2.12(a)   A cross-section of the western Atlantic Ocean illustrating the vertical distribution of salinity. This general pattern is typical of all oceans, although the details will vary from ocean to ocean.

(b) Salinity-depth profiles along lines A and B in (a).

The problem of osmotic balance is particularly acute in those marine invertebrates and vertebrates that inhabit estuarine areas with rapidly changing salinities (see Section 8.5), and in those fish that migrate between freshwater and seawater (see Section 6.6.1). Species that can tolerate a wide range of salinity are called **euryhaline**, and they may display various means of osmotic control ranging from simple impermeability (for example, closing of mollusc shells) to complex forms of active transport as described above. Those animals that can only tolerate a narrow salinity range are called **stenohaline**.

QUESTION 2.5   How could the ambient salinity affect the buoyancy of pelagic organisms?

2.4 DENSITY

The **density** (mass per unit volume) of seawater is governed by temperature and salinity (and, to a lesser extent, by hydrostatic pressure). As salinity increases, the density increases; as the temperature increases, the density decreases. Salinity and temperature are physically independent variables but, as we have seen, they are not randomly distributed in the ocean. Global climate establishes the temperature and salinity distribution in the surface layers of the ocean. Distinctive combinations of these variables are thus developed in large volumes of water, and these easily measured temperature–salinity characteristics can be used to define particular **water masses**. Each of these bodies of water thus forms a different type of environment, and each supports distinctive communities of organisms.

Figure 2.13 shows the major upper-layer water masses of the world that extend in depth to about the base of the thermocline. The definitive temperature and salinity characteristics of these water masses are acquired at the surface, but once water is out of contact with the atmosphere, its physical characteristics will only change very gradually and very slowly through mixing with adjacent waters of different characteristics. This means that even though water masses move both horizontally and vertically in the ocean, each can be traced over long distances by its definitive combination of temperature and salinity. New water masses also are eventually formed by mixing of waters of different origins, and these too develop their own temperature–salinity signatures that indicate the amount of mixing.

The upper water layers are moved horizontally by surface currents generated by wind systems (see Section 2.6). Vertical movements of water are controlled in part by temperature and salinity variations that change the density of seawater. Figure 2.14 illustrates the relationship between temperature, salinity, and density. Note that density itself cannot be used to define a water mass because different combinations of temperature and salinity may produce the same density.

Figure 2.13   The global distribution of major upper water masses.

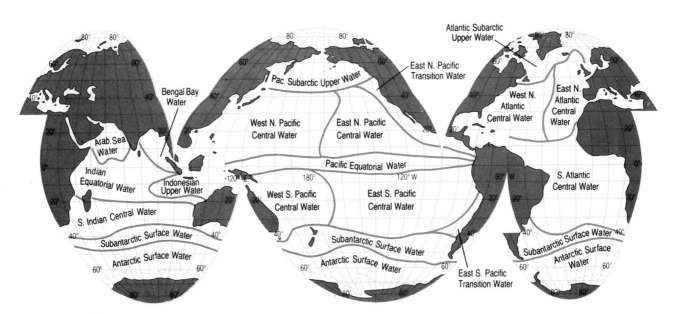

QUESTION 2.6 (a) What is the density of seawater having a temperature of 9°C and a salinity of 33.5? (b) What is the density of seawater with a temperature of 20°C and a salinity of 36.5? Refer to Figure 2.14.

Water that is less dense than underlying layers will remain at the surface (e.g. water in equatorial latitudes). Surface water masses that increase in density will sink to depths determined by their densities relative to the vertical density structure of the surrounding waters. Figure 2.15 shows the water masses lying between about 550 m and 1500 m depth; these masses are denser than the waters above 500 m. The densest water masses occupy depths from below 1500 m to the seafloor.

Salinity also has the important effects of lowering the temperature at which maximum density occurs, and depressing the freezing point of seawater. These changes are shown in Figure 2.16; note that the temperature of maximum density and the freezing point are the same at a salinity of about 25. As the oceans are generally more saline than this (average salinity = 35), the density of seawater continues to increase with decreasing temperature all the way to the freezing point (about −1.9°C at a salinity of 35). In contrast, the temperature of maximum density of freshwater (salinity = 0) is 4°C, and water becomes less dense as the temperature falls to 0°C, its freezing point. This is a profound difference between freshwater and seawater, and it has important ramifications on oceanic circulation and on marine life.

The densest and deepest water masses originate primarily around Antarctica, or in the vicinity of Greenland and Iceland (Figure 2.17). During winter in high latitudes, surface waters become colder and, because seawater density continues to increase to the freezing point, there is a continual sinking of water until that point is reached. As sea-ice forms, it is less saline than the seawater so the salinity of the water is elevated, and the density further increased. This very dense polar water sinks and flows toward the Equator (Figure 2.17) at intermediate depths (Antarctic Intermediate Water and North Atlantic Deep Water) or along the seafloor (Antarctic Bottom Water). Antarctic Bottom Water in particular penetrates far into the northern parts of

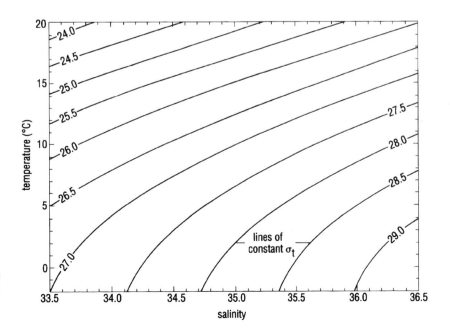

Figure 2.14 A T−S diagram showing the relationship between temperature (T), salinity (S) and density. For convenience, the density contours are shown as lines of equal values of σ_t (sigma-t) where $\sigma_t = (\text{density} - 1) \times 1000$. Therefore a density of 1.02781 g cm^{-3} has a $\sigma_t = 27.81$.

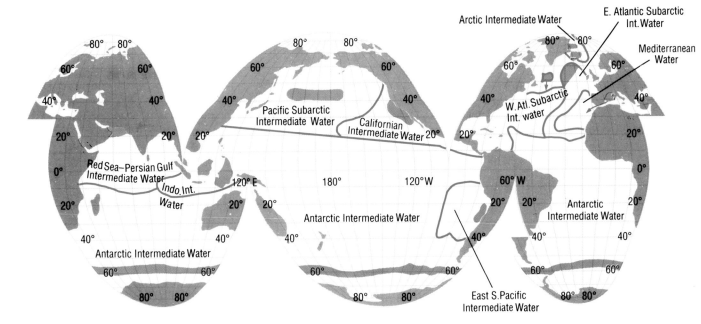

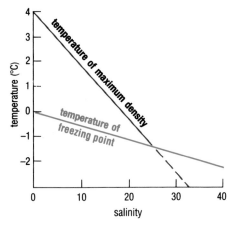

Figure 2.15 The global distribution of intermediate water masses lying between about 550 and 1500 m depth. The source regions of the water masses are indicated by dark blue.

Figure 2.16 Temperatures of freezing and maximum density of water as a function of salinity.

the Atlantic and Pacific oceans. Deep water will eventually be returned to the surface by wind-driven mixing, and thus there is a continuous, but very slow (on the order of several hundred to a thousand years), cycling between surface and deep waters.

In temperate latitudes during winter, cooling surface water over the deep open ocean continues to sink and never reaches the freezing point. Thus the ocean surface remains ice-free, except in shallow marginal seas like the Gulf of St. Lawrence off eastern Canada. By contrast, the water in freshwater lakes and ponds may cool to 4°C (its temperature of maximum density) from top to bottom. With additional cooling, the surface layer in lakes becomes lighter and floats, and vertical circulation ceases. Then the surface water can continue to cool to the freezing point, with ice formation closing the surface relatively easily.

QUESTION 2.7 How do you think the relatively low salinity (< 34.5) of Arctic surface water would affect freezing? Refer to Figure 2.16.

When dense water sinks from the surface, water moves horizontally into the region where sinking is occurring, and elsewhere water rises to complete the cycle. The horizontal or vertical movement of water is referred to generally as **advection**. Because water is a fixed quantity in the oceans, it cannot be accumulated or removed at given locations without movement of water between these regions. The sinking of water is called **downwelling**; upward movements of water are called **upwelling**. Downwelling transports oxygen-rich surface water to depth; upwelling returns essential nutrients (e.g. nitrate, phosphate) to the euphotic zone where they can be utilized by plants to produce organic materials. Because upwelling is so important for marine productivity, it is discussed in more detail in Section 3.5.

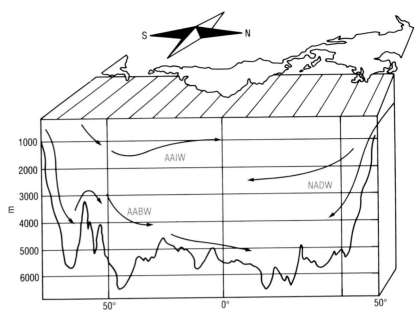

Figure 2.17   A cross-section of the Atlantic Ocean showing the formation and movement of intermediate and deep-water masses from polar regions. AABW, Antarctic Bottom Water; AAIW, Antarctic Intermediate Water; and NADW, North Atlantic Deep Water.

2.5   PRESSURE

Hydrostatic pressure is another physical environmental factor affecting life in the sea. Pressure is determined by the weight of the overlying water column per unit area at a particular depth. For the purposes of this book, the relationship between pressure and depth is considered to be effectively linear although, in fact, pressure is also influenced by density which increases with depth.

Figure 2.18   The relationship between pressure and depth. Both scales are logarithmic simply to accommodate the range of numbers.

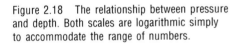
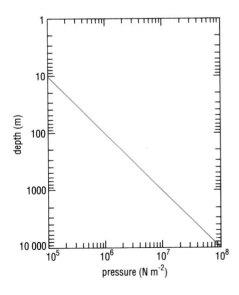

Pressure can be measured in a variety of units. Figure 2.18 expresses pressure in newtons per square metre. At 10 m depth, the pressure is 10^5 N m^{-2}; this is roughly equivalent to 1 atm or 1 bar, which are the units conventionally used by biologists (and scuba divers). It is easiest to remember that, with increasing depth, pressure increases by 1 decibar every metre (= 10^4 N m^{-2}), or by 1 atm every 10 m.

No matter which units of pressure are used, it is apparent that many marine organisms inhabiting deep waters are subject to very high pressures. In the deepest ocean basins, organisms exist at pressures exceeding 1000 atm. There are also marine animals that daily migrate vertically over distances of several hundred metres and thus experience considerable pressure changes.

Because of the difficulties in collecting deep-sea species under pressure and maintaining them at their *in situ* pressures in the laboratory, the biological effects of pressure remain somewhat uncertain. It has also been difficult to separate the effects of hydrostatic pressure on the metabolism of deep-sea animals from those effects due to living in a low temperature and dark environment. It is known, however, that liquids can only be slightly compressed under high pressure, but that gases are highly compressible. This means that animals with gas-filled structures, like the swim bladders of some fish, may be markedly affected by pressure change, whereas animals lacking these structures may be more tolerant of depth change. Air-breathing marine mammals, with gas-filled lungs, have evolved a variety of special anatomical and physiological adaptations that permit them to make deep dives. More generally, experiments have shown that many planktonic organisms are sensitive to pressure and will respond to laboratory-controlled pressure changes by swimming upward or downward. Animals living permanently in the deep sea do not have gas-filled organs and may have special biochemical adaptations to living under high pressures.

Certain animals (both benthic and pelagic species) do inhabit wide depth ranges in the sea and these are considered to be **eurybathic**. Other species are intolerant of pressure change and remain restricted to narrow depth ranges; these are the **stenobathic** species. Indeed, some stenobathic forms are restricted to deep areas and seem to require high pressures for normal development.

2.6 SURFACE CURRENTS

The major surface currents in the ocean (Figure 2.19) are primarily wind-driven and thus closely related to the major wind systems. However, the eastward rotation of the Earth modifies the direction of water movement by deflecting currents to the right in the Northern Hemisphere, producing a tendency to clockwise circulation patterns. In the Southern Hemisphere, the deflection is to the left, and major currents move counterclockwise.

Figure 2.19 illustrates the major clockwise gyres in both the North Atlantic and North Pacific oceans. North of the Equator, persistent north-east trade winds force water westward to form a North Equatorial Current in both oceans. When this water reaches continental land masses, it turns northward as the Gulf Stream in the Atlantic and as the Kuroshio in the Pacific. At about 40°N, dominant westerly winds assist returning eastward-flowing currents. The circuits are completed by water flowing southward to form the Canaries Current in the Atlantic, and the California Current in the Pacific.

The oceanic gyres flow counterclockwise in the Southern Hemisphere, forming mirror images of their northern counterparts. South-east trade winds generate westward-flowing South Equatorial Currents, most of which are deflected southward (left) along eastern South America (Brazil Current) and Australia (East Australia Current). Water flows northward along western Africa in the Atlantic and along Chile and Peru in the Pacific before rejoining the South Equatorial Currents.

In all of these gyres, currents are narrower, deeper, and faster along the western edges of the oceans compared with those along the eastern margins. For example, the Gulf Stream and Kuroshio are western boundary currents flowing at velocities up to 200 cm s^{-1}, roughly ten times faster than eastern boundary currents (< 20 cm s^{-1} for the Canaries or California currents). As these large volumes of water circulate, they mix with other water bodies and their characteristics change gradually. Note, for example, the joining and

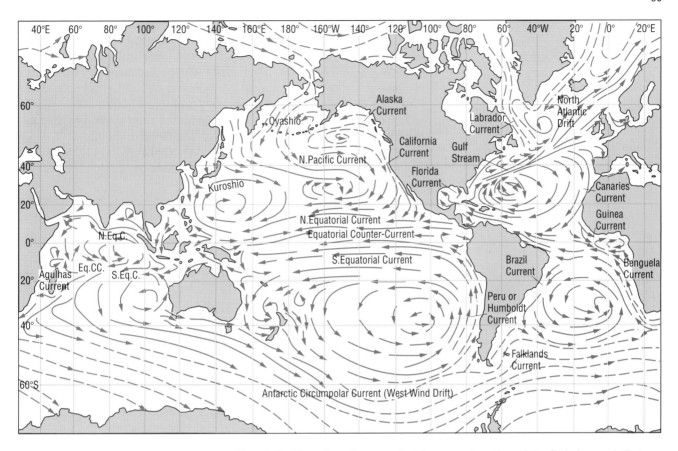

Figure 2.19   The major surface currents of the oceans in northern winter. Dashed arrows indicate cool currents; solid arrows show warm currents.

mixture of cold, low-salinity Laborator Current water with the warm, high salinity Gulf Stream (Figure 2.19 and Colour Plate 7).

QUESTION 2.8   Figure 2.19 shows one eastward-flowing surface current that is unimpeded by Land barriers and makes a complete circuit around the world, connecting the Atlantic, Pacific, and Indian oceans. What is the name of this current and where is it located?

2.6.1   BIOLOGICAL SIGNIFICANCE OF CURRENTS

The issue of ocean circulation is the subject of physical oceanography, and the description of generating forces for surface currents given here is far from complete. However, the basic pattern of circulation is presented here because water movements greatly influence biological productivity. As currents meet and mix, or meet continental land masses or major rivers, or move over shallower depths, various types of vertical circulation patterns are generated that affect the distribution of nutrients available to phytoplankton. Section 3.5 explores the mechanisms that produce these geographical differences in production. The patterns of ocean currents also influence the geographical distributions of both pelagic and benthic marine species.

The oceans form a dynamic, fluid environment moving over the surface of the Earth, and this creates one of the most difficult problems for biological oceanographers because it is impossible to follow the same population or community of pelagic organisms for any appreciable time period. A body of water sampled at a specific locality will not be the same body of water 1 hour later at the same position. Even if a patch of surface water is marked by a floating buoy or drogue, the resident animals in the underlying water column will change as they vary their depth positions. By doing so, they enter water moving at different speeds and in different directions relative to the surface. This is why many biological processes that require sampling over longer time spans, such as growth rates of zooplankton, are measured in the laboratory using captive animals, or are inferred using indirect techniques.

2.7 SUMMARY OF CHAPTER 2

1   The amount of sunlight arriving at the sea surface varies with time of day, season, and weather. Approximately 50% of the solar radiation penetrating the sea surface is within the visible spectrum (about 400–700 nm), and these are approximately the same wavelengths used in plant photosynthesis. The intensity of photosynthetically active radiation (PAR) at the sea surface ranges from zero (in darkness) to about $2000\ \mu E\ m^{-2}\ s^{-1}$ with the Sun directly overhead.

2   Different wavelengths of light are absorbed and scattered at different depths in water, and they have different extinction coefficients, with red light being attenuated most rapidly and blue light penetrating deepest in clear water. The depth to which any wavelength penetrates depends partly on the amount of suspended particles and chlorophyll in the water.

3   Three ecological zones have been defined, based on the penetration of light in seawater. The euphotic zone is that region where light is sufficient for the growth of plants, and it extends from the surface to a maximum of about 150 m in the clearest oceanic water. The lower boundary is defined by the compensation light depth, where only enough light is present for photosynthesis to balance plant respiration over 24 hours. The disphotic zone is dimly lighted; there is sufficient light for vision, but too little for plant production. The deepest and largest zone is the aphotic zone, a region of darkness extending to the seafloor where the only light emanates from the bioluminescence of certain animals.

4   Infrared wavelengths are absorbed within the first few metres of the sea surface and are the primary heat source of the oceans. Sea surface temperatures vary with latitude and fluctuate seasonally but remain within a moderate range of about 40°C to −1.9°C, the freezing point of water with a salinity of 35.

5   In many parts of the ocean, there is thermal stratification consisting of an upper mixed layer of water of almost homogeneous temperature; a region of rapid temperature decrease known as the permanent thermocline; and an underlying cold deep layer of water formed originally at the surface in polar regions.

6   In mid-latitudes where seasons are pronounced, seasonal thermoclines are formed in the surface layer during spring and summer. These zones of steep temperature change are established because increased solar radiation elevates surface temperatures at a time when lessened winds reduce the amount of mixing in the water.

7   The average salinity of the open ocean is about 35 parts per thousand by weight, with ten major ions making up about 99.99% of all the dissolved substances in the oceans. In inshore or isolated areas with little water exchange, salinity may vary from about 5 to 25 in brackish waters, to more than 40 in such hypersaline areas as the Red Sea and some shallow lagoons. Variations in salinity are primarily caused by evaporation (which elevates salinity) and precipitation (which decreases salinity).

8   Whereas total salinity is variable, the major dissolved ions are not significantly affected by biological or chemical reactions and the relative proportions of these dissolved constituents remain constant.

9   The combined properties of salinity and temperature are used to define water masses. Each of these large bodies of water has a discrete origin and forms a distinctive environment, supporting a distinctive community of pelagic organisms.

10   Salinity, temperature, and pressure establish the density of seawater. Changes at the sea surface that result in higher density will lead to downwelling of that water. Very dense water formed at high latitudes sinks to form the bottom water masses of the oceans, and this process is important in maintaining oxygen levels at all depths. Upwelling of water is partly caused by wind-driven mixing and is of importance in returning biologically essential elements to surface waters, where they are used by plants in photosynthesis.

11   The salt content of the sea lowers the temperature of maximum density and depresses the freezing point of seawater relative to freshwater. This not only results in the winter downwelling of polar water (see 10 above), but it also prevents sea-ice formation except in polar areas and in shallow high-latitude marginal seas.

12   Oceanic surface currents are generated by global wind systems, and their direction is modified by the Earth's rotation. This results in large clockwise-moving gyres in the northern oceans and anticlockwise gyres in the Southern Hemisphere. The patterns of movement and mixing of these currents produces geographic regions of differing biological productivity. Horizontal transport of water also establishes the geographic distribution of many marine species.

13   Hydrostatic pressure effectively increases linearly with depth, at a rate of 0.1 atm m^{-1}. In the deepest areas, organisms live at pressures exceeding 1000 atm.

Now try the following questions to consolidate your understanding of this Chapter.

QUESTION 2.9   In Section 2.1.2, compensation light intensities (I_C) for phytoplankton are given as ranging between 1 and 10 μE m^{-2} s^{-1}. What are these values in watts m^{-2}? (Refer to Appendix 2 for conversion factors.)

QUESTION 2.10   The majority of marine animals (both pelagic and benthic) are poikilothermic, whereas many land animals (birds, mammals) are homoiothermic species. Can you think of a reason to explain this difference?

QUESTION 2.11 Refer to the global ranges of surface salinity shown in Figure 2.11. (a) Explain the low salinity value (34.5) at the Equator. (b) Why is salinity higher in surface waters of the Antarctic (ca. 57° S latitude) compared to the Arctic (57° N)?

QUESTION 2.12 Refer to Figure 2.14. Which combination of high or low temperature and high or low salinity would produce water of greatest density?

QUESTION 2.13 Review what you have learned about abiotic environmental factors in this Chapter and describe the deep-sea environment below 2000 m in terms of light, salinity, temperature, pressure, and relative density.

The great majority of the plants in the ocean are various types of planktonic, unicellular algae, collectively called phytoplankton. Although some phytoplankton are large enough to be collected in fine-mesh nets, many of these microscopic plants can only be collected by filtering or centrifuging sizable volumes of seawater. There are also macroscopic floating algae in some oceanic areas, *Sargassum* in the Sargasso Sea being a well known example, but they are relatively restricted in locality. Similarly, the benthic species of algae, including attached macroscopic seaweeds, are limited in distribution to coastal, shallow areas because of the rapid attenuation of light with depth. In contrast, phytoplankton are present throughout the lighted regions of all seas, including under ice in polar areas. Because the phytoplankton are the dominant plants in the ocean, their role in the marine food chain is of paramount importance.

Table 3.1 A taxonomic survey of the marine phytoplankton.

Class	Common name	Area(s) of predominance	Common genera
Cyanophyceae (Cyanobacteria)	Blue-green algae (or blue-green bacteria)	Tropical	*Oscillatoria* *Synechococcus*
Rhodophyceae	Red algae	Cold temperate	*Rhodella*
Cryptophyceae	Cryptomonads	Coastal	*Cryptomonas*
Chrysophyceae	Chrysomonads	Coastal	*Aureococcus*
	Silicoflagellates	Cold waters	*Dictyocha*
Bacillariophyceae (Diatomophyceae)	Diatoms	All waters, esp. coastal	*Coscinodiscus* *Chaetoceros* *Rhizosolenia*
Raphidophyceae	Chloromonads	Brackish	*Heterosigma*
Xanthophyceae	Yellow-green algae	—	Very rare
Eustigmatophyceae	—	Estuarine	Very rare
Prymnesiophyceae	Coccolithophorids Prymnesiomonads	Oceanic Coastal	*Emiliania* *Isochrysis* *Prymnesium*
Euglenophyceae	Euglenoids	Coastal	*Eutreptiella*
Prasinophyceae	Prasinomonads	All waters	*Tetrasalmis* *Micromonas*
Chlorophyceae	Green algae	Coastal	Rare
Pyrrophyceae (Dinophyceae)	Dinoflagellates	All waters, esp. warm	*Ceratium* *Gonyaulax* *Protoperidinium*

This table is included for completeness and information, but it is not necessary to remember all details. It is important to note the diversity shown among the phytoplankton.

3.1 SYSTEMATIC TREATMENT

Approximately 4000 species of marine phytoplankton have been described, and new species are continually being added to this total. A taxonomic list of the major types of phytoplankton is given in Table 3.1, but only the better known groups are considered in some detail below.

3.1.1 DIATOMS

Diatoms (Figure 3.1; Colour Plates 1 and 2) belong to a class of algae called the Bacillariophyceae. They are among the best studied of the planktonic algae and are often the dominant phytoplankton in temperate and high latitudes. Diatoms are unicellular, with cell size ranging from about 2 μm to over 1000 μm, and some species form larger chains or other forms of aggregates in which individual cells are held together by mucilaginous threads or spines. All species have an external skeleton, or frustule, made of silica and fundamentally composed of two valves. Silica in the skeleton

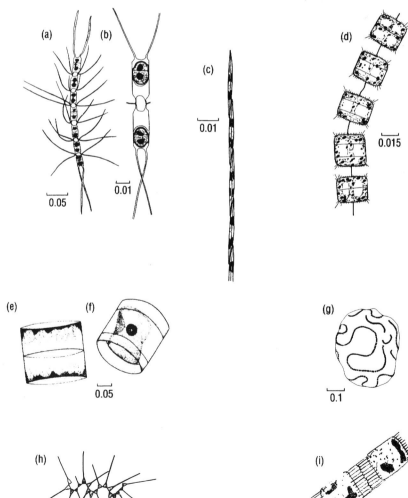

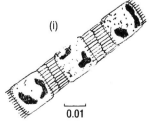

Figure 3.1 Diatoms. (a) a typical chain of *Chaetoceros laciniosus*; (b) *C. laciniosus* chain with resting spores; (c) *Nitzschia pungens* chain of dividing cells; (d) *Thalassiosira gravida* chain; (e) *Coscinodiscus* showing the two valves of the frustule; (f) *Coscinodiscus wailesii*, lateral view; (g) *Chaetoceros socialis* chains in a gelatinous colony formation; (h) a chain of *Asterionella japonica*; and (i) *Skeletonema costatum*. (scales in mm)

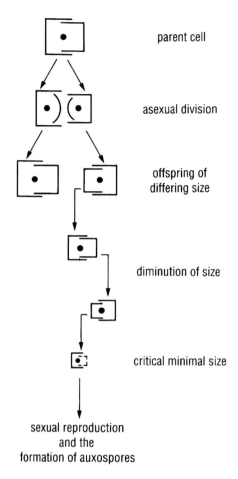

parent cell

asexual division

offspring of
differing size

diminution of size

critical minimal size

sexual reproduction
and the
formation of auxospores

Figure 3.2   The life cycle of diatoms.

makes up 4–50% of the dry weight of the cell. The frustule is usually sculptured into patterns of spines, pores, channels, and/or ribs which are distinctive to individual species. Diatoms have been abundant in the seas since the Cretaceous (about 100 million years ago) and, over geological time, sedimented frustules have formed seafloor deposits called **diatomaceous ooze**.

Two types of diatoms are recognized: the pennate and centric forms. Pennate diatoms have elongate shapes and are mostly benthic, but the few planktonic genera such as *Nitzschia* (Figure 3.1c) may be abundant in some regions. Centric diatoms have valves that are arranged radially or concentrically around a point, and they are much more common in the plankton, with somewhat over 1000 species. *Chaetoceros*, *Coscinodiscus*, *Skeletonema*, and *Thalassiosira* are all common centric genera, some of which are illustrated in Figure 3.1.

Planktonic diatoms do not have any locomotor structures and are usually incapable of independent movement. Because it is essential for diatoms and other phytoplankton to remain in lighted surface waters in order to carry out photosynthesis, these algae exhibit a variety of mechanisms which retard sinking. These include their small size and general morphology, as the ratio of cell surface area to volume determines frictional drag in the water. Colony or chain formation also increases surface area and slows sinking. Most species carry out ionic regulation, in which the internal concentration of ions is reduced relative to their concentration in seawater. Diatoms also produce and store oil, and this metabolic by-product further reduces cell density. In experimental conditions, living cells tend to sink at rates ranging from 0 to 30 m day^{-1}, but dead cells may sink more than twice as fast. In nature, turbulence of surface waters is also important in maintaining phytoplankton near the surface where they receive abundant sunlight.

The usual method of reproduction in diatoms is by a simple asexual division in which the cell forms two nuclei, the two halves of the frustule separate, and each resulting daughter cell grows a new inner valve of the frustule (Figure 3.2). This can result in the two new cells being of slightly unequal size, the one receiving the inner half of the original frustule being slightly smaller than the cell formed from the outer valve. Asexual division can lead to very rapid population growth under optimal conditions. However, with repeated divisions, there may be a diminution in size of some of the progeny.

When a diatom reaches a certain critical minimal size, it undergoes sexual reproduction by forming a cell that lacks a siliceous skeleton and contains only half of the genetic material. Such cells fuse to form a zygote, and this swells to produce an **auxospore**. Subsequently, a larger cell is formed that ultimately produces a frustule of the normal shape and size. Sexual reproduction in diatoms does not necessarily require a reduction in size of the cell, however.

Some diatoms, particularly neritic species living in relatively shallow water, produce **resting spores** (Figure 3.1b) under adverse environmental conditions. These form when the protoplasm of a normal cell becomes concentrated and surrounded by a hard shell. This heavy spore sinks to the bottom and remains dormant until favourable conditions are restored, in which case it is capable of becoming a normal planktonic cell.

Motile = capable of motion.

3.1.2 DINOFLAGELLATES

The second most abundant phytoplankton group following the diatoms is composed of algae belonging to the Pyrrophyceae, and commonly referred to as **dinoflagellates** (Figure 3.3; Colour Plate 3). Most of these unicellular algae exist singly; only a few species form chains. Unlike the diatoms, dinoflagellates possess two flagella, or whiplike appendages, and are therefore motile.

Different species of dinoflagellates utilize different energy sources. Only some dinoflagellates are strictly **autotrophic**, building organic materials and obtaining all their energy from photosynthesis. Other species carry out **heterotrophic production**; that is, they meet their energy needs by feeding on phytoplankton and small zooplankton. Indeed, about 50% of the dinoflagellates are strict heterotrophs that lack chloroplasts and are incapable

Figure 3.3   Dinoflagellates. (a) Two views of *Prorocentrum marinum*; (b) *Prorocentrum micans*; (c) *P. micans* dividing; (d) *Protoperidinium crassipes*; (e) *Gymnodinium abbreviatum*; (f) *Dinophysis acuta*, and (g) *Gonyaulax fragilis*. (All scale bars represent 0.02 mm.)

DESMOPHYCEAE

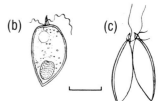

DINOPHYCEAE

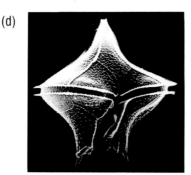

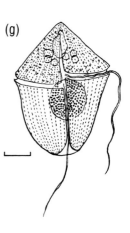

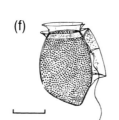

of carrying out photosynthesis; these species form part of the zooplankton and are considered in Section 4.2. Some dinoflagellates are **mixotrophic**, and are capable of both autotrophic and heterotrophic production (see also Section 4.2). Still other dinoflagellate species are parasitic or symbiotic (e.g. see Section 8.6.4). There are an estimated 1500 to 1800 species of free-living, planktonic dinoflagellates.

Conventionally, dinoflagellates are divided into thecate species, which have a relatively thick cellulose cell wall called a theca, and naked forms which lack this structure. Taxonomically, the dinoflagellates are separated into the Desmophyceae and the Dinophyceae. The former is a small group in which the species are characterized by having both flagella arising from the anterior end of the cell (Figure 3.3a, b). The cell wall is composed of two longitudinal valves that separate during asexual division to form two new cells of equal size (Figure 3.3c). *Prorocentrum* is a common planktonic genus belonging to the Desmophyceae.

The majority of planktonic dinoflagellate species form the Dinophyceae (Figure 3.3d–g; Colour Plate 3), and the majority of these are thecate. In all of them, the cell is divided into an anterior and posterior half by a transverse groove known as a girdle. The flagella are so arranged that one extends posteriorly from the cell, and the other wraps transversely around the cell in the girdle region. In those species with a theca, the cell wall is divided into a number of separate cellulose plates that are ornamented with pores and/or small spines. Common thecate genera include *Ceratium*, *Protoperidinium*, *Gonyaulax*, and *Dinophysis*. *Gymnodinium* is a common naked form belonging to the Dinophyceae.

Reproduction in dinoflagellates is normally by simple asexual division, with the cell dividing obliquely to form two cells of equal size. The theca may divide, with each new cell forming a new half, or the theca may be lost before division, in which case each new cell forms an entirely new cell wall. Asexual division can lead to rapid population development when conditions favour these algae. Dinoflagellates often become abundant in summer or autumn, following blooms of diatoms, as they are better adapted at living under lower light conditions and in nutrient-impoverished water. This is partly because dinoflagellates are capable of moving vertically in the water column; during the day they can carry out photosynthesis in sunlit surface waters that have been stripped of nutrients by fast-growing algae, and at night they may move deeper to take advantage of higher nutrient concentrations. For the same reason, dinoflagellates are usually the most numerous of the phytoplankton in stratified, nutrient-poor tropical and subtropical waters (see Section 2.2.2 and Figure 3.9).

Sexual reproduction also occurs in at least some species of dinoflagellates. This may lead to the formation of thick-walled, dormant **cysts** that settle on the seafloor, where they can survive for years. When triggered by environmental change, the cysts germinate to produce swimming cells.

Phytoplankton blooms develop when a species suddenly increases greatly in numbers under favourable conditions. In some circumstances, the rapid reproduction of dinoflagellates results in such high densities of organisms that their reddish-brown pigment visibly colours the water, producing so-called **red tides** (Colour Plate 4) (see also Section 3.1.3, Cyanophyceae). This red water may be caused by very high concentrations of innocuous species of dinoflagellates, or of species that contain potent toxins (see below). In any case, red tides begin with a sudden increase in numbers of the dinoflagellate. The water becomes noticeably coloured when concentrations reach about 200 000 to 500 000 cells l^{-1} and, as the bloom develops, concentrations may exceed 10^8 cells l^{-1}. When essential nutrients are exhausted by the dinoflagellates and the bloom decays, the bacterial decomposition of large amounts of organic material depletes the available oxygen and fish may die as a result of the lowered oxygen concentrations. The development of anoxic conditions is not exclusively a property of dinoflagellate blooms; such conditions can also occur following large blooms of other types of phytoplankton.

Some red tides are caused by certain species of *Alexandrium*, *Pyrodinium*, and *Gymnodinium* which produce a variety of neurotoxins collectively referred to as **saxitoxin**, which is 50 times more lethal than strychnine and 10 000 times more deadly than cyanide. Even when present in concentrations too low to colour the water, these dinoflagellates can be poisonous to certain animals and to humans. While the dinoflagellates are growing and reproducing, they build up saxitoxin in their cells and some of this is released into the water. The toxic dinoflagellates are also ingested by some zooplankton and by filter-feeding shellfish like clams, mussels, scallops, and oysters. Zooplankton and shellfish accumulate and concentrate saxitoxin in their own tissues, where it may be retained for considerable periods without harmful effects. However, vertebrates, such as fish, are sensitive to saxitoxin and may die from eating contaminated zooplankton. In serious outbreaks, seabirds and even dolphins and whales may also perish by accumulating saxitoxin from their food.

The minimum lethal dose of saxitoxin for humans is 7 to 16 μg kg^{-1} of body weight, and eating a single contaminated clam may be enough to cause death from **paralytic shellfish poisoning** (or PSP). Saxitoxin is heat stable, so cooking of the shellfish does not destroy the potency of this neurotoxin. In North America within historic medical times, about 1000 cases of shellfish poisoning have been recorded, with about one-quarter of these resulting in death. One of the earliest recorded cases occurred off the west coast of Canada on 15 June 1793, when one death and four illnesses resulted from crew on Captain George Vancouver's ship eating toxic mussels. In 1799, 100 men on a Russian expedition off Alaska died from eating mussels. Paralytic shellfish poisoning remains a problem on both coasts of North America, in Central America, and the Philippines; it also occurs in Europe, Australia, South Africa, and Japan. In 1987, three human fatalities and 105 cases of acute poisoning were reported in eastern Canada as the result of consuming toxic mussels. In this case, the neurotoxic compound was identified as **domoic acid** that originated in a diatom, *Pseudonitzschia*, which had previously been considered harmless. Developed countries typically have monitoring programmes that permit the closure of contaminated shellfish beds (natural or cultivated) when toxins are detected in the water or in shellfish tissues; the incidence of sickness and fatalities from algal-derived shellfish poisoning is higher in coastal developing countries.

A related health problem, **ciguatera fish poisoning** (or CFP) is found in tropical and subtropical countries, where certain species of toxic dinoflagellates live attached to seaweeds. Fish that feed on seaweeds also ingest the dinoflagellates and accumulate toxin in their tissues, and this is passed on through the food web to carnivorous fish, and eventually to humans who consume contaminated fish. Symptoms range from headache and nausea in mild cases to convulsions, paralysis, and even death in severe cases. It is estimated that CFP causes more human illness than any other kind of toxicity originating in seafood, with 10 000 to 50 000 individuals being affected each year.

3.1.3 OTHER PHYTOPLANKTON

Coccolithophorids (Colour Plate 5) are unicellular phytoplankton that form part of the nanoplankton (refer to Figure 1.2), with most of the 150 or so species being smaller than 20 μm. Their outstanding characteristic is an external shell composed of a large number of calcareous plates called coccoliths. The shape and arrangement of the coccoliths can be used to identify species. The coccoliths accumulate in bottom sediments, and they are the major constituent of the uplifted sediments known as chalk, which forms the famous White Cliffs of Dover. Like the dinoflagellates, coccolithophorids possess two flagella, although they may have a life cycle which includes an alternation with a non-motile stage lacking flagella. Although coccolithophorids can be found in neritic as well as in oceanic waters (see Figure 1.1), and at times are near the surface, the majority of species occur in warmer seas and thrive in reduced light intensities; some species reach maximum abundance at depths of about 100 m in clear, tropical, oceanic water. However, *Emiliania huxleyi* is probably the most widespread coccolithophorid in the sea, and it is present in all oceans except the polar seas. *Emiliania* sometimes forms enormous blooms, one having been measured to cover approximately 1000 km by 500 km of sea surface in the North Atlantic Ocean — or an area roughly the size of Great Britain. Reproduction in coccolithophorids is by longitudinal division, with the shell being divided and afterwards reformed into a whole by each new cell. However, life histories in this group are complex and may involve several different types of stages.

Allied with coccolithophorids in the algal group Prymnesiophyceae are several other important phytoplankton which lack coccoliths and are superficially very different in appearance. These include such unicellular and motile genera as *Isochrysis*, a small alga commonly cultured in the laboratory, and *Phaeocystis*, which forms large gelatinous colonies that can foul fish nets and also beaches when washed ashore. *Prymnesium* is characteristic of low salinity water and can be a major cause of mortality in farmed salmon along the Norwegian coast because it interferes with gas exchange across the gills of the fish.

The best known marine forms of the Chrysophyceae, or golden-brown algae, are the **silicoflagellates** with an internal skeleton formed of siliceous spicules (Colour Plate 6). These uniflagellate organisms are small (10–250 μm) and contain very numerous yellow-brown chloroplasts. Only a few species of silicoflagellates are known, and these are usually most abundant in colder waters.

Numerous species of small, naked, flagellated phytoplankton also make up other taxonomic divisions (Table 3.1). Some of these species are truly rare, but many remain poorly known because of the difficulties in collecting and preserving very small cells (including picoplankton of 0.2–2 μm) which do not have rigid skeletal structures. Some flagellates that survive collection will disintegrate during filtration or when placed in preservatives. Nevertheless, some of these minute phytoplankton can be very abundant and important in ecological cycles.

Some of the smallest, and also some of the largest, species of phytoplankton belong to the **Cyanophyceae** or **Cyanobacteria** (also known as **blue-green algae**, or **blue-green bacteria**). A single genus, *Oscillatoria* (formerly called *Trichodesmium*) is well known, and is important in the tropical open ocean. At times this alga exists in single long filaments formed by chains of cells; at other times, the filaments clump together to form macroscopic bundles of several millimetres in diameter. Interest has been directed toward this genus as the species are capable of utilizing and fixing dissolved gaseous nitrogen (N_2), unlike other phytoplankton which can only utilize combined forms of nitrogen such as nitrate, nitrite, and ammonia. The attribute of nitrogen fixation may explain the relative success of *Oscillatoria* in tropical waters which typically have low concentrations of the nitrogen sources normally utilized by other algae. Nitrogen fixation does not seem to be a physiological feature of *Synechococcus*, another genus of marine cyanobacteria. *Synechococcus* is of picoplankton size (refer to Figure 1.2); it occurs abundantly in the euphotic zone of both coastal and oceanic waters of temperate and tropical oceans. Concentrations of *Synechococcus* may reach up to 10^6 cells ml^{-1} and, at such high concentrations and in the absence of larger phytoplankton, this single genus can play a major role in the primary productivity of the sea. Recently scientists have discovered even smaller (0.6–0.8 μm diameter) photosynthetic organisms called **prochlorophytes**, which are closely related to the cyanobacteria and occur in both coastal and oceanic waters. Although few ecological studies have been made of these organisms, the genus *Prochlorococcus* apparently contributes to a significant fraction of the total primary production in the oceanic equatorial Pacific.

QUESTION 3.1 Assuming a spherical shape, how many *Synechococcus* cells of 1 μm diameter are equivalent in volume to a single dinoflagellate cell of 50 μm diameter?

3.2 PHOTOSYNTHESIS AND PRIMARY PRODUCTION

Phytoplankton are the dominant **primary producers** of the pelagic realm converting inorganic materials (e.g. nitrate, phosphate) into new organic compounds (e.g. lipids, proteins) by the process of **photosynthesis** and thereby starting the marine food chain. The amount of plant tissue build up by photosynthesis over time is generally referred to as **primary production**, so called because photosynthetic production is the basis of most of marine production. As we will see later in Sections 5.5 and 8.9, there are other types of primary production that are carried out by bacteria capable of building organic materials through chemosynthetic mechanisms, but these are of minor importance in the oceans as a whole.

Although a number of steps are involved, the chemical reactions for photosynthesis can be very generally summarized as:

photosynthesis

(requiring sunlight)

$$6CO_2 + 6H_2O \rightleftharpoons C_6H_{12}O_6 + 6O_2$$

carbon dioxide   water   carbohydrate   oxygen

respiration

(requiring metabolic energy)

Carbon dioxide utilized by the algae can be free dissolved CO_2, or CO_2 bound as bicarbonate or carbonate ions (see also Section 5.5.2). The total carbon dioxide (all three forms) is about 90 mg CO_2 l^{-1} in oceanic waters, and this concentration is sufficiently high so that it does not limit the amount of photosynthesis by phytoplankton. This type of production, involving a reduction of carbon dioxide to produce high-energy organic substances, is also called **autotrophic production**; autotrophic organisms do not require organic materials as an energy source. Note that this process not only results in the production of plant carbohydrate, but it also produces free oxygen (which is derived from the water molecule, not from the carbon dioxide). The reverse process is **respiration**, in which there is an oxidative reaction that breaks the high-energy bonds of the carbohydrates and thus releases energy needed for metabolism. All organisms, including plants, carry out respiration. Whereas photosynthesis can proceed only during periods of daylight, respiration is carried out during both light and dark periods.

Solar energy is used to drive the process of photosynthesis, and the conversion of radiant energy to chemical energy depends upon special photosynthetic pigments that are usually contained in chloroplasts of the algae. The dominant pigment is **chlorophyll** a, but chlorophylls b, c, and d plus **accessory pigments** (carotenes, xanthophylls, and phycobilins) are also present in many species and some of these pigments can also be involved in this conversion. All of these photosynthetically active pigments absorb light of wavelengths within the range of about 400–700 nm (PAR), but each shows a different absorption spectrum. Figure 3.4a gives the absorption spectrum of chlorophyll a, the most commonly occurring pigment; maximum absorption takes place in the red (650–700 nm) and blue-violet (450 nm) range. Figure 3.4b shows the absorption spectra of several accessory pigments. It is often these accessory pigments that dominate over the green colour of chlorophyll, and therefore many phytoplankton appear to be brown, golden, or even red in colour.

QUESTION 3.2 Some planktonic (and benthic) algae contain large amounts of accessory pigments as well as chlorophyll. Refer to Figures 3.4a, b and Figure 2.4 and suggest how these pigments may be ecologically important for the algae concerned.

When chlorophyll or other photosynthetically active pigments absorb light, the electrons in the pigments molecule acquire a higher energy level. This energy in the electrons is then transferred in a series of reactions in which ADP (adenosine diphosphate) is changed to higher energy ATP (adenosine triphosphate), and a compound called nicotinamide adenine dinucleotide phosphate (or $NADPH_2$) is formed. These reactions, which are entirely dependent on light energy and involve the conversion of radiant energy to chemical energy, are called the **light reactions** of photosynthesis.

The light reactions are inextricably linked with a series of reactions that do not require light and which are referred to as the **dark reactions** of photosynthesis. They involve the reduction of CO_2 by $NADPH_2$ and require the chemical energy of ATP to produce the end products of high-energy carbohydrates (usually polysaccharides) and other organic compounds such as lipids. Additionally, the reduction of nitrate (NO_3^-) yields amino acids and proteins.

Note that in the reactions of photosynthesis, compounds are formed that contain nitrogen and phosphorus as well as the elements supplied by carbon

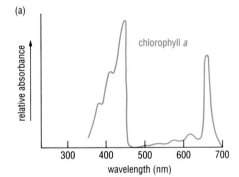

Figure 3.4(a)   The absorption spectrum of chlorophyll a.

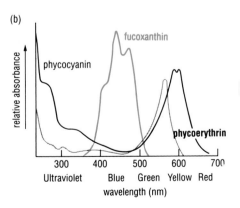

(b) The absorption spectra of the accessory pigments fucoxanthin (a xanthophyll) and phycocyanin and phycoerythrin (phycobilins).

dioxide and water. As with all plants, phytoplankton have absolute minimum requirements for these elements. Nitrogen is usually taken up by the phytoplankton cell in the form of dissolved nitrate, nitrite or ammonia; phosphorus is normally taken up in dissolved inorganic form (orthophosphate ions), or sometimes as dissolved organic phosphorus. Other elements may be required as well. Dissolved silicon, for example, is an absolute requirement for diatoms in producing the frustule. In addition, vitamins and certain trace elements may also be required, with types and amounts depending upon the species of phytoplankton. When photosynthetic species require vitamins or other organic growth factors, the production is termed **auxotrophic**. In seawater, all of the compounds referred to here are present in relatively low concentrations that vary according to the rates of photosynthesis and respiration and other biological activities, such as excretion by animals or bacterial decomposition. Therefore the concentrations of these essential elements or substances may at times become so low as to limit the amount of primary production. These considerations are discussed in Section 3.4.

3.2.1 METHODS OF MEASURING BIOMASS AND PRIMARY PRODUCTIVITY

Standing stock refers to the number of organisms per unit area or per unit volume of water at the moment of sampling. For phytoplankton, this can be measured by microscopic cell counts of preserved phytoplankton filtered from seawater samples, and the standing stock is given in number of cells per volume of water. However, because phytoplankton vary greatly in size, total numbers are not as ecologically meaningful as estimates of their biomass. **Biomass** is defined as the total weight (total numbers × average weight) of all organisms in a given area or volume. It is possible to count numbers and measure volumes of phytoplankton electronically, and this method attempts to provide an estimate of phytoplankton biomass, although cell volume may not always accurately reflect cell weight. Biomass is then expressed as the total volume (total numbers × volumes = mm^3) of phytoplankton cells per unit volume of water. The distinction between standing stock and biomass is not always made evident, however, and often the terms are used synonymously.

Another laboratory method that attempts to estimate phytoplankton biomass determines the quantity of chlorophyll a in seawater. This method is often used because chlorophyll a is universally present in all species of phytoplankton, can be easily measured, and its relative abundance enables estimates to be made of the productive capacity of the phytoplankton community. A known volume of water is filtered, and plant pigments are extracted in acetone from the organisms retained on the filter. The concentration of chlorophyll a is then estimated by placing the sample in a fluorometer to measure fluorescence, or in a spectrophotometer which measures the extinction of different wavelengths in a beam of light shining through the sample. The biomass is expressed as the amount of chlorophyll a per volume of water, or as the amount contained in the water column under a square metre of water surface.

However, the *rate* at which plant material is produced, or the **primary productivity**, is of more ecological interest than instantaneous measures of standing stock or biomass. The most popular method of measuring productivity in the sea is the ^{14}C **method**. In this method, a small measured amount of radioactive bicarbonate (HCO_3^-) is added to two bottles of seawater containing phytoplankton. One bottle is exposed to light and permits photosynthesis and respiration; the other is shielded from all light so that only respiration takes place. The amount of radioactive carbon taken up per unit time is later measured on the phytoplankton when they are filtered out of the original samples. This radioactivity is measured using a scintillation counter, and primary productivity (in mg C m^{-3} h^{-1}) is calculated from:

$$\text{rate of production} = \frac{(R_L - R_D) \times W}{R \times t}$$

(3.1)

where R is the total radioactivity added to a sample, t is the number of hours of incubation, R_L is the radioactive count in the 'light' bottle sample, and R_D is the count of the 'dark' sample. W is the total weight of all forms of carbon dioxide in the sample (in mg C m^{-3}), and this is determined independently, either by titration or from assuming a specific carbon dioxide content related to the salinity of the sample. The productivity is expressed as the amount (in mg) of carbon fixed in new organic material per volume of water (m^{-3}) per unit time (h^{-1}); it varies between zero and as much as about 80 mg C m^{-3} h^{-1}. This method is applied to water samples taken from a series of depths. In order to calculate production throughout the euphotic zone and to facilitate comparisons, the results obtained at different depths may be integrated to give production in terms of the amount of carbon fixed in the water column under a square metre of surface per day (g C m^{-2} day^{-1}). If the amount of carbon fixed per unit time is coupled with chlorophyll a measurements of biomass, one obtains a measure of growth rate in units of time (mg C per mg chlorophyll a per hour); this measure of productivity is sometimes called the **assimilation index** (see Table 3.2).

The carbon-14 method described above can be made very precise by careful experimental techniques but, at the same time, there is reason to question its accuracy. For example, the uptake in the dark bottle (R_D) is assumed to represent a blank with which to correct the uptake in the light bottle (R_L). This assumes that, except for photosynthesis, the same biological activities go on in both the light and dark bottles; but this may not be quite true. Also, any soluble organic material that is lost by the phytoplankton (a process known as **exudation**) during the period of photosynthesis will not be measured as it is not retained during filtration. Therefore, although the ^{14}C method is the most practical measurement of photosynthesis in the sea, it may sometimes lead to errors.

Other techniques have been developed for measuring chlorophyll concentration and thus relative phytoplankton abundance over large expanses of sea. A fluorometer that produces a certain wavelength of ultraviolet light will cause chlorophyll to emit a red fluorescence, and this device can then estimate the amount of chlorophyll in a volume of water. The method is very sensitive, and a fluorometer towed from a research vessel (see Figure 4.2) can rapidly record changes in chlorophyll concentration over large distances of sea surface. Remote sensing by aircraft or satellites provides even broader spatial coverage of phytoplankton abundance. This technique is based on the fact that the radiance reflected from the sea surface in the visible (or PAR) spectrum (400–700 nm) is related to the concentration of chlorophyll. Because chlorophyll is green, and water colour changes from blue to green as chlorophyll concentration increases, the relative colour differences can be used as a measure of chlorophyll concentration (see Colour Plate 8). Satellite measurements are not as sensitive as others and have restrictions of limited depth penetration, but they provide useful patterns of relative plant production on a global scale.

3.3 RADIATION AND PHOTOSYNTHESIS

The amount of light (or solar radiation) strongly affects both the amount and rate of photosynthesis. Thus the photosynthesis occurring in a water sample is proportional to the light intensity, as shown in Figure 3.5 where photosynthesis increases with increasing light intensity up to some maximal value (P_{max}). At still higher light intensities, there may be a significant decrease in photosynthesis (called **photoinhibition**) that is caused by a number of physiological reactions such as shrinkage of chloroplasts in bright light.

The point on the curve in Figure 3.5 at which the amount of respiration exactly balances the amount of photosynthesis is called the compensation point, and this occurs at a compensation light intensity (I_C) which was defined earlier (Section 2.1.2) as marking the lower boundary of the euphotic zone. The term **gross primary productivity** (P_g) is used to describe the total photosynthesis, and **net primary productivity** (P_n) denotes gross photosynthesis minus plant respiration.

The curve in Figure 3.5 can be described by mathematical equations that closely approximate two independent series of reactions, one series (shown by the initial slope $\Delta P/\Delta I$) being the light-dependent reactions of photosynthesis and the other (P_{max}) being the dark reactions, both of which were defined in Section 3.2. The simplest equations which describe the curve up to P_{max} (i.e. with no photoinhibition) are:

$$P_g = \frac{P_{max}[I]}{K_I + [I]}$$

(3.2)

Figure 3.5   The response of photosynthesis (P) to changes in light intensity (I). I_C, compensation light intensity; K_I, the half-saturation constant, or the light intensity when photosynthesis equals 1/2 of maximal photosynthesis (P_{max}); P_g, gross photosynthesis; and P_n, net photosynthesis. Absolute units not shown because all units are species specific.

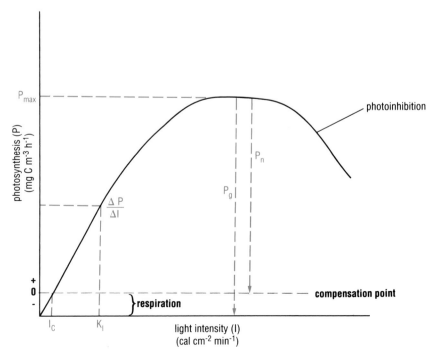

and

$$P_n = \frac{P_{\max}[I - I_C]}{K_I + [I - I_C]} \tag{3.3}$$

where P_g and P_n are gross and net productivity, respectively, as defined above; and K_I is the half-saturation constant, or the light intensity when $P = P_{\max}/2$. K_I values range from about 10 to 50 μE m^{-2} s^{-1}. [I] is the amount of ambient PAR light, and $[I - I_C]$ is the ambient PAR light less the compensation light intensity, I_C.

In the above equations, it is implied that there is a defined light response for all algae growing under constant physiological conditions and that this response can be described by two constants, $P_{\max}$ and K_I. In fact, different species have different values of $P_{\max}$ and K_I and, even within the same species, the photosynthetic response of a cell to light can change over time (e.g. over the course of a day from bright light near the surface to shade adaptation deeper in the water column). In general, the initial slope of the curve ($\Delta P/\Delta I$) in Figure 3.5 will respond to physiological changes in the photosynthetic biochemistry of a cell (i.e. the light-dependent reactions). The upper limit of the curve ($P_{\max}$) will respond to changes in environmental parameters, such as nutrient concentration and temperature, which affect the dark reactions of photosynthesis. Because different species of phytoplankton respond differently to changes in surface radiation and *in situ* light intensity, changing environmental conditions will favour different species at different times and lead to a succession of different dominant species in the community. Values for $P_{\max}$ and $\Delta P/\Delta I$ are given in Table 3.2. Note that $P_{\max}$ is generally increased at higher temperatures and under high nutrient conditions, but the initial slope of the photosynthetic curve ($\Delta P/\Delta I$) is more dependent on cellular properties; e.g. picoplankton generally have higher $\Delta P/\Delta I$ values than larger phytoplankton. Consequently, picoplankton can grow deeper in the water column where there is less light.

Table 3.2 Representative values of $P_{\max}$ and $\Delta P/\Delta I$. $\Delta P/\Delta I$ is the initial slope of the curve in Figure 3.5 and is given in terms of productivity divided by solar radiation; $P_{\max}$ is given as the maximum value of the assimilation index (see Section 3.2.1).

$P_{\max}$ (assimilation index) (mg C mg^{-1} Chl a h^{-1})	Comment
2–14	General range
2–3.5	Low temperatures, 2–4°C
6–10	High temperatures, 8–18°C
0.2–1.0	Low nutrients (e.g. in the Kuroshio current)
9–17	High nutrients and high temperatures (e.g. in tropical coastal waters)

$\Delta P/\Delta I$ (initial slope) (mg C mg^{-1} Chl a h^{-1})/(μE m^{-2} s^{-1})	Comment
0.01–0.02	Temperate ocean
0.005–0.01	Subtropical waters
0.02–0.06	Picoplankton (< 1.0 μm)
0.006–0.13	Annual range for temperate
(annual average, 0.045)	coastal waters

$$P_g \frac{P_{max}(I)}{KI + (I)}$$

$P_{max} = 2$

$I = 50$

$KI = 10$

$$\frac{2 \times 50}{10 + 50} = 1.6 \, mg \, (mg^{-1}$$
$$chl \, a \, h^{-1}$$

$P_{max} = 6$

$I = 50$

$K_o = 20$

$$\frac{6 \times 50}{10 + 50} = 4.3.$$

sample 2 grows
faster.

QUESTION 3.3 Using equation 3.2 and assuming a P_{max} value of 2 mg C mg^{-1} Chl a h^{-1} and a K_I value of 10 μE m^{-2} s^{-1} for one species of phytoplankton and respective P_{max} and K_I values of 6 mg C mg^{-1} Chl a h^{-1} and 20 μE m^{-2} s^{-1} for a second species, which species will be growing faster at a PAR light intensity of 50 μE m^{-2} s^{-1}?

In an earlier section (2.1.2), we considered how to calculate the extinction of light in water and the compensation depth of light. Equations 2.1 and 2.2 can now be extended to deal with the problem of phytoplankton being mixed vertically in the water column. When phytoplankton are being mixed up and down in the surface layers of the sea, it is useful to know the *average* amount of light ($\bar{I}_D$) in the euphotic zone. This is given by the expression:

$$\bar{I}_D = \frac{\bar{I}_0}{kD}(1 - e^{-kD}) \tag{3.4}$$

where I_0 is the surface radiation, k is the extinction coefficient, and D is the depth over which the light intensity is averaged.

A useful application of equation 3.4 is to consider how far down a population of phytoplankton cells can be mixed until photosynthetic gain is balanced by respiratory losses (i.e. where $P_w = R_w$, Figure 3.6). This depth is called the **critical depth** (D_{cr}). If equation 3.4 is rearranged and I_C, the compensation light intensity, is substituted for $\bar{I}_D$, we get the following expression to calculate the critical depth:

$$D_{cr} = \frac{I_0}{kI_C}(1 - e^{-kD_{cr}}) \tag{3.5}$$

If $kD_{cr} >> 0$, then equation 3.5 can be simplified to:

$$D_{cr} = \frac{I_0}{kI_C} \tag{3.6}$$

The importance of the critical depth is illustrated in Figure 3.6. This figure indicates that if the amount of plant material used up in respiration (in the area bounded by ABCD) is matched against the amount gained by photosynthesis (area ACE), then diagrammatically one arrives at the same depth as calculated in equation 3.6, that is, the critical depth. If phytoplankton cells are mixed downward below this depth by intensive storm action, there can be no net photosynthesis. However, as long as the depth of mixing is above the critical depth, positive net photosynthesis can occur. Thus by using a simple formula based on the amount of radiation at the surface (I_0), the extinction coefficient (k), and a known compensation light intensity (I_C), it is possible to estimate when the spring production of phytoplankton can start in temperate latitudes.

QUESTION 3.4 The surface radiation is 500 μE m^{-2} sec^{-1} of which 50% is PAR, the compensation light intensity of the phytoplankton is 10 μE m^{-2} sec^{-1}, and the depth of mixing in the water column is 100 m. Using the extinction coefficient obtained in Question 2.2, is there any net positive photosynthesis in this water column?

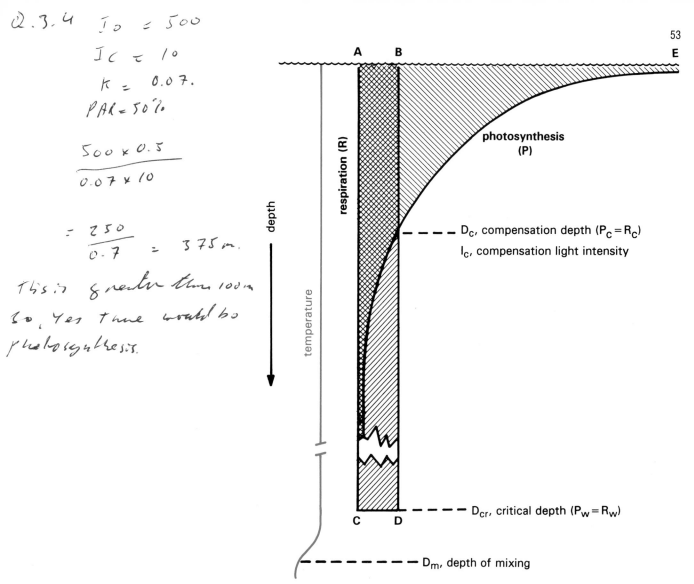

Q.3.4 $I_0 = 500$
$I_C = 10$
$k = 0.07$
$PAR = 50\%$

$$\frac{500 \times 0.5}{0.07 \times 10}$$

$$= \frac{250}{0.7} = 375 \text{ m}.$$

This is greater than 100 m so, yes there would be photosynthesis.

Figure 3.6   An illustration of the relationships among the compensation light depth, the critical depth, and the depth of mixing. At the compensation depth (D_C), the light intensity (I_C) is such that the photosynthesis of a single cell (P_C) is equal to its respiration (R_C); above this depth there is a net gain from photosynthesis ($P_C > R_C$) and below it there is a net loss ($P_C < R_C$). As phytoplankton cells are mixed above and below the compensation depth, they experience an average light intensity (I_D) in the water column. The depth at which I_D equals I_C is the critical depth (D_{cr}) where photosynthesis throughout the water column (P_w) equals phytoplankton respiration throughout the water column (R_w). The area bounded by points A, B, C and D represents phytoplankton respiration, and the area bounded by points A, C and E represents photosynthesis; these two areas are equal at the critical depth. When the critical depth is less than the depth of mixing (D_M) (as illustrated in this figure), no net production takes place because $P_w < R_w$. Net production of the phytoplankton ($P_w > R_w$) only occurs when the critical depth lies below the depth of mixing.

3.4   THE EFFECT OF NUTRIENTS ON GROWTH RATE

In Section 3.2.1, productivity was represented as the amount of carbon fixed per unit time. This is a convenient convention because it is what the ecologist actually measures. It was also pointed out that productivity can be represented by the assimilation index, in which growth is expressed as mg of carbon produced per mg of chlorophyll a per hour. This value is useful for

Area A

$$\frac{20}{2} = 10\ C\ mg^{-1}\ chl.a\ h^{-1}$$

Area B.

$$\frac{50}{25} = 2\ C\ mg^{-1}\ chl.a\ h^{-1}$$

B - less nutrient
end of a bloom.

comparing photosynthesis from different areas because it normalizes all measurements to a unit of chlorophyll *a*.

QUESTION 3.5   In comparing two different areas of the ocean, we find that the photosynthetic production is 20 mg C m^{-3} h^{-1} in area A and 50 mg C m^{-3} h^{-1} in area B. The standing stocks of phytoplankton are 2 mg Chl *a* m^{-3} and 25 mg Chl *a* m^{-3} in areas A and B, respectively. (a) In which area are the phytoplankton most photosynthetically active? (Use assimilation indices to determine the answer.) (b) What could cause this difference in activity?

Another useful way of comparing growth rates of phytoplankton is to express growth as an increase in cell numbers. For unicellular organisms, this is an exponential function:

$$(X_0 + \Delta X) = X_0 e^{\mu t} \tag{3.7}$$

where X_0 is the population of cells at the beginning of the experiment, ΔX is the number produced during time t, and μ is the growth constant of the population per unit of time. If ΔX has been measured in units of photosynthetic carbon, then X_0 must be expressed as the total standing stock of phytoplankton carbon instead of in terms of cell numbers.

One additional expression that can be obtained from equation 3.7 is the **doubling time**, which is defined as the time taken for a population to increase by 100%. Doubling times for phytoplankton can be derived from:

$$X_t = X_0 e^{\mu t} \tag{3.8}$$

where X_t is $(X_0 + \Delta X)$ in equation 3.7. The time required for X_0 to double (d) is given as,

$$\frac{X_t}{X_0} = 2 = e^{\mu d}, \tag{3.9}$$

and then doubling time (d) can be calculated from:

$$d = \frac{\log_e 2}{\mu} = \frac{0.69}{\mu} \tag{3.10}$$

The reciprocal of doubling time (or $1/d$ when d is in days) gives the **generation time** as number of generations produced per day.

The effect of nutrient concentration on the growth constant, μ, can be described by the same expression that was used for photosynthesis (equations 3.2 and 3.3). Hence,

$$\mu = \frac{\mu_{max}[N]}{K_N + [N]} \tag{3.11}$$

where μ is the growth rate (time^{-1}) at a specific nutrient concentration $[N]$ which is usually expressed in micromoles (μM) per litre, μ_{max} is the maximum growth rate of the phytoplankton, and K_N (given in μM) is a half-saturation constant for nutrient uptake that is equal to the concentration of nutrients at $1/2\ \mu_{max}$.

Equation 3.11 is valid when the growth rate of phytoplankton is controlled by the nutrient concentration in seawater. However, in some surface waters with extremely low concentrations of nutrients, some larger photosynthetic dinoflagellates (See Section 3.1.2) can migrate to deeper layers where nutrients are more abundant. The zone where nutrient concentrations increase rapidly with depth is the **nutricline**, and this may be below the euphotic zone. After taking nutrients such as nitrate into the cell, these flagellates can return to sunlit waters to carry out photosynthesis. In such cases, the (photosynthetic) growth rate of the phytoplankton is proportional to the nutrients *within* the cell, and not to the external nutrient concentration.

Among the principal nutrients in the sea that are required for phytoplankton growth, only certain elements may be in short supply. In general, the quantities of magnesium, calcium, potassium, sodium, sulphate, chloride, etc. (Table 2.1) are all in sufficient quantities for plant growth. Carbon dioxide, which may be limiting in lake waters, is present in excessive quantities in seawater. However, some essential inorganic substances, like nitrate, phosphate, silicate, iron, and manganese, may be present in concentrations that are low enough to be limiting to plant production. There may also be synergistic effects between essential nutrients. For example, the concentration of iron in a metabolizable form governs the ability of phytoplankton to utilize inorganic nitrogen. This is because iron is required in the enzymes nitrite reductase and nitrate reductase, and these enzymes are necessary for the reduction of nitrite and nitrate to ammonium, which is used to make amino acids. Large diatoms may be affected by iron limitation, but small flagellates usually are not because they can take up iron at lower concentrations. Ocean areas that are limited by iron are characterized by having high nitrate but low chlorophyll concentrations, and they are referred to as **HNLC areas**. They include the subarctic North Pacific, Equatorial Pacific, and parts of the Antarctic Ocean. In addition, certain organic substances (e.g. vitamin B_{12}, thiamine, and biotin) are required for auxotrophic growth of some phytoplankton, and these substances may also be in short supply in seawater and thus limiting to growth.

Many different species of phytoplankton can be found living in the same water mass. What factors allow the coexistence of so many species, all of which have the same basic requirements for sunlight, carbon dioxide, and nutrients, and all of which may compete for requirements that are in limited supply?

Each phytoplankton species has a specific half-saturation concentration (K_N in equation 3.11) for the uptake of each of the limiting nutrients, and each species has a different maximum growth rate (μ_{max}). These species-specific differences in growth rates and responses to nutrients allow a great variety of phytoplankton to grow in what seems to be a very uniform environment. This is illustrated in part by Figure 3.7 which shows changes in growth rates (μ) of different hypothetical species of phytoplankton having different values of K_N and μ_{max} and responding to variations in the ambient concentration of one nutrient (see equation 3.11). In the first example (a) of this figure, species 1 has a higher maximum growth rate than species 2. Because K_N is the same for both species, they grow at the same rate to a certain level of nutrient, beyond which species 1 continues to its higher maximum growth rate. In a second example (b), two different species have the same value for μ_{max}, but they achieve this at different nutrient concentrations. The value of K_N is lower for species 1, so it reaches the maximum growth rate at a lower nutrient concentration. In the last example (c), two other competing species have differing values for both μ_{max} and K_N, and the competitive advantage

(a)
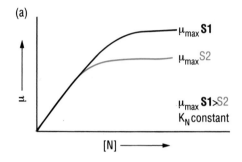

(b)
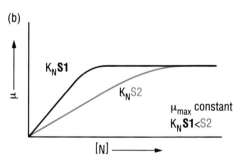

(c)
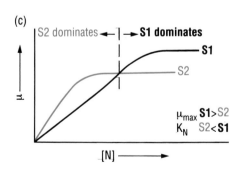

Figure 3.7 Three possible variations (a, b, and c) in the nutrient–growth curves of competing pairs (species 1 and 2) of hypothetical species of phytoplankton. μ is the specific growth rate; μ_{max} is the maximum rate of growth; K_N is a half-saturation constant for nutrient uptake; [N] is ambient nutrient concentration; S1 is species 1; and S2 is species 2. All units are arbitrary. See text for discussion of the differences in results between competing species.

shifts between the species as the nutrient concentration changes. At lower nutrient concentrations, species 2 dominates because it grows faster; but at higher nutrient concentrations, species 1 becomes dominant because it achieves a higher maximum rate of growth.

If one considers further that two, three, or many growth-rate limiting nutrients may occur in any body of water, and that there are also differences in light and other physical properties such as temperature and salinity, it is obvious that there is a constantly changing mosaic of rate-limiting factors governing the growth of phytoplankton. Since each species responds differently to the mosaic, and since growth cannot be limited by more than one process at any one time, the physical/chemical restrictions on phytoplankton growth allow for the coexistence of many species in the same body of water, with successive changes in the relative abundance of the component species.

Figure 3.8 further explores how several nutrients and several species can interact to produce the diverse phytoplankton populations discussed above. In Figure 3.8a, a single species (1) of phytoplankton is considered in relation to two potentially limiting resources (e.g. nutrients such as nitrate and phosphate). The species requires a certain minimum concentration of each nutrient (R_{I^*} and R_{II^*} values). If the concentrations drop below these levels, species 1 cannot exist even without competition from another species; above these minimal nutrient levels, species 1 can survive and grow. If a second

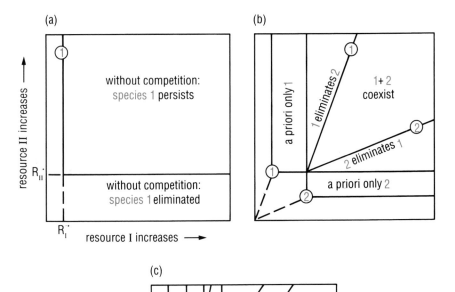

Figure 3.8   Coexistence of phytoplankton species when limited by two resources. (a) One species (1) which is limited only by the two lower concentrations of each resource. (b) Two species (1 and 2) showing the region of coexistence on each rate-limiting resource. (c) Four species showing regions of coexistence based on each resource: 1 is the best competitor for resource 1; 2 is the second best; 3, third; and 4, fourth; the competitive rank order for resource ii is reversed. A circled number indicates that only that species can exist under the given circumstances.

phytoplankton species (2) having different nutrient-concentration requirements is introduced (Figure 3.8b), the situation becomes more complex. In this hypothetical example, each species can be limited by a different nutrient concentration. Species 1 is a superior competitor for resource 1; it will be the only species capable of existing at very low levels of resource 1, and it will outcompete and exclude species 2 at slightly higher concentrations of this resource. Conversely, species 2 is the superior member and only survivor when resource 2 is in low concentrations. Above these minimal nutrient levels, there is a region where both species can coexist. As more species, each with its own nutrient requirements, are added to the community (Figure 3.8c), there are more possibilities of establishing coexistence.

With reference to the examples in Figure 3.8, if both nutrients are abundant and phytoplankton growth is not limited by any of the K_N values for nutrient uptake, then species dominance will be determined by the μ_{max} values of the species and the fastest growing phytoplankton species will dominate. In the extreme case of very high nutrient concentrations, there will be a single species, the one with the highest μ_{max}. Thus at both extremes of the nutrient field, very low and very high nutrient levels tend to lead to a low diversity of phytoplankton species. Very low nutrient concentrations can lead to dominance of the community by a single species with the lowest K_N, and very high nutrient conditions can lead to dominance by a single species having the highest μ_{max}. If the simple illustrations in Figure 3.8 are expanded to include additional resources and other species, an almost infinite combination of physical/chemical backgrounds is produced in which many phytoplankton species can grow.

The physico-chemical environmental mosaic itself is not constant. The light and temperature background changes daily and seasonally, and nutrient concentrations vary. Sometimes it is the change itself that affects different phytoplankton responses. For example, nutrient concentrations may change sporadically by pulsing inputs resulting from diel upwelling of deeper water with high nutrient levels. Such fluctuating changes in nutrient concentration will have a different effect on phytoplankton species composition than when nutrients are maintained at relatively constant levels through, for example, sustained upwelling. On the other hand, toxic pollutants will work in the opposite direction to nutrient resources; at higher concentrations, they will selectively inhibit the growth of certain phytoplankton species, so that eventually diversity is reduced to only the most pollutant-resistant forms. It must also be added that selective grazing by herbivorous zooplankton can alter the relative abundance of phytoplankton species.

Some values for growth rates (μ_{max}) and half-saturation constants (K_N) for phytoplankton are given in Table 3.3. The relative availability of nutrients for phytoplankton (particularly of nitrate and ammonium which are most often present in limiting quantities) can be used to classify aquatic environments. Regions that have low concentrations of essential nutrients, and therefore low primary productivity, are called **oligotrophic**. Such areas typically have chlorophyll concentrations ranging from <0.05 μg l^{-1} at the surface to a maximum of $0.1–0.5$ μg l^{-1} at depths of $100–150$ m. **Eutrophic** waters contain nutrients in high concentrations; high phytoplankton densities are manifested by chlorophyll concentrations of 1 to 10 μg l^{-1} in surface layers. **Mesotrophic** is a term that is sometimes applied to waters of intermediate character. Eutrophic waters tend to be dominated by one or two fast-growing, r-selected phytoplankton species (see Table 1.1). In contrast, oligotrophic waters tend to have many competing

Table 3.3 Maximum growth rates (μ_{max}) and half-saturation constants (K_N) for some phytoplankton.

Maximum growth rates (μ_{max}) (in generations day^{-1})		Comments
	0.1–0.2	Oligotrophic, tropical waters
	0.4–1.0	Temperate, eutrophic coastal waters
	1.0–3.0	Tropical upwelling; and picoplankton under eutrophic conditions and high temperatures
Half-saturation constants (K_N) (in μM)		
Nitrate or ammonium		
	0.01–0.1	Oligotrophic waters
	0.5–2.0	Eutrophic oceanic waters
	2.0–10.0	Eutrophic coastal waters
Silicate		
	0.5–5.0	General range for diatoms
Phosphate		
	0.02–0.5	General range for oligotrophic to eutrophic waters

k-selected species, each dependent on a different limiting nutrient; the community thus tends to be in equilibrium with the total nutrient supply.

QUESTION 3.6 From Table 3.3, what is the general relationship between the definitions of eutrophic and oligotrophic and the half-saturation constants (K_N) of phytoplankton cells? What does this imply in terms of relative nutrient uptake by the phytoplankton living in eutrophic or oligotrophic waters?

3.5 PHYSICAL CONTROLS OF PRIMARY PRODUCTION

Light is one of the two major physical factors controlling phytoplankton production in the sea. The second includes those physical forces which bring nutrients up from deep water, where they accumulate, into the euphotic zone. These two features together largely determine what type of phytoplankton develop and how much primary production occurs in any part of the world's ocean. They are also major factors in determining the amount and type of marine animals that are produced, including fish which are caught commercially.

The amount of light decreases from the Equator towards the poles. On the other hand, the amount of wind mixing, which brings nutrients up to the surface, increases from the tropics (where water is vertically stabilized by solar heating) toward the poles. Thus the abundance of light and the abundance of nutrients in the euphotic zone form an inverse relationship (Figure 3.9) which largely determines the pattern of phytoplankton production in different latitudes. In polar regions, a single pulse of phytoplankton abundance occurs during the summer when light becomes sufficient for a net increase in primary productivity. In temperate latitudes, primary productivity is generally maximal in the spring and again in the autumn when the combination of available light and high nutrient concentrations allows plankton blooms to occur. In the tropics, where intense surface heating produces a permanent thermocline (see Section 2.2.2), the phytoplankton are generally nutrient-limited throughout the year, and there are only small and irregular fluctuations in primary production due to local conditions.

Figure 3.9 is a general representation of the annual cycle of phytoplankton production in the world's ocean. However, there are many physical features that affect nutrient levels in the euphotic zone and thereby greatly modify the general pattern. These include **fronts**, which are relatively narrow regions characterized by large horizontal gradients in variables such as temperature, salinity, and density, and **eddy**-formations such as rings and large-scale gyres, which have characteristic rotational patterns of circulation. These modifying physical features may be thousands of kilometres wide (e.g. gyres) or only a few kilometres long (e.g. tidal and river-plume fronts). The size depends on the topography and ocean climate of any particular

Figure 3.9   The relative abundance of light (unshaded area) and nutrients (shaded area) at the sea surface and the relative seasonal change in primary productivity at three different latitudes. (Productivity expressed in arbitrary vertical scales.)

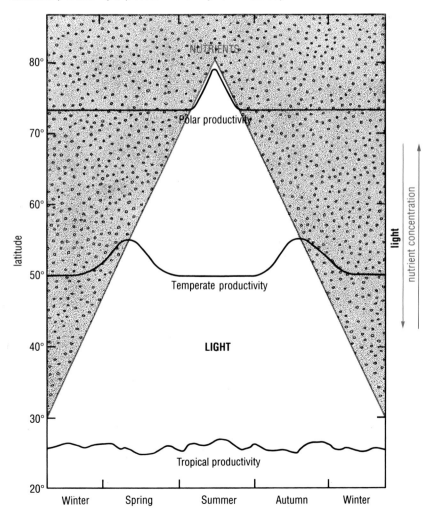

location. The common feature of all these structures is that there is some mechanism involved for bringing nutrients up to the euphotic zone from deeper water, on time scales which may range from days to months. These mechanisms are superimposed on the seasonal wind mixing that partly generates the global pattern of phytoplankton production shown schematically in Figure 3.9. Some of the nutrient-enhancing processes can result in 'oases' of plankton production during periods of the year when the production of phytoplankton would otherwise be low.

3.5.1 OCEANIC GYRES AND RINGS

The general circulation of surface water in the global ocean (discussed in Section 2.6 and shown in Figures 2.19 and 3.10) results in large gyres. In the **anticyclonic gyres**, water flows in a clockwise direction in the Northern Hemisphere, and in an anticlockwise direction in the Southern Hemisphere (see Table 3.4). In the Northern Hemisphere, the clockwise flow results in **convergent gyres** because the direction of water circulation tends to draw surface water in toward the centre. This is illustrated in Figure 3.11b where it can be seen that anticyclonic gyres in the Northern Hemisphere tend to deepen the thermocline due to the convergent tendency of the circulation. In this situation, no new nutrients can come to the surface from the deep water,

Figure 3.10   The location of upwelling zones and coral reefs in the world's ocean.

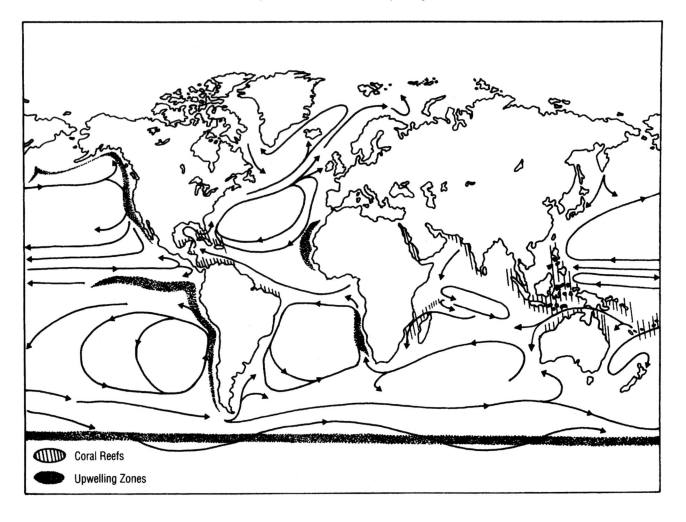

Coral Reefs

Upwelling Zones

Table 3.4 Water flow in gyres and rings in the Northern and Southern Hemispheres.

	Cyclonic gyres or **Cold core rings** divergence leading to high production	**Anticyclonic gyres** or **Warm core rings** convergence leading to low production
Northern Hemisphere	↺	↻
Southern Hemisphere	↻	↺

and convergent gyres like the Sargasso Sea in the North Atlantic are relative 'deserts' of ocean production. In the Southern Hemisphere, the reverse rotational direction of the gyres also reverses the vertical flow of water within the system, so that the anticlockwise circulation also forms convergent gyres with relatively low productivity.

Cyclonic gyres are formed by water circulating in an anticlockwise direction in the Northern Hemisphere and in a clockwise direction in the Southern Hemisphere. These are **divergent gyres**, which tend to draw water up from below the thermocline (Figure 3.11a); this results in a plentiful supply of nutrients at the surface that should make such areas highly productive. The Alaskan Gyre in the Gulf of Alaska is a divergent gyre in which the actual vertical movement of water from below the thermocline is believed to be about 10 m yr^{-1}. Although this would be a highly productive gyre if situated farther south, its location at north of 50° N means that the area is limited by light in winter, and the productivity of the gyre is actually controlled more by seasonal events than by oceanic circulation.

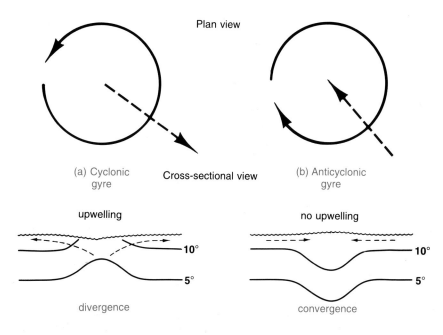

Figure 3.11   Plan and cross-sectional views of a cyclonic (a) and anticyclonic (b) gyre in the Northern Hemisphere. The dashed arrows indicate net transport of water away from and towards the centre, respectively. The same pattern of circulation applies to warm core and cold core rings, but on a smaller scale.

(a)

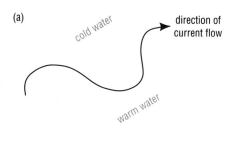

cold water

direction of
current flow

warm water

(b)

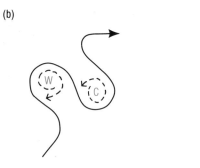

(c)

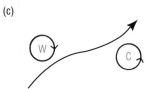

Figure 3.12 The sequential formation of
warm (W) and cold (C) core rings from a major
current system (such as the Gulf Stream) in the
Northern Hemisphere. As the current flows
between water of contrasting temperature, it
begins to develop a meandering pattern (a) with
eddies forming on the edges. As the meanders
become more pronounced (b), the eddies are
eventually pinched off to form independent
circulatory systems called rings (c). Note that
this results in warm water rings being isolated
in areas of cold water, and in cold water being
transferred across the current into an area of
predominantly warm temperatures.

Ring structure in the ocean have the same morphology as gyres, but they are
much smaller, being hundreds, rather than thousands, of kilometres in
diameter. They are formed as eddies that detach from a major current system
such as the Gulf Stream. Such large currents tend to meander and, in so
doing, large eddies or rings spin off as independent bodies of circulating
water which may survive for several years (i.e. long enough to influence the
primary productivity within the ring). The two types of rings shown in
Figures 3.12 and Colour Plate 7 are referred to as **warm core rings**
(anticyclonic) and **cold core rings** (cyclonic). A cross-section of each ring
type would look like the cyclonic and anticyclonic gyres shown in
Figure 3.11, but on a much smaller scale. The rotational circulation of the
rings maintains cooler (cyclonic circulation) or warmer (anticyclonic)
temperatures because of the respective vertical flow of water within the rings
(see Table 3.4 and Figure 3.11). However, although the isotherms in cold
core rings bow up in the middle as they do in cyclonic gyres (Figure 3.11),
this does not necessarily signify upwelling. High productivity within cold
core rings may result because the water which has been captured by the
meander is already nutrient-rich. Similarly, water in the centre of warm core
rings is not necessarily sinking.

3.5.2 CONTINENTAL CONVERGENCE AND DIVERGENCE

Very large frontal zones occur along the edges of continents due to
wind-driven oceanic circulation. Major **divergent continental fronts** are
associated with the Peru Current and California Current in the Pacific, and
with the Canaries Current and Benguela Current in the Atlantic
(Figure 2.19). Currents such as these that flow toward the Equator along the
western coasts of continents are driven away from the coasts due to the
Earth's eastward rotation, and this consequently leads to coastal upwelling
(Figure 3.10). Upwelling of nutrient-rich water in these areas continues for
many months of the year. Further, the location of these currents in latitudes
between 10° and 40° means there is generally enough solar radiation to
allow maximum photosynthesis during most of the year. These four
divergent continental fronts are among the most productive regions in the
ocean. They are characterized by having large populations of fish and birds,
and they have been the subject of much scientific investigation because of
their exploitable resources.

Another divergent continental front exhibiting upwelling and extremely high
production occurs around the continent of Antarctica. Known as the
Antarctic Divergence, this area is the home of huge stocks of krill and
other zooplankton which give rise to abundant stocks of whales, seals, and
seabirds (see Section 5.2 and Figure 5.4).

Contrary to expectations, the west coast of Australia does not support a large
fishery that would be indicative of upwelling. Although water does have a
tendency to upwell on this western coast, the upwelling is suppressed by a
continual strong flow of warm water from the north which covers the area
(Figure 3.10). A similar flow of warm water across the Pacific Ocean can
sometimes suppress the effect of the Peruvian upwelling by greatly
increasing the depth of the thermo-nutricline, an event that has become
known as an El Niño.

An opposite type of **convergent continental front** tends to form on the
eastern sides of continents, where water flows away from the Equator. These

regions are characterized by the accumulation of large quantities of warm, nutrient-poor water. They are usually areas where coral reefs occur in maximum abundance (Figure 3.10); these include the Great Barrier Reef off eastern Australia in the South Pacific, the coral reefs of Madagascar in the Indian Ocean, and the reefs of the Caribbean Sea.

3.5.3 PLANETARY FRONTAL SYSTEMS

The continental frontal systems described above are large enough to be described as planetary fronts, but they have been dealt with separately because of their very special association with continents. Other **planetary frontal systems** are formed by the convergence or divergence of two current systems which often have contrasting properties. Thus the Oyashio off the northern coast of Japan is a cold nutrient-rich current that meets the warm and vertically stable Kuroshio in the western Pacific (Figure 2.19). These two currents join to form the North Pacific Current which flows from Japan to the west coast of North America. Mixing of these waters produces a very large frontal zone that is highly productive for marine life. A similar situation occurs in the North Atlantic where the cold Labrador Current meets the warm Gulf Stream (Figures 2.19 and Colour Plate 7).

A planetary frontal system is also formed around the Antarctic continent at latitudes of about 57°–59°S; here there is a convergence of subtropical water with Antarctic water, forming the **Antarctic Polar Front** (or **Antarctic Convergence**). This convergent zone of sinking water is an important source of cold deep water for the world's ocean (see Section 2.4 and Figure 2.17).

Finally, the last of the fronts which can affect productivity on a planetary scale is the upwelling caused largely by divergent current patterns across the Equator. Equatorial upwelling is particularly pronounced at about 10°N in the Pacific Ocean, where it results in an extension of the Californian and Peruvian continental upwellings out into the Pacific Ocean. It also occurs in the Atlantic and to a lesser extent in the Indian Ocean.

3.5.4 SHELF-BREAK FRONTS

Shelf-break fronts occur along the edges of continental shelves and other banks (which are often undersea 'island' extensions of the continental shelves). A shelf-break front is formed by a combination of the sudden shallowing of water across a continental shelf, and by the change in current speed across the shelf which may be induced by residual oceanic circulation or, especially, by tidal exchange. The process by which a shelf-break front is formed can be analysed by considering the ratio (R) of the potential energy (PE) in maintaining well-mixed conditions to the rate of current energy dissipation (TED) in a water column of unit cross-sectional area:

$$R = \frac{PE}{TED} \tag{3.12}$$

The two forms of energy $(PE$ and $TED)$ can be formulated in terms of a number of parameters, most of which are constant for a defined area where the major form of stratification is a thermocline. Two important terms which are not constant are the average water velocity, $|\overline{U}|$, and water depth, h. These are considered in formulating a stratification index expressed as:

$$S = \log_{10} \frac{h}{C_D |\overline{U}|^3} \text{(in c.g.s. units)} \tag{3.13}$$

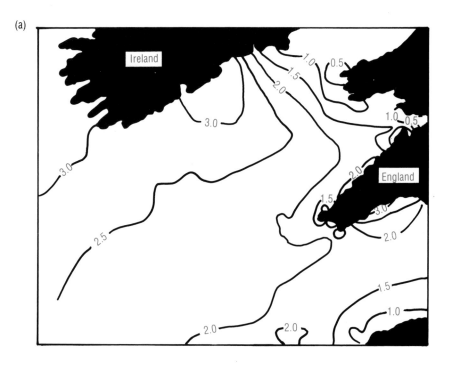

Figure 3.13   (a) Average stratification index values obtained for the Celtic Sea between Ireland and England.

(b) Surface distribution of chlorophyll *a*, in April, for the Celtic Sea.

where C_D is a frictional or drag coefficient that can be approximated as a constant (*ca.* 0.003) for a sandy bottom. The stratification index can be easily calculated for any coastal region, and it usually falls within the range of $+3$ and -2, the former value indicating highly stratified water and the latter, highly turbulent. A value of $S \approx 1.5$ provides the best conditions for phytoplankton production, indicating water that is not too stratified and not too turbulent. Nutrients that are brought to the surface by turbulence as the water velocity increases over a shelf or bank can be utilized by the phytoplankton, resulting in a shelf-break front of high primary productivity. Note, however, that the highest standing stock of phytoplankton (i.e.

$$3.3^3 = 35.937$$

$$35.937 \times 0.003 = 0.107811$$

$$\frac{5000\ cm}{0.107811}$$

$$= 46377.45685$$

$$Log\ 46377.45685$$

$$= 4.67$$

enter log then the answer into the calculator.

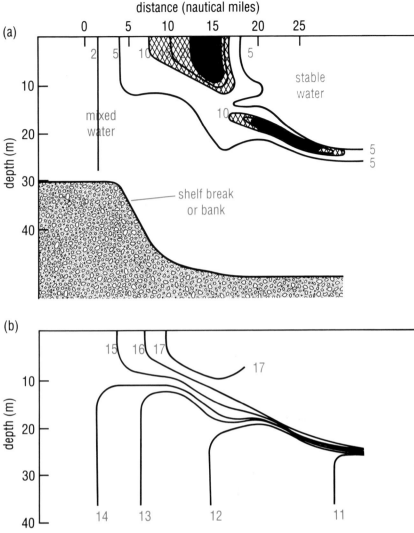

Figure 3.14   (a) Chlorophyll *a* values through a frontal region.
                  (b) Corresponding temperature (°C) through a frontal region.

chlorophyll *a* concentration) will develop over time on the more stable side of the front.

Shelf-break fronts are illustrated in Figures 3.13 and 3.14. Calculations of the stratification index (Figure 3.13a) coincide spatially with a chlorophyll maximum (Figure 3.13b) in the Celtic Sea. In Figure 3.14, the vertical distribution of chlorophyll is shown within a mixed water column on one side of the front and within a stable water column on the other side of the front.

QUESTION 3.7   The mean tidal flow across a shallow bank having a minimum depth of 50 m is 3.3 cm s^{-1}. Assuming a sandy bottom, will this bank produce a frontal zone?

$$S = \log_{10} \frac{50\ m.}{0.003 \times (3.3)^3}$$

$$= l \frac{5000\ cm}{0.11} = ?$$

3.5.5   RIVER-PLUME FRONTS

Rivers entering the sea often carry high nutrients, derived either from natural sources or from agricultural fertilizers and sewage. These nutrients enrich

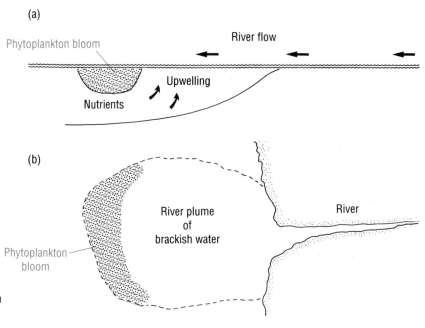

Figure 3.15 Nutrient entrainment at the mouth of a river. (a) Cross-sectional view. (b) Plan view.

coastal waters and increase productivity off the mouth of the river. In addition, estuarine waters are often highly productive because the flow of the river at the sea surface causes nutrients to be entrained from deeper water (Figure 3.15) upwelling into the surface water. Providing the deep waters are rich in phosphates and nitrates, the entrainment of nutrients also contributes to phytoplankton blooms off the river mouth. The exact position of a bloom in the river plume (or the location of the front) is a function of many factors including the quantity of nutrients introduced and/or entrained, the settling out of river silt which allows light to penetrate deeper, the depth of the mixed layer, grazing by zooplankton, etc. A phytoplankton bloom may also be disrupted or enhanced by the prevailing oceanic climate affecting the estuary.

3.5.6 ISLAND MASS EFFECT AND LANGMUIR FRONTAL ZONES

In addition to the five major physical processes that bring nutrients up to the euphotic zone as discussed above (Sections 3.5.1-3.5.5), there are many additional minor effects that form smaller frontal zones by physically altering nutrient concentrations in surface waters. Among these is the **island mass effect** (also known as island wake effect). This is a disturbance in the flow of water caused by the presence of an island (or an undersea mountain), resulting in upwelling from below the thermocline and subsequent nutrient enrichment of surface waters. First described from enhanced phytoplankton biomass and production around Hawaii, it is now known in many localities. For example, a plume of high production (> 4 mg Chl a m^{-3}) extends west of the Scilly Isles (off south-west England) for about 50 km into water that otherwise contains less than 0.5 mg Chl a m^{-3}. Similar upwelling and enhanced production can result as currents pass headlands and bays on a rugged coastline.

A different process affecting production on a smaller scale is **Langmuir circulation**. This pattern of circulation is set up when wind blows steadily across the surface of relatively calm seas. As a result, vortices of several

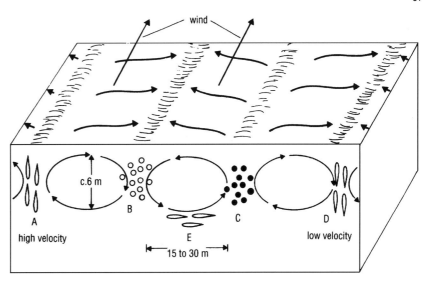

wind

Figure 3.16   Langmuir vortices and plankton distributions. Neutrally buoyant particles are randomly distributed, but downward swimming organisms are aggregated in high velocity upwellings (A); particles that tend to float are aggregated in downwellings (B); particles that tend to sink are aggregated in upwellings (C); upward swimming organisms are aggregated in low-velocity downwellings (D); and horizontally swimming organisms are aggregated where there is less relative current velocity than within the vortices (E).

metres in diameter start to revolve around horizontal axes and lead to both upwelling and downwelling of water (Figure 3.16). The vertical scale of this interaction is not large enough to bring nutrients up from deep water, but it is sufficient to concentrate plankton and this will enhance feeding interactions and result in a faster regeneration of nutrients. The phenomenon of Langmuir circulation is often visibly apparent as a series of parallel foam lines extending for great distances. In the Sargasso Sea, the seaweed *Sargassum* lines up in windrows in response to this type of circulation.

3.6   GLOBAL PHYTOPLANKTON PRODUCTIVITY

The primary productivity of phytoplankton in various areas of the global ocean varies with season and location. The highest productivity values of >1 g C m^{-2} day^{-1} are encountered in upwelling areas (see Section 3.5.2), and the lowest values (<0.1 g C m^{-2} day^{-1}) occur in the subtropical convergent gyres (see Section 3.5.1). During the summer in subarctic latitudes of the Pacific and Atlantic oceans, daily primary productivity may be >0.5 g C m^{-2}, but during the winter there may be no net primary productivity for several months. Integration of these different values on an annual basis gives the range of primary productivity values shown in Table 3.5. These differences in relative production can also be seen in Colour Plate 8 which shows relative chlorophyll concentrations in surface waters of the global ocean as detected by remote sensing from satellites. In total, the primary productivity of the world's ocean is about 40 × 10^9 tonnes of carbon per year. This figure is the same order of magnitude as for photosynthetic production by terrestrial plants, but the pattern of production is very different.

QUESTION 3.8   Much of the Indian Ocean between latitudes 0° and 40° S has a low primary productivity of less than 150 mg C m^{-2} day^{-1}. What feature(s) limit production in this ocean?   *Anticyclonic.*

In the terrestrial ecosystem, very high productivities occur in relatively small areas and the range of values for production are very great. For example, the

Table 3.5 The range of annual primary productivity in different regions of the global ocean.

Location	Mean annual primary productivity $(g\ C\ m^{-2}\ year^{-1})$
Continental upwelling (e.g. Peru Current, Benguela Current)	500–600
Continental shelf-breaks (e.g. European shelf, Grand Banks, Patagonia shelf)	300–500
Subarctic oceans (e.g. North Atlantic, North Pacific)	150–300
Anticyclonic gyres (e.g. Sargasso Sea, subtropical Pacific)	50–150
Arctic Ocean (ice-covered)	<50

estimated primary productivity of a rainforest is 3500 g C m^{-2} year^{-1}, or about six times the highest phytoplankton productivity. On the other hand, much of the terrestrial land mass is desert with little or no photosynthetic production. In contrast, marine productivity occurs virtually everywhere in the euphotic zone of the oceans (covering > 70% of the planet's surface), even under polar ice. It is the accumulative effect of the marine primary productivity throughout the world ocean that adds up to a total annual production of photosynthetic carbon approximately equivalent to that on land.

Latitudinal and seasonal differences in marine productivity result from differences in light and nutrient availability (see Figure 3.9). These physical forces largely determine the maximal phytoplankton production that is possible in any marine area. There are also biological processes that modify regional primary production levels. As algae grow, they reduce nutrient concentrations in the euphotic zone, and their own increasing numbers create **self-shading** by reducing the penetration of light, thus causing the euphotic zone to become shallower. Balancing these effects are the grazing activities of herbivorous zooplankton which remove part of the production, and there are regional differences in how the phytoplankton community is utilized by these animals.

When primary productivity increases, it is generally accompanied by a measurable increase in the standing stock of phytoplankton. During a bloom in coastal areas, the standing stock of chlorophyll *a* may increase from less than 1 mg m^{-3} to more than 20 mg m^{-3} over a period of several days. In some areas, however, the zooplankton may graze the phytoplankton as fast as it is produced, with the result that the increase in primary productivity does not show any discernible increase in the standing stock of phytoplankton. This situation is found in the North Pacific Ocean at about 50° N (Figure 3.17). Here, outside of coastal influences, there is virtually no change in the standing stock of phytoplankton throughout the year; it remains constant at about 0.5 mg chlorophyll *a* m^{-3}. However, primary productivity in this area increases from winter values of less than 50 mg C m^{-2} day^{-1} to more than 250 mg C m^{-2} day^{-1} in July. The excess primary productivity is grazed by the indigenous zooplankton which increase their biomass as

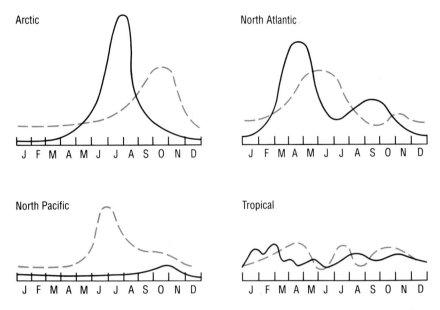

Figure 3.17 Summary of annual cycles in plankton communities in different regions. The solid black lines represent changes in phytoplankton biomass; the dashed blue lines indicate changes in zooplankton biomass (arbitrary units).

indicated in Figure 3.17. The close phasing between phytoplankton and zooplankton also has implications for deep-sea benthos in the North Pacific as there is little uneaten phytoplankton sinking into deep water to serve as a food supply for benthic animals (see Section 8.8.4).

In contrast, in the Atlantic Ocean at the same latitude, the spring bloom is characterized by a ten-fold increase in chlorophyll a from 0.1 to about 1.0 mg m^{-3}. Primary productivity increases as in the Pacific Ocean, but the zooplankton are less efficient at keeping pace with increases in primary production. Because only a fraction of the phytoplankton are eaten, there is an increase in the standing stock as measured by chlorophyll a. In much of the North Atlantic there is also an autumn bloom of phytoplankton, shown in Figure 3.17 as a second peak in phytoplankton and zooplankton biomass. As much of the phytoplankton is not eaten in North Atlantic waters, decaying blooms sink into deep water and become a food source for animals on the seafloor.

Two other annual cycles of phytoplankton and zooplankton are shown in Figure 3.17. One shows the pattern in the Arctic Ocean, where a single pulse of phytoplankton occurs soon after the disappearance of the ice and is followed somewhat slowly by a single pulse in the biomass of zooplankton. The lag in response time of the zooplankton to increased food is due to relatively slow growth rates in cold water. In tropical environments, the biomass of phytoplankton and zooplankton shows no substantial change throughout the year. However, storm activities can disrupt this otherwise very stable environment so that small pulses in plankton biomass may occur irregularly throughout the year. In warm tropical water, any increase in phytoplankton standing stock is quickly tracked by the fast-growing zooplankton.

Primary productivity varies with depth, and the vertical distribution of phytoplankton may change seasonally. This is illustrated over a time

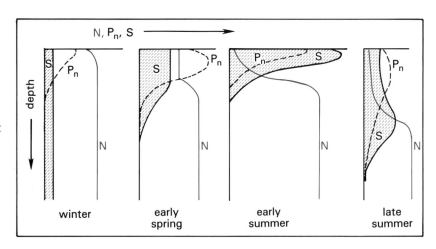

Figure 3.18   Schematic seasonal depth changes in phytoplankton biomass (S), daily net photosynthetic rate (P_n), and nutrient concentration (N) in stratified temperate water. S (shaded area), usually expressed in mg chl a m^{-3}, P_n (broken line), usually expressed as mg C per mg Chl a per day, N (blue line), usually expressed as μM nitrate. The figure omits any changes caused by significant zooplankton grazing.

sequence in Figure 3.18. In temperate latitudes, phytoplankton will be well mixed in the surface layer during the winter months and any photosynthesis will follow a light attenuation curve (Figure 2.5) except for some photoinhibition near the surface. As the spring progresses, primary productivity will increase near the surface, and this may be accompanied by an increase in the standing stock of phytoplankton. In late summer as the system runs out of nutrients near the surface, maximum primary productivity will shift deeper in the water column, resulting in a chlorophyll maximum at depth.

In stable water masses (i.e. most tropical and subtropical oceans), the vertical distribution of nutrients, primary productivity, and chlorophyll a resembles that shown for late summer in Figure 3.18 and is characteristic of the water mass throughout the year. Chlorophyll maxima in such waters can be found anywhere from 20 m to over 100 m depending on the long-term stability of the water. Under these conditions, the euphotic zone is really divided vertically into two communities. The top community is nutrient-limited and largely governed by biological and chemical processes that regenerate nutrients within the zone. The bottom community is light-limited, but it is located at the nutricline, where the maximum change in nutrient concentration occurs, and thus additional nutrients enter the system from deeper water. Since some zooplankton and fish migrate vertically through both communities, there is a degree of biological transport between the two vertically separated environments.

3.7   SUMMARY OF CHAPTER 3

1   The marine phytoplankton community is composed of several diverse groups of algae that carry out autotrophic production and begin the pelagic marine food chain. Photosynthesis results in the production of high-energy organic materials from carbon dioxide and water plus inorganic nutrients.

2   Photosynthesis involves a series of interrelated chemical reactions. The light reactions depend upon chlorophyll and accessory pigments capturing photons of light, so that radiant energy is converted to chemical energy. The dark reactions do not require light; they reduce the carbon dioxide and produce high-energy carbohydrates as end products. Respiration in plants and

animals is the reverse process of photosynthesis, whereby oxygen is used to release the energy contained in carbohydrates and carbon dioxide is liberated.

3   All phytoplankton species require certain inorganic substances to carry out photosynthesis, including sources of nitrogen, phosphorus, and iron (also silica for diatoms) which may be in concentrations that are low enough to be limiting to plant production. Some species also require certain organic substances (e.g. vitamins) for auxotrophic growth, and these also may be present in limiting concentrations.

4   Estimates of the total phytoplankton crop (standing stock or biomass) in a particular locality can be determined by measurements of cell numbers, total volume, or most commonly, by quantity of chlorophyll a. The rate of primary production is most often measured by following the uptake of radioactive ^{14} C in samples of seawater containing phytoplankton.

5   The amount of photosynthesis increases with light intensity up to a maximum value known as P_{max} which is specific for each species. When light intensity increases beyond this value, the rate of photosynthesis declines due to photoinhibition. The light intensity at which plant photosynthesis (production) exactly equals plant respiration (losses) is the compensation intensity. Gross photosynthesis describes total photosynthesis; net photosynthesis is gross photosynthesis less respiratory losses.

6   Photosynthetic responses of phytoplankton species to light can be described by a series of equations based on values for P_{max} and K_I. P_{max} values are generally higher at warmer temperatures and in eutrophic waters.

7   Phytoplankton are exposed to differing light intensities as light changes over the course of a day and as the algae are mixed vertically in the surface layers of the sea. At the critical depth, photosynthetic gains throughout the water column are just balanced by respiratory losses in the phytoplankton. If the depth of water mixing is greater than the critical depth, no net primary production can take place. Net production occurs only when the critical depth exceeds the depth of mixing.

8   Growth rates of phytoplankton are also controlled by the concentrations of essential nutrients in seawater. Oligotrophic regions have low concentrations of essential nutrients and therefore low productivity. Eutrophic waters contain high nutrients and support high numbers of phytoplankton.

9   Each species of phytoplankton has a particular response to different concentrations of limiting nutrients, and each has a different maximum growth rate. These differences and the species-specific responses to different light intensities, temperatures, salinities and other parameters, mean that heterogeneous and fluctuating environmental conditions favour different species at different times and allow many species to coexist in the same body of water. Thus phytoplankton species diversity can be high in what appears superficially to be a homogeneous aqueous environment.

10   Solar radiation and essential nutrient availability are the dominant physical factors controlling phytoplankton production in the sea. The amount of light varies with latitude, and the amount of nutrients contained in the euphotic zone is largely determined by physical factors controlling vertical mixing of water.

11   Despite year-round high light intensity, tropical regions are generally low in productivity because solar heating stabilizes the water column and

nutrients remain at low concentrations within the euphotic zone. Conversely, polar regions are generally high in nutrients but low in solar radiation except for a brief period in the summer. Maximum annual productivities are generally found in temperate latitudes where light and nutrients are both reasonably abundant.

12   The general latitudinal patterns of primary productivity are altered by a number of different physical processes that lead to nutrients being redistributed in the water column in discrete areas. These processes occur on scales varying from very large (e.g. gyres and continental upwelling), to smaller (e.g. tidal fronts and rings), to the very small scales of Langmuir circulation in which only the top few metres of the water column are mixed.

13   The standing stock of phytoplankton in the surface layers of the sea ranges from less than 1 mg chlorophyll a m^{-3} to about 20 mg m^{-3} during a phytoplankton bloom. Regional oceanic primary productivity ranges from <50 to >600 g C m^{-2} year^{-1}, with coastal upwelling regions having the highest values. Total primary productivity of the world ocean is about 40×10^9 tonnes of carbon per year, a figure that is approximately equivalent to terrestrial plant production.

14   Zooplankton grazing removes different proportions of the phytoplankton production in different marine areas. Much of the plant production is consumed in areas where growth rates and generation times of the zooplankton permit tight coupling with any phytoplankton increase (e.g. tropical waters). Where there is a lag in the development of zooplankton relative to increases in phytoplankton biomass, then some of the algal community dies and sinks to become a food source for deeper-living pelagic or benthic animals (e.g. North Atlantic).

15   The vertical profile of phytoplankton production changes with season and with latitude. High surface productivities generally occur in temperate latitudes in spring and autumn, whereas chlorophyll and productivity maxima occur considerably deeper in tropical waters.

Now try the following questions to consolidate your understanding of this Chapter.

QUESTION 3.9   It is generally considered that a high diversity of species is found in spatially heterogeneous environments such as rainforests and coral reefs. What are the reasons for the great diversity of phytoplankton (see Table 3.1) found in the pelagic environment?

QUESTION 3.10   If a rate of photosynthesis is measured at 0.2 mg C m^{-3} h^{-1} and the standing stock of phytoplankton is 2.5 mg C m^{-3}, what is the doubling time of the phytoplankton in the sample? Refer to equations 3.7 and 3.10.

QUESTION 3.11   Refer to Table 3.3 (a) Is the growth rate of the population in Question 3.10 rapid or slow? (b) Where might you find such a population growing?

QUESTION 3.12   If the half-saturation constants (K_N) for nitrate uptake were 0.1 μM for species A and 0.5 μM for species B and the maximum growth rates (μ_{max}) of A and B were 1 and 2 doublings per day respectively, which species would dominate at a nitrate concentration of 0.4 μM? This can best be shown by drawing a graph.

QUESTION 3.13   Which term in the equation (3.11) for nutrient uptake by phytoplankton is more important in determining high species diversity, K_N or μ_{max}?   *K_N*

QUESTION 3.14   Could you control the type of phytoplankton that grow in the sea through the introduction of artificial nutrient media?   *No*

QUESTION 3.15   What features would affect the growth rates and type of phytoplankton living under ice in the Arctic Ocean?

light.
temperture.

QUESTION 3.16   In what ways can phytoplankton produce harmful effects on marine animals and humans?

Toxic blooms.
axoxic condition.

The animals making up the zooplankton are taxonomically and structurally diverse. They range in size from microscopic, unicellular organisms to jellyfish several metres in diameter (refer to Figure 1.2). Although all zooplankton are capable of movement, by definition none are capable of making their way against a current. By definition also, all zooplankton — indeed all animals and some micro-organisms — are **heterotrophic**. That is, they require organic substrates (as opposed to inorganic ones) as sources of chemical energy in order to synthesize body materials. Unlike plants, which carry out autotrophic production by utilizing solar energy to reduce carbon dioxide, animals obtain carbon and other essential chemicals by ingesting organic materials. Animal species differ in how their energy is obtained: some species are **herbivores** which consume plants; others are **carnivores** which are capable of eating only other animals; and some species are predominantly **detritivores** which consume dead organic material. Many animals, however, are **omnivores** with mixed diets of plant and animal material. Different types of zooplankton often are placed in categories which describe their diets.

In addition to size categories and positions in food chains, zooplankton can be subdivided into classifications based on habitat (oceanic vs. neritic species; see Section 1.2) and taxonomy. They also form two categories depending upon the length of residency in the pelagic environment; **holoplankton** (or permanent plankton) spend their entire life cycles in the water column, whereas **meroplankton** are temporary residents of the plankton community. The meroplankton includes fish eggs and fish larvae (the adults are nektonic), as well as the swimming larval stages of many benthic invertebrates such as clams, snails, barnacles, and starfish. The more common types of holoplankton and meroplankton are described below in Sections 4.2 and 4.3, respectively.

4.1 COLLECTION METHODS

Zooplankton larger than 200 μm (refer to Figure 1.2) traditionally have been collected by towing relatively fine-mesh nets through the water column. Plankton nets vary in size, shape, and mesh size (Figure 4.1), but all are designed to capture drifting or relatively slow-moving animals that are retained by the mesh. The simplest nets are conical in shape, with the wide mouth opening attached to a metal ring and the narrow tapered end fastened to a collecting jar known as the **cod end**. This type of net can vary in length and diameter and in mesh size, and it can be towed vertically, horizontally, or obliquely through the desired sampling depths. Such a net will filter water and collect animals during an entire towing period. More sophisticated nets are equipped to be opened and closed at selected depth intervals, and a series of such nets may be attached to a single frame to allow sampling of different discrete depths during a single towing operation. Analyses of the collected samples permit a more detailed picture of the vertical distribution of zooplankton within a particular area. Because many zooplankton migrate vertically during each 24-hour period, the time of sampling is also critical in tracking these changes in depth distribution.

Figure 4.1   A 'Bongo' zooplankton sampler, consisting of duplicate plankton nets, being retrieved on board a research vessel.

The selection of a particular net depends upon the type of organisms desired and the characteristics of the water being sampled. For example, a fine-mesh net (with a mesh opening of *ca.* 100–200 μm) obviously must be used to collect small mesozooplankton. However, the same net is not suitable for sampling larger, relatively fast-swimming zooplankton, like fish larvae, because it clogs quickly and must be towed slowly to avoid tearing. In deep water, where zooplankton tend to be larger and are less abundant, it is common to use a very large coarse-mesh net. All nets can be equipped with a flowmeter that estimates the total volume of water filtered during a tow; this permits a quantitative representation of the zooplankton collected. The newest towed samplers, such as the *Batfish* shown in Figure 4.2, simultaneously measure salinity, temperature, depth, and chlorophyll *a* concentration while counting and sizing zooplankton that pass through an optical sensor. Such devices can be set to undulate over a set depth range, thus making it possible to obtain samples at several depths.

No single sampler is capable of capturing all zooplankton within its path. Zooplankton smaller than 200 μm (nano- and microplankton) cannot be satisfactorily sampled in nets; instead, a known volume of water is collected in sampling bottles or by pumps from defined depths and the smallest zooplankton are concentrated and removed by filtration, centrifuging, or settling and sedimentation. Both planktonic protozoans and phytoplankton can be counted in the concentrated water samples.

Figure 4.2   A towed Batfish plankton sampler which simultaneously estimates phytoplankton and zooplankton abundance while recording environmental parameters. F, fluorometer for detecting chlorophyll *a*; L, light sensor; OPC, optical zooplankton counter; PI, intake for zooplankton sampling; SB, stabilizer; STD, salinity–temperature–depth sensor; T, towing arm.

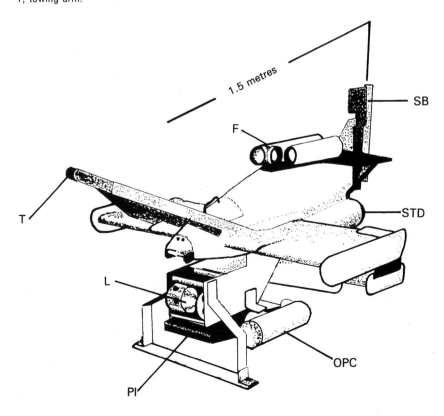

Figure 4.3 A scuba diver counting the number of zooplankton (larvacea) contained within a frame of known dimensions.

Further, it is now recognized that some zooplankton are capable of avoiding towed samplers, which they detect either visually or by sensing turbulence created in advance of the moving gear. In addition, certain gelatinous plankton are so fragile that they are impossible to collect intact in nets; others disintegrate rapidly in the preservatives that are routinely used to store collections. Crustaceans usually comprise the majority of zooplankton in net collections because most are too small to avoid capture, and because their hard exoskeletons protect them from damage and distortion in nets and preservatives. Thus the numbers of Crustacea relative to other types of zooplankton may be overestimated, and net-collected plankton may not provide a true representation of the natural plankton community in many areas. The numerical dominance and biomass contribution of crustaceans, especially copepods, needs to be reassessed in many localities.

Zooplankton can also be observed directly in the field, either by scuba diving down to depths of about 30 m (Figure 4.3) or, in deeper waters, by using manned submersibles or ROVs (remotely operated vehicles) which are tethered to ships and coupled with underwater video cameras of high resolution. These techniques have resulted in discoveries of new species, particularly of fragile forms; in an awareness of the problem of underestimating numbers and biomass of these animals from net collections; and in new behavioural observations of many species. Bioacoustic methods, developed from the use of sonar to locate fish schools, are also being applied to locate and estimate densities of larger zooplankton, such as euphausiids, which form dense aggregations.

4.2 HOLOPLANKTON: SYSTEMATICS AND BIOLOGY

There are approximately 5000 described species of holoplanktonic zooplankton (excluding protozoans) representing many different taxonomic groups of invertebrates (Table 4.1). Those groups that are commonly found in the sea and that form significant fractions of the plankton community are described below. In addition to providing descriptive anatomical accounts, particular attention is given to describing the food and feeding mechanisms of each group as these are important in the discussions of food webs and energy transfer which follow.

The smallest of the zooplankton are certain unicellular **protists** (Table 4.1). Included are many species of **dinoflagellates** that are partly or wholly heterotrophic (see Section 3.1.2 for a discussion of autotrophic species). These heterotrophic dinoflagellates feed on bacteria, diatoms, other flagellates, and ciliate protozoans that are either drawn to the predator by flagella-generated water currents, or that are trapped in cytoplasmic extensions of the dinoflagellate. Some species are only capable of functioning as heterotrophs; others that contain chloroplasts may also function as autotrophs part of the time. The best known heterotrophic dinoflagellate is *Noctiluca scintillans* (Figure 4.4), which has the form of a gelatinous sphere 1 mm or more in diameter. *Noctiluca* often occurs in dense swarms near coasts, and it feeds on small zooplankton (including fish eggs) as well as on diatoms and other phytoplankton.

A taxonomically diverse group of flagellated protists, commonly called **zooflagellates**, includes all those species that are colourless and strictly heterotrophic. All of the organisms in this group lack chloroplasts and plant

Table 4.1 Major taxonomic groups and representatives of holoplanktonic zooplankton.

Phylum	Subgroups	Common genera
Protozoa	Dinoflagellates	*Noctiluca*
(= heterotrophic	Zooflagellates	*Bodo*
protists)	Foraminifera	*Globigerina*
	Radiolaria	*Aulacantha*
	Ciliates	*Strombidium; Favella*
Cnidaria	Medusae	*Aglantha; Cyanea*
(formerly	Siphonophores	*Physalia; Nanomia*
Coelenterata)		
Ctenophora	Tentaculata	*Pleurobrachia*
	Nuda	*Beroe*
Chaetognatha		*Sagitta*
Annelida	Polychaetes	*Tomopteris*
Mollusca	Heteropods	*Atlanta*
	Thecosomes	*Limacina; Clio*
	Gymnosomes	*Clione*
Arthropoda		
(Class Crustacea)	Cladocera	*Evadne; Podon*
	Ostracods	*Conchoecia*
	Copepods	*Calanus; Oithona*
	Mysids	*Neomysis*
	Amphipods	*Parathemisto*
	Euphausiids	*Euphausia*
	Decapods	*Sergestes; Lucifer*
Chordata	Appendicularia	*Oikopleura*
	Salps	*Salpa; Pyrosoma*

pigments, and many feed on bacteria and detritus. Although they are very small (typically 2–5 μm), they have potentially high reproductive rates and therefore can become exceedingly abundant under favourable environmental conditions. Heterotrophic flagellates account for 20–80% of the nanoplankton by cell number, and thus they may be an important food for zooplankton that feed on small organisms.

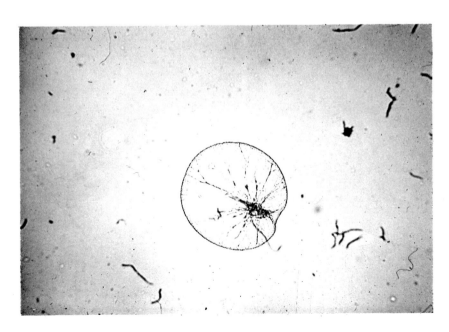

Figure 4.4 *Noctiluca*, a heterotrophic dinoflagellate. (Diameter, 1 mm.)

Figure 4.5 *Hastigerina pelagica*, a planktonic foraminiferan. A bubble capsule of cytoplasm surrounds a central shell with numerous radiating spines. Fine strands of cytoplasm, known as rhizopodia, project along the spines. The rhizopodia have sticky surfaces and are used to snare prey such as small copepods and a variety of microzooplankton. (Diameter is approximately 3 mm.)

Marine amoebae include the **Foraminifera** (Figure 4.5), which are characterized by having a calcareous perforated shell, or test, that is usually composed of a series of chambers. The size range of planktonic species is about 30 μm to a few millimetres. Food, consisting of bacteria, phytoplankton or small zooplankton, is captured by specialized slender pseudopodia (called rhizopodia) that project through the pores of the test. Although there are less than 40 known planktonic species (but *ca.* 4000 benthic species), these holoplanktonic foraminifera are very abundant, particularly between 40°N and 40°S where they generally inhabit the top 1000 m of the water column. After death, the shells of these protozoans sink and accumulate in large quantities on the seafloor, forming a sediment known as **foraminiferan ooze**.

The **Radiolaria** (Figures 1.7, 4.6 and Colour Plate 9) are spherical, amoeboid protozoans with a central, perforated capsule composed of silica. Most are omnivorous, and they have branched pseudopodia (called axopodia) for food capture; prey includes bacteria, other protists, and tiny crustaceans as well as phytoplankton (especially diatoms). The size of individual organisms ranges from about 50 μm to as much as several millimetres in diameter; some species form gelatinous colonies composed of many individuals and up to a metre or so in length. Radiolaria occur in all oceanic regions but are especially common in cold waters, and many are deep-sea species. A sediment composed of the siliceous remains of these protozoans is called **radiolarian ooze**.

Planktonic **ciliates** (Colour Plate 10) are present in all marine regions and are often extremely abundant. All use cilia for locomotion, and some have modified oral cilia used for food capture. Ciliates can feed on small phyto- and zooflagellates, small diatoms, and bacteria. **Tintinnids** (Colour Plate 11) make up one large subgroup (>1000 species) of marine ciliates. They are noted for their vase-like external shells that are composed of protein; because this substance is biodegradable, the shells are not present in sediments. Despite their small size (about 20–640 μm), tintinnids are of considerable ecological significance as they are widely distributed in both

Figure 4.6   A large radiolarian with a central spherical skeleton composed of silica. Numerous axopodia radiate from the central capsule of cytoplasm that lies within the sponge-like skeleton. Prey such as tintinnids, small copepods and other microplankton are captured by the sticky surfaces of the axopodia. Many of the white spots in the photo are algae that typically live in association with this radiolarian. (Size is approximately 1 mm.)

open seas and coastal waters, where they feed primarily on nanoplanktonic diatoms and photosynthetic flagellates. In coastal waters, tintinnids may consume 4–60% of the phytoplankton production. In turn, they are prey for a wide variety of mesozooplankton.

Jellyfish, or **medusae** (Colour Plate 12), are conspicuous and common inhabitants of both the open sea and coastal waters. Some species are holoplanktonic, but others have an asexual benthic stage in their life cycle and thus their medusae are part of the meroplankton. Although jellyfish belong to several different taxonomic groups within the Phylum Cnidaria, all are characterized by a primitive structural organization, and all are carnivorous, capturing a variety of zooplanktonic prey by tentacles equipped with stinging cells called nematocysts. They range in diameter from just a few millimetres to 2 m for *Cyanea capillata*, a common northern species with 800 or more 30–60 m-long tentacles. Some well-known pelagic Cnidaria are colonial forms, like **siphonophores** (Figure 4.7 and Colour Plates 13 and 22), in which many individuals with specialized functions are united to form the whole organism. *Physalia*, or the Portuguese man-of-war, is a tropical siphonophore that floats at the surface with its tentacles extending as far as 10 m below; it is capable of capturing sizeable fish, and its stings can be painful to swimmers. However, the medusae known as box jellyfish are much more dangerous. *Chironex fleckeri* of tropical Australia is the most venomous animal on Earth, and this 'sea wasp' has caused at least 65 human fatalities in the last century. The stings of a large individual, with up to 60 tentacles stretching some 5 m, can cause death within four minutes. In nature, *Chironex* uses this potent venom to quickly kill prey such as shrimp.

(a)

(b)

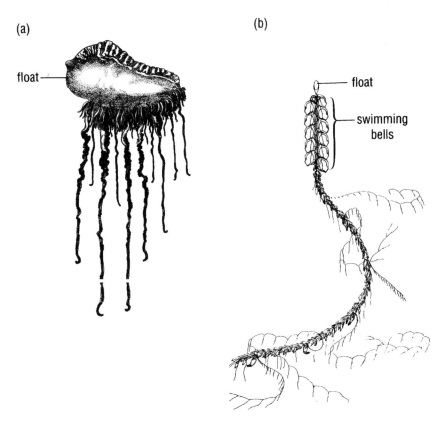

float

float

swimming
bells

Figure 4.7   Siphonophores: (a) a surface floating species, *Physalia physalis*, the Portuguese man-of-war, with tentacles up to 10 m long; (b) a swimming species, *Nanomia* sp., *ca.* 10 cm long. Both species use their long trailing tentacles to capture prey.

Ctenophores are closely related to jellyfish, but their structure is sufficiently different to warrant their being placed in a separate phylum. These are transparent animals that swim by means of fused cilia arranged in eight rows (called comb plates). Like the Cnidaria, ctenophores are carnivores, but they lack the nematocysts of their close relatives. Certain of the ctenophores like *Pleurobrachia* (Colour Plate 14) have long paired tentacles with adhesive cells that are used to capture prey; other species (e.g. *Bolinopsis*) capture food in large ciliated oral lobes. These ctenophores can have significant impacts on fish populations as they feed directly on fish eggs and fish larvae, and they also compete with young fish for smaller zooplankton prey such as copepods. Some ctenophores like *Beroe* (Colour Plate 15) lack tentacles but have large mouths; they engulf their prey, which consists principally of tentaculate ctenophore species.

Chaetognaths, or arrow worms, (Figure 4.8), are one of the best known and most abundant carnivorous planktonic groups. These hermaphroditic animals are found only in the sea, down to depths of several thousand metres. They have transparent, elongate and streamlined bodies, and most are less than 4 cm long. They often remain motionless in the water, but are capable of swift darting motions when in pursuit of prey. Food, consisting of smaller zooplankton, is captured by clusters of chitinous hooks located around the mouth of the predator. Chaetognaths do not seem to be selective in prey type; often the type of food eaten reflects local relative abundance of suitable prey. There are a few other holoplanktonic worms belonging to different phyla but, in most regions, these are generally found in very low numbers.

Figure 4.8   The chaetognaths (a) *Sagitta pulchra* and (b) *Sagitta ferox.*

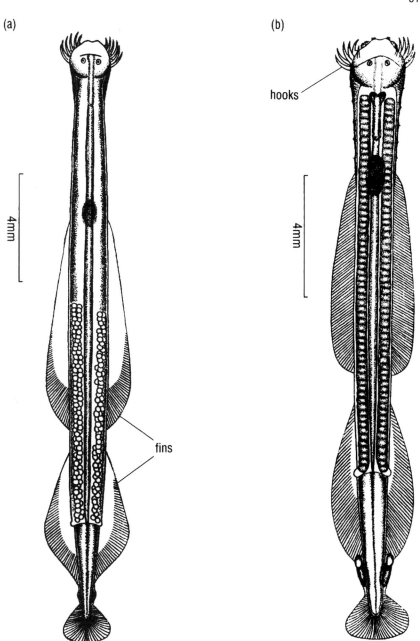

(a)

4mm

fins

(b)

hooks

4mm

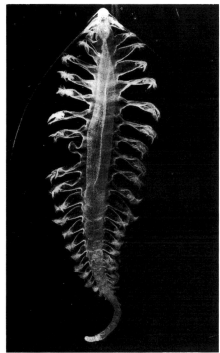

Figure 4.9   The planktonic polychaete *Tomopteris helgolandica* with multiple swimming legs and long, slender, paired antennae. Length, 45 mm.

One exception is the **polychaete** (Phylum Annelida) genus *Tomopteris* (Figure 4.9) with about 40 species (all predators) distributed throughout the world ocean.

QUESTION 4.1   Chaetognaths, ctenophores, some jellyfish, and many other zooplankton living in the upper layers of the sea are transparent. Can you suggest why transparency is a useful characteristic for animals living in the euphotic zone?   *camoflarge.*

Only a few species belonging to the Phylum Mollusca have become holoplanktonic. **Heteropods** are a small group (*ca.* 30 species) of molluscs that are closely related to snails, but these pelagic forms swim by undulating motions of a single fin developed from the creeping foot of their benthic ancestors. Some of the species can completely withdraw into a small (<10 mm) external, spirally-coiled shell (Colour Plate 16); others have reduced shells or lack shells entirely and are highly transparent animals that may attain lengths of up to 50 cm (Colour Plate 17). Despite these external differences, all heteropods have remarkably well-developed eyes and are visual predators, feeding on other planktonic molluscs, or on copepods, chaetognaths, salps, or siphonophores. Prey are actively pursued and captured by large chitinous teeth which can be protruded from the mouth. Heteropods are generally found in oceanic warm water areas, but, like many carnivorous plankton, they are not very abundant in any given locality. Their calcium carbonate shells are sometimes found in sediments on the ocean floor.

The shelled **pteropods**, or **thecosomes**, are also holoplanktonic snails. Most of them have a thin, external, calcareous shell measuring from a few to about 30 mm in largest dimension. The shell is spirally coiled in primitive species but assumes a variety of shapes in more advanced members (Colour Plates 18, 19). One subgroup (called pseudothecosomes) (Colour Plate 20) is made up of larger animals (>30 mm) that lack a true shell; instead, there is a cartilaginous, internal, skeletal structure. All thecosomes swim by means of paired wings or a fused wingplate, structures that developed from the foot of benthic molluscan ancestors. Despite considerable structural diversity, all thecosomes are suspension feeders. They produce large, external, mucous webs that are held in the water while the animal remains motionless below. As the web fills with organisms entangled in the sticky strands of mucus, it is withdrawn and ingested. Food consists of phytoplankton as well as small zooplankton and detrital material. Some of the shelled pteropods can be very abundant in epipelagic areas, particularly those species that inhabit polar seas. Some thecosomes are an important food source for pelagic fish, including commercially important species like mackerel, herring, and salmon. The shelled pteropods have an unusual reproductive pattern in which an animal is first a male that mates with another male; the sperm is stored until the animal changes into a female that lays fertilized eggs in mucoid floating masses. The carbonate shells of dead animals eventually sink and accumulate in certain areas to form a type of sediment known as **pteropod ooze**.

Thecosomes are preyed upon by another group of planktonic gastropod molluscs, the naked pteropods or **gymnosomes**. These animals lack shells as adults, but they too swim by means of paired wings. Of the approximately 50 species, *Clione limacina* (Colour Plate 21) is the largest (to 85 mm long), most abundant, and best known. It lives in polar and subpolar regions of the Northern Hemisphere, where it feeds only on several species of the thecosome genus *Limacina* (Colour Plate 18). Other gymnosomes also are predators that feed exclusively on specific shelled pteropods. All capture prey with special tentacles and chitinous hooks, and they remove the soft parts of the prey from its shell before swallowing it.

The segmented Crustacea are represented in the sea by several different groups, but **copepods** are the predominant forms. Some of the most abundant and best known marine zooplankton belong to the Order **Calanoida** (Figure 4.10a-c) which comprises about 1850 species. These free-living calanoid copepods are present in all marine regions and usually make up 70% or more of all net-collected plankton. All species have three distinctive body regions: the head and first segment of the body are fused and bear two pairs of antennae and four pairs of mouthparts; the segmented mid-body has paired swimming legs; and the narrow posterior section lacks

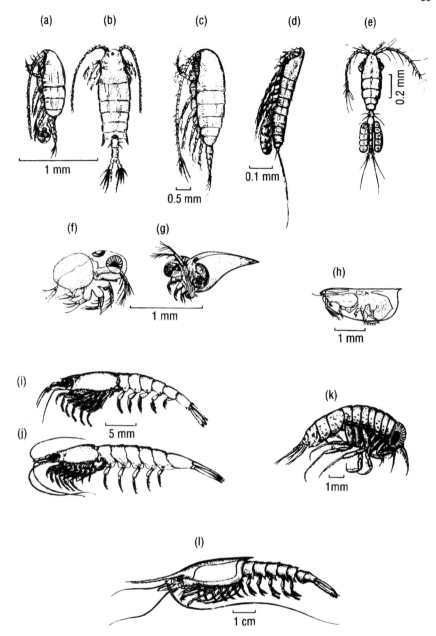

Figure 4.10 Planktonic Crustacea. Calanoid copepods: (a), *Pseudocalanus elongatus*; (b), *Centropages typicus*; (c), *Calanus finmarchicus*. (d), *Microsetella norvegica*, a harpacticoid copepod. (e), *Oithona similis*, a cyclopoid copepod. Cladocera: (f), *Podon leuckarti*; (g), *Evadne nordmanni*. (h), *Conchoecia elegans*, an ostracod. Euphausiids: (i), *Thysanoessa inermis*; (j), *Meganyctiphanes norvegica*. (k), *Themisto abyssorum*, an amphipod. (l), *Gnathophausia zoea*, a mysid.

appendages. Total body length is usually less than 6 mm, but some exceptional species exceed 10 mm in length. Many of the species feed by capturing phytoplankton, especially diatoms, in currents generated by movements of the swimming legs and mouthparts. Some calanoids, however, are omnivorous or carnivorous and feed on small zooplankton. The sexes are separate, and fertilized eggs may be laid freely in the water or may be retained in an external cluster by the female. Development involves twelve different stages, each separated by moulting, or casting off of the exoskeleton, and marked by the appearance of new segments and additional appendages. The first six stages are **nauplius** (plural, nauplii) larvae (designated NI to NVI); the last six are **copepodite** stages (CI to CVI), with CVI being the sexually mature adult.

Another copepod group, the Order **Cyclopoida** (Figure 4.10e), differs in that members have relatively shortened antennae and more segments in the posterior third of the body. The order contains over 1000 species, but the majority live among benthic algae or in bottom sediments; only about 250 species are planktonic. Small species belonging to the planktonic genera *Oithona* and *Oncaea* can, however, be very abundant. Some of the cyclopoid copepods have specially modified antennae for capturing individual microzooplankton.

The majority of copepods belonging to the Order **Harpacticoida** (Figure 4.10d) are coastal or live in association with the sea bottom. Approximately 20 species are holoplanktonic, and these are characterized by usually being very small (<1 mm long) and without distinct divisions between body regions. Although some species are widely distributed and may be seasonally or locally abundant, their ecological importance in the plankton community does not seem to be great.

The **euphausiids** (Figure 4.10i-j) form another important group of marine Crustacea with 86 species. These shrimp-like animals are generally of relatively large size, with many species attaining a length between 15 mm and 20 mm and with some exceeding 100 mm. *Euphausia superba* is the **krill** of the Antarctic Ocean, where this abundant species is a major component of the diet of many larger animals and is itself harvested commercially (see Sections 5.2 and 6.1). Euphausiids are fast-swimming and are usually undersampled by large nets because of their visual perception and avoidance capabilities. But it is known that euphausiids form major fractions of the zooplankton biomass in the open ocean of the North Pacific and North Atlantic and in the Arctic, and they are important food for fish (e.g. herring, mackerel, salmon, sardines, and tuna) and some seabirds in these areas. Euphausiids are generally omnivorous with food consisting of detritus, phytoplankton, and a variety of smaller zooplankton. Larger species are also capable of feeding on fish larvae. Euphausiids, like copepods, have a series of anatomically distinct larval stages separated by moulting and growth.

Amphipods (Figure 4.10k) are distinguished from other Crustacea by having laterally compressed bodies. They usually constitute only a small fraction of the total zooplankton. *Parathemisto gaudichaudi* is a common pelagic species with a wide distribution at relatively high latitudes in both hemispheres. The adults of this species are free-living carnivores, feeding on copepods, chaetognaths, euphausiids, and fish larvae. Many pelagic amphipods, however, are commonly found attached to siphonophores (Colour Plate 22), medusae, ctenophores, or salps, and the amphipods either feed as predators on these animals or act as parasites. In contrast to copepods and euphausiids, amphipods have direct development; the young are released from a brood pouch and look like miniature adults.

Ostracods (Figure 4.10h) are usually minor components of the zooplankton community. These crustaceans have a unique, hinged, bivalved exoskeleton into which the animal can withdraw. Most species are rather small although *Gigantocypris*, a deep-water inhabitant, reaches more than 20 mm in diameter. Little work has been done on feeding habits in this group, but some species are regarded as scavengers.

Although there are over 400 species of freshwater **Cladocera** (including *Daphnia*, the common water flea), there are only about eight marine species (Figure 4.10f-g) in this primitive group of crustaceans. The marine cladocerans are primarily of interest in coastal and brackish water, although there are species that become seasonally abundant for brief periods in the open ocean. Because they are capable of producing cloned offspring by **parthenogenesis** (i.e. reproduction without males and without fertilization), Cladocera are able to rapidly increase their numbers when environmental conditions are favourable.

Mysids (Figure 4.10*l*) are listed in Table 4.1 and mentioned here for completeness, but they seldom are important components of the plankton community. Many of these shrimp-like animals spend part of the time on the seafloor, but rise into the overlying water at night or when forming breeding swarms. A few oceanic species are residents of near-surface waters, but most live in deeper zones. The most abundant and best known species are estuarine or inshore residents, and some of these are harvested commercially in parts of Asia.

The most advanced Crustacea are **decapods**, a group that encompasses shrimp, lobsters, and crabs. Most are benthic, but some are holoplanktonic or nektonic. Pelagic species include about 210 species of shrimps (Colour Plate 23) that typically measure 10–100 mm or more in length and thus constitute some of the larger zooplankton. They are strong swimmers and avoid capture by ordinary plankton nets. Most live below 150 m depth in daytime. They are usually omnivores or predators, and copepods, euphausiids and other planktonic Crustacea are their predominant foods. Densities of pelagic shrimp in oceanic areas are typically on the order of 1 individual per 200–2000 m^3 of water, but they can be important prey for various fish, including albacore tuna, and for dolphins and whales.

Two groups of chordates are important members of marine zooplankton communities. **Appendicularians** (Colour Plate 24) are closely related to benthic tunicates or sea squirts (see Section 7.2.1). Because they closely resemble the larval stages of these bottom-dwelling relatives, they are also known as **larvaceans**. The body of appendicularians looks like a tadpole; it consists of a large rounded trunk containing all the major organs and a longer, muscular tail. Most species secrete a spherical balloon of mucus, called a house, in which they reside. The body is generally only a few millimetres long, whereas houses range from about 5 mm to 40 mm long. Movements of the animal's tail create a current of water that enters the mucoid house through mesh-covered filters which remove larger particles of suspended material. As water flows through the house, it passes another feeding filter where nanoplankton and bacteria are collected and transported to the mouth. Periodically the filters become clogged with particles and the house must be abandoned, an activity that can be repeated up to a dozen times per day; new houses can be secreted within a few minutes. Discarded houses can reach densities of more than 1000 m^{-3}, and they contribute to the formation of **marine snow**, a term applied to macroscopic aggregates of amorphous particulate material derived from living organisms. Abandoned houses represent rich sources of food and surfaces for attachment by other organisms in the water column and, for these reasons, they are rapidly colonized by bacteria and protozoans. Larvaceans grow rapidly, and have short generation times of 1–3 weeks. They are among the most common members of the zooplankton, being especially abundant in coastal waters and over continental shelves where densities may reach 5000 m^{-3}. The 70 or so species are distributed in all the oceans.

Salps (Colour Plates 25, 26) constitute another class of chordates, but these animals are commonly found only in warm surface or near-surface waters. Each individual salp has a cylindrical, gelatinous body with openings at each

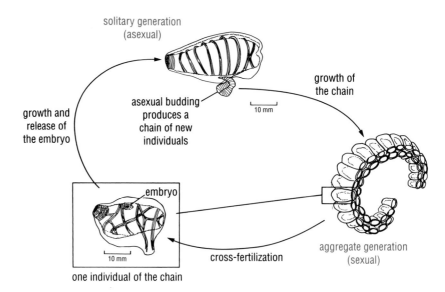

Figure 4.11   The life cycle of salps.

end. Locomotion is achieved by muscular pumping which propels water through the body. This water current also brings food particles into contact with an internal net of mucus that is continuously secreted by the animal. Cilia transport food entrapped in mucus to the esophagus where it is ingested. Food consists primarily of phytoplankton and bacteria, ranging in size from about 1 μm to 1 mm. Because salps often form dense swarms and have high feeding rates, their feeding activities may significantly reduce the concentrations of small-sized organisms in the surrounding water. Salps have an unusual life cycle (Figure 4.11) in which sexual reproduction alternates with asexual budding. Each species of salp has two different forms: the asexual form is a **solitary** individual (1–30 cm long) that buds to produce a chain of up to several hundred individuals and as long as 15 m; each **aggregate** in the released chain is an hermaphrodite and will produce both sperm and a single egg. Self-fertilization does not usually occur because the egg and sperm ripen at different times. Following cross-fertilization, the single embryo grows within the parent and eventually breaks through the parental body wall to become a young, free-swimming, solitary individual which once again will continue the cycle with asexual budding. Salps and the related appendicularians are good examples of *r*-selected species (see Section 1.3.1) with extremely rapid growth rates and short life spans; thus they can quickly respond to favourable environmental conditions by producing large populations.

QUESTION 4.2   Can you think of any reasons why salps may have developed a complex life cycle that involves two different reproductive patterns?

4.3   MEROPLANKTON

Some benthic marine invertebrates have no free-swimming larval stage. Their young hatch as miniature adults from eggs attached to the sea bottom, or emerge directly from the parent. But approximately 70% of benthic species release eggs or embryos into the water column, and the resulting larvae become part of the plankton community. Depending on the species, these meroplanktonic larvae may spend from a few minutes to several months (or even years in exceptional cases) in the plankton before they settle onto a substrate and metamorphose into the adult form. During this time, the larvae drift in currents and may be dispersed away from the parent population.

Some of the more common types of meroplanktonic larvae of benthic invertebrates are illustrated in Figure 4.12. Benthic snails and clams produce a shelled **veliger** larva that has a distinctive ciliated membrane (called a velum) that is used for locomotion as well as food collection. Sessile barnacles have free-swimming **nauplius** stages, usually six, which are similar to the nauplii of copepods and other planktonic Crustacea, but with characteristic pointed projections on the anterior edges of the exoskeleton; these naupliar stages are succeeded by a **cypris** which attaches to a substrate and metamorphoses to the adult. Starfish, sea urchins, sea cucumbers, and

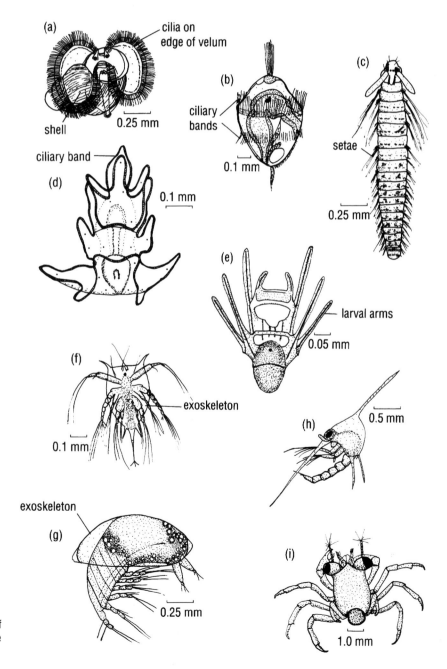

Figure 4.12   Meroplanktonic larvae of benthic invertebrates. (a) snail veliger; (b) polychaete trochophore; (c) late larva of a polychaete; (d) bipinnaria of a starfish; (e) echinopluteus of a sea urchin; (f) barnacle nauplius; (g) barnacle cypris; (h) crab zoea; (i) crab megalopa.

other benthic echinoderms have various types of meroplanktonic larvae, some of which are shown in Figure 4.12. Benthic worms belonging to different phyla also have distinctive larvae: polychaetes, for example, produce a **trochophore** larva, with several bands of cilia, that eventually develops a segmented body and appendages before settlement. Benthic decapods, like crabs, typically have a succession of different planktonic larval stages that are separated by moulting. Crabs usually hatch as a spiny **zoea**, which eventually changes before settlement to a **megalopa** that resembles a miniature adult. These examples are but a few of the more common types of larvae that appear temporarily in the plankton community.

Benthic invertebrates living in shallower zones often produce planktonic larval stages, but deep-sea species commonly lack a planktonic stage and instead have direct development or brood protection of young. This may be related to a lack of suitable and abundant suspended food for planktonic larvae in deep water. In temperate and cold-water inshore regions, meroplanktonic larvae of benthic invertebrates typically appear seasonally in response to warmer temperatures and increased phytoplankton. In tropical waters, the reproduction of benthic invertebrates may be more or less continuous but with peaks of reproductive activity tied to other environmental events, such as rainfall; in such areas, meroplanktonic larvae may be present throughout the year, but in differing abundances.

Fish eggs and fish larvae (Figure 4.13) also form an important part of the meroplankton; they are referred to as the **ichthyoplankton**. Some fish

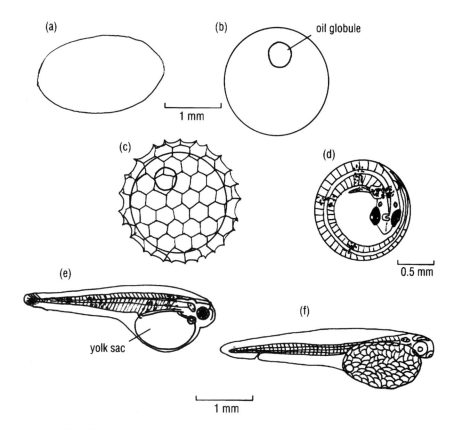

Figure 4.13   Meroplanktonic fish eggs and fish larvae. (a) anchovy egg; (b) mackerel egg; (c) myctophid (lantern-fish) egg; (d) cod embryo developing in egg; (e) newly hatched cod larva; (f) newly-hatched pilchard.

species attach their eggs to substrates; salmonids, for example, typically place eggs in the gravel of streams and herring deposit eggs on seaweeds or directly on the seafloor. Many marine fish, however, release free-floating, planktonic eggs; these include sardines, anchovy, tuna, and many other commercially harvested species. These planktonic eggs are typically spherical, transparent, and small, usually of the order of 1–2 mm in diameter. The eggs contain varying amounts of clear yolk which is the food of the developing embryos and newly hatched larvae. Some eggs also contain oil globules, apparently to aid in flotation.

As with the meroplanktonic larvae of benthic invertebrates, the appearance of fish eggs in the plankton is dependent on the spawning cycles of the adults, and these are often linked with environmental change. Rate of development within the egg is species-specific but is also closely tied to ambient sea temperatures, with hatching being delayed in colder waters. Hatching generally occurs within a few days to a few weeks after the eggs are spawned. The numbers of eggs produced may be very large. Each female plaice, for example, produces about 250 000 eggs; haddock lay about 500 000; and cod lay over one million. If these figures are multiplied by the number of spawning fish, the annual production of eggs may be enormous. Numbers of pilchard eggs in the English Channel alone have been estimated to exceed 4×10^{14}. Obviously few of the young from these eggs survive to adulthood; many form an important food source for the holoplankton and also for adult fish.

In general, for both fish and benthic invertebrates, there is an inverse correlation between egg size and the numbers of eggs produced. Is there any selective advantage to producing large numbers of small eggs or, alternatively, to producing small numbers of large eggs?

Two life patterns are operable with regard to egg size. Some species produce large eggs, but in small numbers because of size and energy restrictions. Larger eggs contain more yolk and the newly hatched young will be larger relative to young from smaller eggs. Larger young tend to have higher survival rates, probably because they are already too large to be eaten by some pelagic predators, and possibly because they are also more active and better able to evade potential predators. On the other hand, many species produce enormous numbers of small eggs which contain little or no nutritive material for the developing embryos. These young hatch at a small size; they are vulnerable to many predators, and it is critical that they immediately begin feeding. Mortality is much higher for these young compared to those from large eggs, but the high mortality is compensated for by larger numbers.

For the first few days after hatching, fish larvae retain remnants of the yolk in a sac carried under their body, and they continue to rely on this yolk sac as a food source until the mouth and gut develop. When the yolk is exhausted and the larvae begin to feed, they are totally dependent upon suitable food being plentifully available in the plankton. This dependency lasts throughout their planktonic life, for up to several months, until they are large enough to be classified as nekton and can actively seek feeding areas independently of current drift. At the same time, the young are vulnerable to pelagic predators, both larger zooplankton and nekton. Mortality is high during the planktonic stages; usually only a small fraction of fish survive to adulthood. For example, it is estimated that early life mortality in cod is as much as 99.999%.

4.4 VERTICAL DISTRIBUTION

We have previously classified zooplankton according to size, habitat, taxonomic position, and length of residency in the plankton community. Zooplankton are also grouped according to their depth position in the water column.

Those species that live permanently at the sea surface and whose bodies project partly into the air are called **pleuston**. They are sometimes considered to form a special category because they are passively transported by wind instead of by currents. The **neuston** includes those species that inhabit the uppermost few to tens of millimetres of the surface water. Ecologically, it is difficult to separate these categories, and here they are discussed together as organisms living in the uppermost zone of the ocean. This community is richly developed in tropical waters, and most of the following examples are typically warm water species.

Examples of pleustonic species include the colonial cnidarians *Physalia* (Figure 4.7a) and *Velella* (Colour Plate 27) and their relatives, all of which have gas-filled floats that project above the water surface. The long trailing tentacles of *Physalia* enable it to capture zooplankton and small fish well below the sea surface. *Velella* has short tentacles and its food (copepods, larval crustaceans, and eggs of fish and euphausiids) is captured from very near the surface. Despite their stinging tentacles, sea turtles and several surface-dwelling molluscs feed on these Cnidaria. *Janthina* (Colour Plate 28) is a snail that builds a raft of air bubbles encased in mucus; the animal hangs suspended, upside-down, from this float at the sea surface. It feeds on both *Physalia* and *Velella* as well as on other neustonic animals. *Glaucus* (Colour Plate 29) is a nudibranch (or sea slug) that floats upside-down at the surface by ingesting air which it then stores in special sacs of the digestive tract. It too feeds on both *Physalia* and *Velella*, and it ingests the nematocysts along with other tissues. *Glaucus* has the remarkable ability to absorb the stinging cells without causing them to discharge, and the nematocysts are then conveyed by cilia to special sacs in the tips of external papillae on the nudibranch's dorsal surface. There, they can be employed as a defence against predators, and they will discharge if the animal is disturbed. Bathers who come in contact with *Glaucus* which have washed ashore report painful stings lasting several hours. Other permanent members of the neuston community include the only insects found in the open ocean; they belong to the insect order Hemiptera and the genus *Halobates*. These wingless water striders cannot survive immersion in seawater and thus are restricted to eating other organisms living at the immediate surface, including floating cnidarians and neustonic species of copepods.

Smaller organisms are also present in the immediate surface layers of the sea. Bubbles produced by breaking waves accumulate organic material in a thin surface scum, and this provides a rich substrate for bacteria which often concentrate in films. This very thin surface region of only a few millimetres in depth may contain from 10 to 1000 times more bacteria than in the water immediately below, and consequently these organisms may be an important part of the neuston, forming food for large numbers of protozoans such as tintinnids. The high light intensities inhibit photosynthesis and because few phytoplankton are found in this zone, the bacteria and protozoa are probably important components in the diets of grazing copepods. Many types of meroplanktonic invertebrate larvae and the eggs and larvae of certain fish species are also commonly present in this zone. Indeed, the eggs of some fish (e.g. anchovy, mullet) are extremely buoyant, adhering to the surface film, and the upper few centimetres of the sea support many different types of fish larvae. Some of these fish remain at all times near the surface, but others — as well as some other types of zooplankton — are transient members of the neuston, usually moving to the surface zone at night to feed. The accumulation of organisms at the sea-air interface creates an important feeding zone for oceanic birds (e.g. petrels, skimmers, fulmars), many of which have bills adapted for skimming this layer (see Section 6.5 and Figure 6.2).

What special environmental features are present at the sea air interface to which neustonic species must adapt?

This region receives very high levels of infrared and ultraviolet light, the latter being detrimental to many organisms. The high light intensities also inhibit the production of phytoplankton, and consequently few, if any, of the zooplankton residing in this zone are strict herbivores. In calm weather, this region may experience the most extreme 24-hour changes in temperature and salinity, although mixing usually produces a temperature-salinity regime that is indistinguishable from that of the upper 2–3 m. Permanent inhabitants of this zone are also exposed to extreme wave action during storms. And neustonic animals, like intertidal species, are exposed to both marine and aerial predators.

Neustonic animals live in a region of high light intensity. What types of defences might they have against predators that hunt by sight?

Some of the neuston, especially crab and fish larvae, are highly transparent and therefore difficult to discern by sight. But many tropical neustonic animals (e.g. *Velella, Halobates, Janthina, Glaucus*) are strikingly coloured with violet or blue pigments. It has been suggested that these colours may provide protection from the high levels of ultraviolet light, but such coloration also may be effective as camouflage from predators because it blends closely with the blue colour of tropical oceanic water. *Janthina* and *Glaucus* also exhibit **counter-shading;** that is, those body surfaces that are directed downward in the water are lighter in colour than areas close to the surface, and this too may be a predator defence. Marine predators like fish, which approach from below, would view the lighter undersides of these molluscs against a background of lightly coloured sky whereas, to aerial predators, the darker blue upper surfaces of prey would blend with the dark blue colour of the water. Cryptic coloration is not the only defence mechanism of neuston. For example, some neustonic copepods have developed the ability to jump out of the water in response to predators.

A special surface community has developed in the Sargasso Sea. *Sargassum*, a floating seaweed, forms an extensive habitat for an unique association of more than 50 species of animals. The total wet weight of *Sargassum* in the community has been estimated at between 4 and 10 million tonnes. Many of the animals found living in the seaweed are primarily benthic, and these include hydroids, sea anemones, crabs, shrimp, and other Crustacea. Some of the **endemic** species (that is, those restricted to this particular habitat), including some of the crabs and fish, have developed special camouflage protection by coming to resemble *Sargassum* weed both in shape and colour.

In all areas, the region immediately below the sea surface and extending to 200 or 300 m is referred to as the **epipelagic** zone (Figure 1.1). Many zooplankton are permanent residents of this zone, others migrate into this region at night. Only the zooplankton that live in depths shallower than 300 m during the daytime are regarded as truly epipelagic. The epipelagic zone coincides with the euphotic and disphotic zones (see Section 2.1.2 and Figure 2.5), and it supports a great diversity and abundance of life. Many herbivores and omnivores inhabit this region; these include smaller crustaceans (such as copepods), the thecosomatous pteropods, salps, larvaceans, and meroplanktonic larvae. Many of the species are relatively small, and many are transparent.

The **mesopelagic** zone (Figure 1.1) lies between the bottom of the epipelagic region and a depth of approximately 1000 m, and the animals that live here in the daytime are called mesopelagic species. Many mesopelagic zooplankton tend to be larger than their epipelagic relatives. In this deep nonturbulent water, even the delicate-bodied, transparent, gelatinous zooplankton become more diverse and increase in size. For example, *in situ* observations have revealed a deep-sea larvacean (*Bathochordaeus*) that makes a house of 2 m in size, and siphonophores have been seen that can extend to 40 m in length, making them among the longest animals known. Some of the animals of this region, such as euphausiids, are at least partly herbivorous, and they move upward to the epipelagic zone at night to feed on phytoplankton. Many of the residents, however, are carnivores or detritus feeders, feeding on larger particles.

Particulate organic material sinking from above, especially faecal pellets, tends to accumulate at depths of about 400–800 m because of the density gradient (pycnocline) that is associated with the permanent thermocline. This rich food source is decomposed by bacteria, but is also consumed by zooplankton that tend to congregate in these regions, and the decomposition and animal respiration result in high levels of oxygen utilization. These biological activities contribute to the formation of **oxygen minimum layers**, or zones where oxygen concentrations may fall from the normal range of 4–6 mg l^{-1} to less than 2 mg l^{-1}, even approaching anoxic conditions in some areas. Physical factors also are involved in the formation of oxygen minimum layers. Oxygen is replenished at the sea surface through contact with atmospheric oxygen, and is carried into deep areas through the sinking of oxygen-laden surface water. The oxygen minimum layer thus marks the intermediate depth of minimal physical replenishment, as well as a zone of high respiration. Certain species such as the 'vampire squid', *Vampyroteuthis infernalis*, seem to have become uniquely adapted to live permanently within the low-oxygen zones; other species migrate in and out of the oxygen minimum layers.

Many mesopelagic animals have developed red or black coloration. For example, all pelagic shrimp living below 500–700 m by day are uniformly bright red (Colour Plate 23), whereas those living in shallower depths are transparent or semi-transparent. Many mesopelagic zooplankton (and fish) also have larger eyes and increased sensitivity to blue-green wavelengths of light; these are the deepest penetrating wavelengths of solar radiation and also the spectrum of most bioluminescent light.

QUESTION 4.3 What is the adaptive significance of red coloration in deep-sea zooplankton?

Hand to see in Blue-green light.

Bioluminescence refers to light produced and emitted by organisms themselves, and it is known in marine species of bacteria, dinoflagellates, many invertebrates (both pelagic and benthic), and some fish. No amphibians, reptiles, birds, or mammals possess this property, and only one freshwater invertebrate is known to be luminous. Although the phenomenon occurs in many shallow-living marine species, bioluminescence becomes increasingly important in the deep sea where it is the only source of light below about 1000 m. In the midwater disphotic zone, more than 90% of the resident species of crustaceans, gelatinous zooplankton, fish, and squid emit light.

The biochemistry of bioluminescence is not completely understood and apparently varies within different species. But, in general, biologically generated light results from the oxidation of organic compounds known as luciferins in the presence of the enzyme luciferase. The chemicals involved in the reaction are synthesized by living cells. However, midshipman fish (*Porichthys*) are known to acquire luciferin from their diet, by eating luminous crustaceans; in regions where these crustaceans do not occur, the fish are nonluminous even though they possess the enzyme luciferase. In the chemical reactions of bioluminescence, the energy, instead of being released as heat (as occurs in most chemical reactions), is used for the excitation of a product molecule, called oxyluciferin. This 'excited' compound (indicated by an asterisk below) then releases the energy as a photon, producing light.

$$\text{luciferin} + O_2 \xrightarrow{\text{luciferase}} \text{oxyluciferin}^* \rightarrow \text{oxyluciferin} + \text{light}$$

The reaction shown above can be carried out in special cells called **photocytes**, or in complex organs known as **photophores**.

Bioluminescence may be used for various types of communication in the sea, but in many species the behavioural or ecological role of the light signals remains unknown. In some planktonic species, bioluminescent displays result when the organisms are disturbed and are therefore thought to be employed as a predator defence. When disturbed, some medusae, siphonophores, ctenophores, ostracods, and deep-sea squid shed luminescing tentacles or produce clouds of luminous material as apparent decoys for predators while the darkened animal itself swims away. In some pelagic species, bioluminescence may serve as a type of camouflage by acting as a counter-illumination system which eliminates an animal's silhouette against downward penetrating daylight. Animals with eyes may use bioluminescent signals to communicate with other individuals of their own species. Euphausiids, for example, may respond to luminescence in other individuals to form densely crowded schools when chased by large predators, or to aggregate for breeding purposes. Some siphonophores and deep-sea fish employ bioluminescent lures to attract prey within close range, thus eliminating the need to expend energy in hunting for food. The ability to produce light obviously has evolved independently in many organisms, and it clearly serves a variety of roles.

Bioluminescence also occurs in some of the **bathypelagic** species that inhabit the dark water layers from 1000 to 3000 or 4000 m, and in some of the **abyssopelagic** species living below these depths (Figure 1.1). In these zones, many of the zooplankton and fish tend to be deep red or black in colour and many have smaller eyes than the mesopelagic species. Away from the productive surface waters, there are fewer species and fewer individuals.

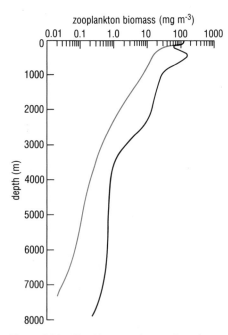

Figure 4.14   The biomass of net-collected zooplankton (excluding Cnidaria and salps) from the surface to 8000 m depth in the Northwest Pacific Ocean (45°N) (black line) and in the tropical Pacific (6°S) (blue line) during July.

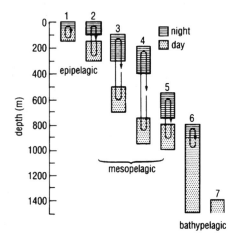

Figure 4.15   A schematic illustration of the diel migration patterns of pelagic shrimps living in different vertical zones. Dotted and hatched areas indicate the depth ranges of the main day and night concentrations, respectively, of the different groups. Each group (1–7) is a composite of different species that occupy similar depth ranges.

The general decrease in zooplankton biomass with depth is shown in Figure 4.14 for several regions in the Pacific Ocean. On average, the biomass of plankton collected in nets decreases by 1–1.5 orders of magnitude from the surface to 1000 m, and decreases by another order of magnitude between 1000 m and 4000 m. This exponential decrease in biomass with depth is correlated generally with longer generation times and lower fecundity in deeper-dwelling species.

One other category of vertical distribution has been established to encompass those pelagic species that live either close to the seafloor, or that are temporarily in direct contact with the sea bottom. These animals are referred to either as **epibenthic** or **demersal**. This category includes many Crustacea, especially shrimp and mysids, and it also applies to bottom-dwelling fish, like sole and plaice. Where the water is not very deep, these species too may move away from the seafloor at night.

The depth classification system given here is based on recognition that environmental conditions change with increasing depth in the sea, and that the animals living at different depths have generally evolved different life strategies. However, it should be stressed that the vertical ecological zones described here are arbitrary regions, and that individuals and species are not distributed uniformly within these depths (see Figure 4.22). Many animals move between the different zones, and the deeper meso-, bathy-, and abyssopelagic zones are not clearly distinguishable from each other. Further, the depth distribution of some species may change with latitude. This is particularly true of those cold-water species that are tolerant of a great range of hydrostatic pressure. The chaetognath *Eukrohnia hamata*, for example, lives near the surface in polar areas, but is found only in deep, cold water in low latitudes.

4.5   DIEL VERTICAL MIGRATION

One of the most characteristic behavioural features of plankton is a vertical migration that occurs with a 24-hour periodicity. This has often been referred to as diurnal vertical migration. However, diurnal refers to events that occur during daytime; it is the opposite of nocturnal. Diel refers to events that occur with a 24-hour rhythm. **Diel vertical migration** (or DVM) is usually marked by the upward migration of organisms towards the surface at night, and a downward movement to deeper waters in the daytime. This phenomenon has been known since the time of the *Challenger* Expedition (Section 1.4), but even now we do not have entirely satisfactory explanations for the widespread occurrence and ecological significance of this 24-hour rhythmical movement. Diel vertical migration occurs in at least some species of all the major groups of zooplankton (freshwater species as well), and it is known in dinoflagellates and in many nektonic species, including both cephalopods and fish. Diel vertical migration occurs in many (but not all) epipelagic and mesopelagic species and, although few studies have been done on deeper-living plankton, it is known in some bathypelagic shrimp (Figure 4.15).

Because of diel vertical migrations, a comparison of day and night plankton tows taken in the same area at the same depths will always show differences in species composition and total biomass. This can be seen, for example, in

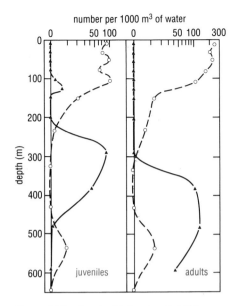

Figure 4.16   Day (solid lines) and night (dashed lines) distributions of the juveniles and adults of *Euphausia hemigibba* at one station in the California Current.

Figure 4.16 which compares the day and night-time distributions of juvenile and adult euphausiids off the coast of California.

QUESTION 4.4   What is the general pattern of vertical migration displayed by euphausiids in Figure 4.15, and how do the migrations of adults and juveniles differ?

Each species has its own preferred day and night depth range, and this may vary with the life stage (as illustrated in Figure 4.16) or with sex of an individual (e.g. adult female *Calanus finmarchicus* are strong migrators, whereas males are not). The preferred depth range may also change with season, geographic location, and general weather conditions (e.g. cloudiness, storm turbulence, etc.). In general, however, there are three patterns shown by migrating marine zooplankton:

1   **Nocturnal migration** is characterized by a single daily ascent, usually beginning near sunset, and a single descent from the upper layers which occurs near sunrise. This is the most common pattern displayed by marine zooplankton.

2   **Twilight migration** is marked by two ascents and two descents every 24 hours. There is a sunset rise to a minimum night-time depth, but during the night there is a descent called the midnight sink. At sunrise, the animals again rise toward the surface, then later descend to the daytime depth.

3   **Reverse migration** is the least common pattern. It is characterized by a surface rise during the day and a night-time descent to a maximum depth.

The vertical distance travelled over 24 hours varies, generally being greater among larger species and better swimmers. But even small copepods and small thecosomes may migrate several hundred metres twice in a 24-hour period, and stronger swimmers like euphausiids and pelagic shrimp may travel 800 m or more. Upward swimming speeds of copepods and the larvae of barnacles and crabs have been measured at 10–170 m h^{-1}, and euphausiids swim at rates of 100–200 m h^{-1}. Although the depth range of migration may be inhibited by the presence of a thermocline or pycnocline, this is not necessarily so, and an animal may traverse strong temperature and density gradients, as well as considerable pressure changes, during its migration.

Diel vertical migrations are responsible for the production of moving **deep scattering layers** (or DSLs). These are sound-reflecting layers picked up by sonar traces. They look like false sea bottoms on echograms (Figure 4.17), and they were initially believed to be the result of physical phenomena. However, these layers may move over 24 hours, and their rhythm provided a clue that they were caused by the movements of animals. During the day, as many as five scattering layers may be recorded at depths of about 100–750 m. At night, the layers rise almost to the surface and diffuse, or they merge into a broad band extending down to about 150 m. These deep scattering layers seem to be most frequently caused by the movements of larger crustaceans (e.g. euphausiids, shrimp) and small fish that possess sound-reflecting air bladders (e.g. myctophids), but other zooplankton such as heteropods and large copepods can occasionally form sound-reflecting layers too.

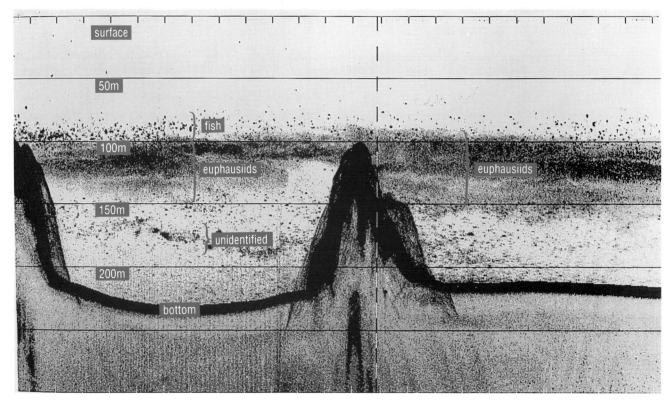

Figure 4.17 An echogram showing day-time deep scattering layers produced by euphausiids (*ca.* 90−150 m), fish (*ca.* 75−100 m) and unidentified animals (*ca.* 175 m) in Saanich Inlet, British Columbia, Canada. Note that the fish show up as discrete dots, whereas the smaller but more abundant euphausiids produce a more even shading pattern. Note also the irregularity of the seafloor, with an abrupt rise from about 225 m to 100 m in the centre of this transect.

The fact that diel vertical migrations are tuned to the natural light:dark cycle suggests that changes in ambient light intensity may be of primary importance as stimuli in initiating and timing the migrations. Light intensity changes can also act as orienting cues for vertically migrating animals, as can gravity and changes in hydrostatic pressure. Natural changes in light intensity which occur seasonally or even daily (e.g. sunny vs. cloudy days; dark nights vs. moonlit nights) can alter the depth ranges inhabited by particular species. Under continuous light in the Arctic summer, migrations may be totally suppressed. Solar eclipses will cause animals to begin an upward migration during the day as light intensity decreases. In the laboratory, the timing of migrations may or may not change to conform with experimental alternations of light and dark periods. Factors other than light may also play a role in initiating the diel migrations; among those suggested as a causal mechanism is hunger, driving animals upward toward the more productive areas under the protective cover of darkness.

While light or other factors may trigger the diel vertical migrations of pelagic organisms, it does not explain why so many species should show this behaviour. What is the adaptive value of vertical migration?

Many hypotheses have been advanced to answer this question, but it may not be realistic to insist on a universal mechanism governing diel vertical migration in all species. It is important to recognize that the hypotheses discussed below may not be mutually exclusive, and that each may be more applicable to some species than others.

1   One hypothesis is that animals which remain in darkness or near-darkness over 24 hours are less vulnerable to visual predators, and this

can be achieved by a daylight descent. The upward migration returns the animals to the surface where food is most abundant.

Considerable evidence supports this hypothesis as being the ultimate cause of DVM in both marine and freshwater species. It has been shown, for example, that diel migrations in several zooplankton species may become more pronounced when predatory fish are more abundant. As well, a 45-year-long study of *Metridia lucens* in the North Atlantic has shown that the length of time this copepod was present near the surface varied seasonally, being shorter in the summer when nights are shorter. However, when food was most abundant during the spring, the animals remained longer at the surface than was predicted from length of daylight; the importance of obtaining food when it is most abundant seems to override the importance of predator avoidance at this time.

2   A second idea is that zooplankton conserve energy by spending non-feeding time in deeper, colder water where metabolic energy demands are less. It has not been proved that the energy saved in a colder environment would offset the amount of energy used in swimming during migration. However, energy required for swimming is very low, generally only a few percent of basic metabolic energy.

3   A third hypothesis recognizes that zooplankton moving vertically in the water column are subjected to currents moving in different directions at different speeds. Thus they encounter a new feeding area each time they ascend. The new feeding area may contain more or less food than the area occupied the night before but, by migrating vertically, small organisms of limited mobility can avoid remaining in an area of little food as well as overgrazing any very productive area. However, experimental manipulations of food concentrations produce conflicting results depending on the species. In some cases, low food levels suppress vertical migration; in other examples, the reverse is true.

Diel vertical migration has several consequences that are biologically and ecologically important. One is that, since all individuals of a species do not migrate at precisely the same time and to the same depths, a population will eventually lose some individuals and gain others. This mixture of individuals from different populations enhances genetic mixing and is especially important in species of limited horizontal mobility.

Another important result of vertical migration is that it increases and hastens the transfer of organic materials produced in the euphotic zone to deeper areas of the sea. The ladder-like series of migrating organisms (Figure 4.15) plays an important role in marine food chains. Each migrating animal removes food from shallower depths during the night; this material is then actively transported to deeper areas in the daytime. Herbivores remove phytoplankton from the euphotic zone, then migrate to deeper areas where they release faecal pellets and other organic debris and where they may be eaten by deeper-living carnivores. The carnivores and scavengers in turn carry out vertical migrations at greater depths. The active vertical transport of organic materials, either in the form of the animals themselves or in their faecal pellets and other wastes is significantly faster than the passive sinking of organic particles.

4.6 SEASONAL VERTICAL MIGRATIONS

In some species, vertical migration patterns change seasonally and may be associated with breeding cycles and changing depth preferences of different stages in the life cycle. In the North Pacific Ocean, the dominant copepods show dramatic changes in their depth patterns. In inshore waters off the western coast of Canada, *Neocalanus plumchrus* adults do not feed, and they overwinter at about 300–450 m depth where the eggs are laid between December and April (Figure 4.18). The eggs float toward the surface, and nauplii (see Section 4.2) hatch and develop at intermediate depths. Nauplii are present in near-surface waters from February to April, and the population matures to the copepodite V stage during March to June when primary productivity is highest. By early June, stage V individuals contain large amounts of lipids accumulated from feeding on phytoplankton, and they begin to migrate to deeper waters where they will subsist on this stored fat reserve. There they mature to the adult stage VI, mate, and lay eggs during the winter. In offshore waters, the life cycle changes somewhat, with spawning in deep water (>250 m) taking place from July to February and early copepodite stages first appearing in the upper 100 m in October. Nevertheless the species continues to show a seasonal migration between surface waters, where larval development takes place, and deeper waters, where mating and spawning occur. A similar pattern of vertical migration associated with different reproductive stages takes place in *Neocalanus cristatus*, a large copepod also common to the North Pacific: adults are present between 500 m and 2000 m, and spawning occurs in deep water; younger stages move upward and live mostly above 250 m.

Figure 4.19 shows both the seasonal and diel vertical migrations in two species of North Atlantic herbivorous copepods, *Calanus helgolandicus* and *C. finmarchicus*. During winter in the Celtic Sea, copepodites V and VI of both species are distributed fairly uniformly from the surface to about 100 m,

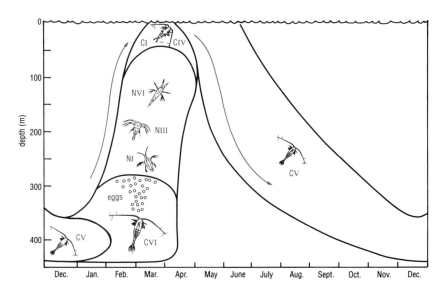

Figure 4.18 A schematic diagram of the life cycle of the copepod *Neocalanus plumchrus* in coastal waters off British Columbia, Canada. The depth distributions of the eggs, larvae (nauplii I–VI and copepodites I–V) and adults (copepodite VI) are shown over the course of one year. C, copepodite; N, nauplius.

(a) *Calanus helgolandicus*

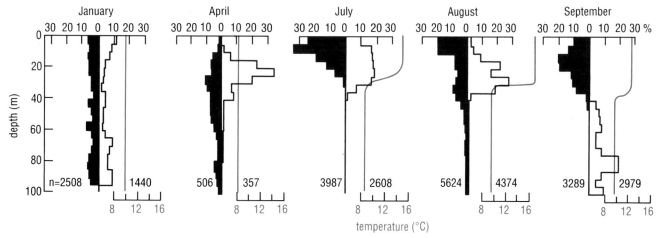

(b) *Calanus finmarchicus*

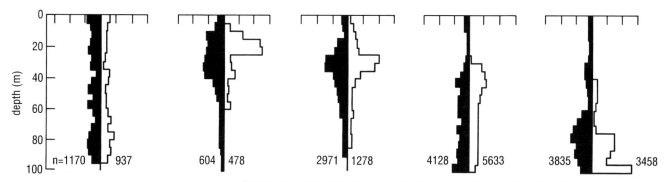

Figure 4.19   Seasonal changes in the day-time (white) and night-time (black) vertical distribution of copepodite stages V and VI of two species of copepods, *Calanus helgolandicus* (a) and *C. finmarchicus* (b), in the Celtic Sea. Numbers in each plankton haul are plotted in 5 m depth intervals as percentages of total numbers (n) present in the haul. Temperature profiles are shown for the day hauls and apply to both species.

and there is little difference between day and night-time distributions. In spring (April), both species begin to concentrate in shallower depths, and they display diel vertical migrations. In July and August, the thermocline becomes well established and the two species show a clear separation in their distributions. *Calanus helgolandicus* continues to develop in the warmer surface zone and to display diel migration, but *C. finmarchicus* moves deeper into cooler water beneath the thermocline and shows little difference in day and night depth preferences. By late September, both species reside in water deeper than 40 m during the day, and *C. helgolandicus* maintains its strong vertical movement toward the surface at night.

QUESTION 4.5   Of what advantage is the seasonal change in vertical migratory patterns to *Calanus helgolandicus* and *C. finmarchicus*?

The Antarctic krill (*Euphausia superba*) also undergoes extensive depth changes during its life cycle. The eggs of the krill are deposited in surface waters but sink rapidly to depths of 500–2000 m, where they hatch. The larvae then gradually float and swim to the surface where development is completed, and juveniles and adults are found at or very near the surface.

The total life span is estimated to be at least 2–4 years. During this time, the vertical migration of the different stages into currents moving in different directions results in transport away from, and back toward, the Antarctic continent. Several common Antarctic species of copepods and chaetognaths also undergo similar seasonal vertical migrations: in the southern summer, the species are present in surface water flowing northward from the area known as the Antarctic Divergence; in the winter, they are found in a southerly-flowing deeper current which rises to the surface at the Divergence.

Extensive seasonal migrations are generally undertaken only by species living in temperate and cold waters, or in upwelling regions. The migrations usually result in young animals being within the productive surface waters at a time when they can obtain sufficient quantities of food for growth. In temperate waters, as production declines in the surface waters during summer and fall, late larval stages or sexually mature adults move to deeper waters. Here, in colder and unproductive waters, they may enter a state called **diapause** in which their metabolism slows and they do not feed. Instead, they subsist on energy reserves built up during their stay in the surface zone.

Both diel and seasonal vertical migrations place migrants in currents that are moving in different directions and at variable speeds. Despite this, populations of marine zooplankton do persist in their own characteristic geographical regions. A distinctive pattern of seasonal vertical migrations in a species may ensure retention within an appropriate habitat (as it does with Antarctic zooplankton), or within a productive upwelling area (as is thought to be the case for several species associated with annual upwelling cycles off the western coasts of North America and Africa). Diel migrations may similarly retain animals in favourable habitats. For example, the diel migrations of estuarine species may be attuned to the inward and outward tidal flows of water to retain the animals within the estuary. In all such cases, natural selection favours those individuals that migrate with appropriate timing; any individuals with behavioural patterns that do not conform to the physical system will tend to be lost from the region. The corollary to this is that at least some species are known to exhibit different vertical migratory patterns in different locations, further suggesting that these patterns can be modified to ensure persistence of populations in particular geographic regions. What appears to be a confusing variability in migratory patterns among and within species may be the result of evolutionary adaptations leading to maintenance of populations in favourable environments of different current regimes.

4.7 ZOOGEOGRAPHY OF THE HOLOPLANKTON

Zoogeographic studies describe the distributions of living organisms and investigate the physiological and ecological causes underlying the patterns. Such studies may also be historical, as the present-day distributions of marine organisms also have been determined by events and changes taking place over geological time.

Compared with terrestrial environments, the pelagic realm has few physical barriers to impede mixing and gene flow between populations of holoplanktonic organisms. But there are hydrographic barriers between different water masses which have distinctive complements of physico-chemical conditions and ecological properties. Some zooplankton have wide distributions encompassing a broad spectrum of environmental determinants, but others are restricted to such narrow limits of temperature, salinity, and other factors that they can be used as **biological indicators** of the particular water-mass types they inhabit. The concept of using indicator species to characterize specific water masses has most frequently been applied to certain species of foraminifera, copepods, and chaetognaths, as they are usually sufficiently abundant to be sampled routinely (see, for example, Figure 6.9).

The sharp north-south temperature gradients set major environmental provinces in the ocean surface waters as described in Section 2.2.1. Even so, roughly 50% of all epipelagic zooplankton species extend from tropical and subtropical waters into at least part of the temperate zone. Only about one-third of the epipelagic holoplankton are restricted to the warm waters of the tropics and subtropics. Fewer species are restricted to cold and/or temperate waters, but included in this category are several species with **bipolar distribution**. Bipolar species are found in both arctic/subarctic water and antarctic/subantarctic regions, but are not present in intervening regions. They include the pteropods *Limacina helicina* and *L. retroversa*, the amphipod *Parathemisto gaudichaudi*, and a siphonophore (*Dimophyes arctica*) as well as several diatoms. There are also Arctic and Antarctic species which are so closely similar that they certainly indicate a common ancestry. An example of these 'species pairs' would be the gymnosomes *Clione limacina* (present in the Northern Hemisphere) and *C. antarctica* (Southern Hemisphere); although morphologically distinguishable, these species occupy the same position in polar food chains where both feed exclusively on *Limacina helicina* and *L. retroversa*. Bipolarity may have arisen from animals being transported in the deep, cold water link between the north and south polar regions. An alternative theory proposes that formerly cosmopolitan species were displaced from lower latitudes by competition with other zooplankton, leaving relict populations in the high latitudes.

There are fewer latitudinal hydrographic barriers in deeper water (see Section 2.4), and meso- and bathypelagic plankton generally have relatively wide distributions. However, they are by no means cosmopolitan; that is, many species are restricted to one of the major oceans. For example, it is believed that approximately half of the bathypelagic fauna of the North Pacific is endemic to that area. About 20% of the deep-living North Pacific copepods are Antarctic forms, which is not unexpected because the Antarctic is the source of much of the deep water in all the oceans (see Section 2.4 and Figure 2.17).

The number of epipelagic species decreases from low to high latitudes, a phenomenon seen also in terrestrial animals. Most of the different zooplankton groups have fewer species in cold waters, and some (e.g. heteropods) have no representatives outside of subtropical boundaries (about 45°N–45°S). As species diversity declines in high latitudes, there is a reverse tendency for cold-water epipelagic species to have greater numbers of individuals. Although many hypotheses have been proposed to explain these latitudinal differences in diversity and abundance, the reasons remain debated.

There are a relatively large number of pelagic marine species that are described as 'circumglobal tropical-subtropical'. These species are present in warm waters of the Atlantic, Pacific and Indian Oceans. They include many of the neustonic animals such as *Janthina* species and *Glaucus atlanticus* (see Section 4.4), as well as representatives of epipelagic euphausiids, chaetognaths, amphipods, and other major groups. This wide present-day distribution can be attributed to the long continuity of the warm water masses of the world via the ancient Tethys Ocean that existed for several hundred million years, from the Paleozoic until the Late Tertiary (see Appendix 1 for dates). Many warm-water species of zooplankton, however, are restricted to one or two of the main oceans, and their distributions may be related to present-day barriers (e.g. the Isthmus of Panama) that now close former routes between major oceans.

Do humans influence the distributions of zooplankton?

Constructions such as the Suez Canal have affected zooplankton distributions. Since this canal was opened in 1869, about 140 species have entered the Mediterranean from the Red Sea. Zooplankton distributions also may be changed by accidental transport in the ballast water of ships (see also Section 9.3). This is believed to be the way in which the ctenophore *Mnemiopsis leidyi* moved from the Atlantic coast of North America into the highly polluted Black Sea. Within two to three years, this predatory species increased to a biomass of about 10^9 tonnes in 1990. Its effect on biological communities and fish stocks in the Black Sea is estimated to be greater than that of all other anthropogenic factors.

The North Atlantic and North Pacific Oceans have been separated by the land mass of North America for approximately 150–200 million years. Yet the temperate zones of these oceans show many similarities in their resident fauna (Table 4.2). The medusa *Aglantha digitale*, the chaetognath *Eukrohnia*

Table 4.2 Numerically dominant net-collected zooplankton species in the epipelagic zones of the northern North Atlantic and northern North Pacific. Species aligned between the two columns are found in both oceans.

Group	North Atlantic Ocean	North Pacific Ocean
Cnidaria	*Aglantha digitale*	
Annelida	*Tomopteris septentrionalis*	
Chaetognatha	*Eukrohnia hamata*	
	Sagitta serratodentata	*Sagitta elegans*
	Sagitta maxima	
Amphipoda	*Parathemisto pacifica*	
Copepoda	*Calanus finmarchicus*	*Calanus pacificus*
	Calanus helgolandicus	*Neocalanus plumchrus*
	Euchaeta norvegica	*Neocalanus cristatus*
	Pleuromamma robusta	*Eucalanus bungii*
	Acartia clausi	*Acartia longiremis*
	Metridia lucens	*Metridia pacifica*
	Oithona spp.	*Oithona similis*
	Oncaea spp.	
	Scolecithricella minor	
	Heterorhabdus norvegicus	*Pseudocalanus minutus*
		Paracalanus parvus
Euphausiacea	*Meganyctiphanes norvegica*	*Euphausia pacifica*
	Thysanoessa longicaudata	*Thysanoessa longipes*
Mollusca	*Limacina helicina*	
	Limacina retroversa	
	Clione limacina	
Salps	*Salpa fusiformis*	

hamata, the pteropods *Limacina helicina* and *Clione limacina* are all examples of species found in both areas, as well as in the Arctic Ocean. On the other hand, the dominant copepods and euphausiids of the North Pacific and North Atlantic are mostly of different species, although some are in the same genera (Table 4.2). Speciation and present-day distributions in these areas have been influenced by geological changes marked by intermittent changes in water flow between the Pacific, Arctic, and Atlantic oceans. At the present time, there is almost no southward flow of water through the Bering Strait, but there is a flow from the Pacific into the Arctic as there presumably has been whenever the Strait has been open. There is a much greater water exchange between the Arctic and the Atlantic. Present-day distributions suggest that the Arctic was the passageway between the Atlantic and Pacific for dispersal of many planktonic species. However, cooling of the Arctic during the Pliocene and Pleistocene periods (see Appendix 1) may have resulted in the many examples of species with discontinuous ranges (found in the northern Atlantic and Pacific but not in the Arctic) and in subsequent evolution into distinct species.

An unbroken circumglobal ocean lies in the Southern Hemisphere between Antarctica and the continents of Africa, Australia, and South America. Zooplankton in this broad oceanic area exhibit continuous, concentric patterns of distribution around the Antarctic continent which conform to concentric isotherms and the general clockwise circulation pattern. However, there are also currents that flow northward from the southward toward Antarctica and, as pointed out in Section 4.6, seasonal vertical migrations in these currents also serve to maintain populations around the continent. The pelagic fauna of the Antarctic Ocean is much richer in numbers of species than that of the Arctic Ocean, and this difference may be related to the higher productivity of the southern ocean.

4.7.1 PATCHINESS

Within the geographical boundaries inhabited by any species, the individuals of that species are not distributed uniformly or randomly, but are usually aggregated into 'patches' of variable size. This **patchiness** is true of both phytoplankton and zooplankton, as well as of other types of marine and terrestrial species. Patchiness in phytoplankton distribution was introduced in Section 3.5, where it was related to physical processes that control nutrient availability and thus plant production on scales ranging from oceanic gyres to Langmuir cells of circulation. Zooplankton patchiness may be correlated with phytoplankton concentrations, or it may be caused by other factors.

Small-scale heterogeneity in the horizontal distribution of zooplankton (also known as microdistribution) is more difficult to detect and study than broad geographic patterns because of the way zooplankton are collected. Nets are generally towed through the water for distances ranging from tens of metres to kilometres, so that the numbers of collected individuals are averaged over distances that will mask any smaller-scale patterns. Specially designed sampling programs have demonstrated microdistributional patterns in zooplankton, as have direct observations by scuba divers and observers in submersibles. Patchy distribution on these smaller scales can be explained in a number of ways, and may be related to physical, chemical, or biological events.

Various types of horizontal and turbulent mixing can result in aggregation or dispersion of planktonic populations. As discussed in Section 3.5, some types of mixing (upwelling) result in elevated surface nutrient concentrations, high primary production, and increased numbers of zooplankton; other forms of mixing (downwelling) have the opposite effect on production and aggregation of organisms. Zones of vertical mixing range in area from very large shelf-break fronts (Section 3.5.4), to smaller scale cold- or warm-core rings (Section 3.5.1), to much smaller Langmuir circulation patterns (Section 3.5.6), all of which can affect zooplankton distribution and abundance on corresponding spatial scales. Differences in scale are shown in Figure 4.20, which illustrates changes in numbers of zooplankton (and phytoplankton) on a scale of kilometres, and in Figure 4.21, which shows smaller-scale patchiness of zooplankton on a scale of metres. These figures illustrate patchiness on a horizontal axis, but zooplankton also form discrete aggregates in the vertical dimension. Figure 4.22 shows the vertical distribution of copepods in the Bering Sea; note the vertical separation of the species within the epipelagic and mesopelagic zones, as well as the discrete depths inhabited by different life stages and by males and females.

What are some of the biological or ecological causes of patchy distribution of zooplankton?

Patchiness may result from interactions between zooplankton and their food. On time scales of months, high primary production may result in high secondary production (as in coastal upwelling) but, on shorter time scales and in smaller areas, dense aggregations of phytoplankton and of herbivorous zooplankton tend to be mutually exclusive (Figure 4.20). This may result from heavy grazing by the herbivores which reduces the numbers of phytoplankton. It also may be the result of differences in growth rates between the algae and the zooplankton; whereas phytoplankton can quickly multiply under favourable light and nutrient concentrations, increases in numbers of zooplankton often lag considerably behind because of their slower generation times. Consequently, when phytoplankton numbers are peaking and nutrients are declining, zooplankton biomass may still be low as the animals begin growing in response to the elevated food supply.

Reproduction may also play a role in causing patchy distribution in some species. Aggregations of zooplankton formed for purposes of breeding will

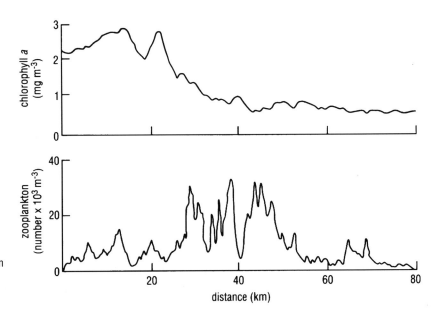

Figure 4.20   Patchiness in phytoplankton (as indicated by chlorophyll *a* concentration) and in zooplankton, on a kilometre scale. Based on night-time data taken from 3 m depth in the northern North Sea, May, 1976.

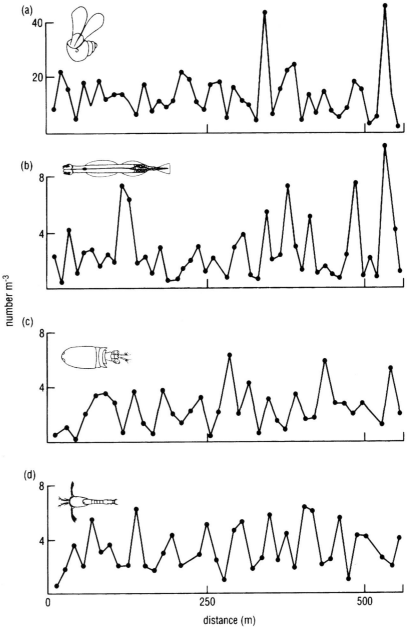

Figure 4.21 Examples of small-scale (in metres) patchy distribution of zooplankton off the California coast. (a) A shelled pteropod. *Limacina*; (b) a chaetognath, *Sagitta*; (c) a copepod, *Corycaeus*; and (d) euphausiid larvae.

cause a small-scale heterogeneous distribution, although the mechanisms in which the members of the swarm unite are not understood. Also, all the progeny hatching from one swarm, or even from one egg mass, tend to remain together for some period of time before they become dispersed.

Considerable attention has been given to the patchy distribution of Antarctic krill because of their abundance and consequent importance for higher trophic levels, and because of the potential for commercial harvesting of this species (see Section 6.1). When feeding, *Euphausia superba* forms swarms in which the individuals are closely packed but move independently of each

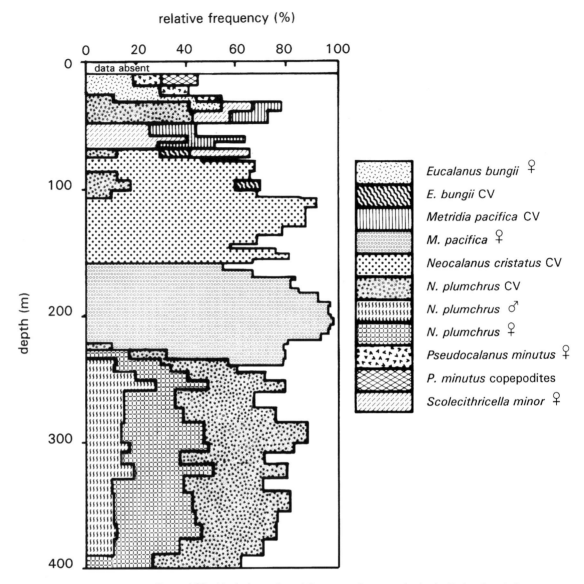

relative frequency (%)

Figure 4.22 Vertical zonation of the copepod community in the Bering Sea during summer. Samples were collected during daylight hours. ♀, females; ♂, males; CV, copepodite stage V (see Section 4.2).

other. At other times, the euphausiids are organized in schools, in which the individuals are uniformly oriented and swim together at a uniform speed. The formation of schools is thought to offer some defence against certain predators but, on the other hand, some predators may force schooling in order to concentrate their prey. For example, some temperate-water species of euphausiids may be driven into tight schools by sharks, or by whales which produce a 'net' of bubbles to encircle and concentrate their prey. In general, extremely high concentrations of predators (e.g. swarms of medusae or ctenophores, or fish schools) will quickly cause local decreases in the numbers of their prey and thus create a patchy distribution of the prey.

Vertical patches of individuals and species, as shown in Figure 4.22, will change over 24 hours as some animals migrate vertically (see Section 4.5). In general, migrators tend to be more dispersed during the night, and to form denser aggregations during the day in deeper water. Vertical separation may be due to physical factors that include the presence of pycnoclines and thermoclines, to light intensity preferences, or to other microenvironmental differences. Vertical aggregations also may result from the distribution of preferred food items, from predation, or from other biological factors.

Table 4.3 summarizes some of the physical and biological processes that cause the patchy distribution of planktonic organisms. As mentioned above, spatial scales vary from thousands of kilometres to very small-scale patches of 10 m or less. The length of time a particular patch of plankton may persist varies according to the cause of the distribution. Very large patches of zooplankton, such as those caught in rings spun off from the Gulf Stream, may persist for months or even years. Mating aggregations of macrozooplankton (e.g. euphausiids) or of nekton (squid, fish) may persist for only a few days, but the planktonic offspring that hatch from spawning aggregations may remain together for many days or months because they will be less independent of water movements than the adults. Patchiness due to Langmuir circulation will persist only as long as wind velocity and direction remain constant; and wave action may cause constantly changing patterns of aggregation and dispersal in near-surface plankton. Thus there is a continuum in size of patches from very large to very small horizontal scales, and in persistence of patches from thousands of days to momentary periods.

Table 4.3   Approximate spatial and temporal scales of some important processes that cause patchy distribution of zooplankton.

Spatial length scale (km)	Physical processes	Biological processes	Persistence time scale (days)
1000+	Gyres (e.g. Sargasso Sea); continental upwelling (e.g. Peru Current); water mass boundaries (e.g. Antarctic Convergence)	Regional ecosystems defined by the water mass	1000+
100	Warm and cold core rings; tidal fronts; seasonal coastal upwelling	Seasonal growth (e.g. spring blooms); differential growth between phyto-and zooplankton	100
		Lunar cycles (e.g. fish spawning)	
10	Turbulence (e.g. estuarine mixing; island wake effects)	Reproductive cycles	10
		Grazing/predation	
1	Tidal mixing		1
		Diel events (e.g. vertical migration)	
0.1	Wind-induced vertical mixing	Physiological adaptation (e.g. buoyancy; light adaptation)	0.1
0.01	Langmuir circulation; wave action	Behavioural adaptation (feeding swarms)	0.01

QUESTION 4.6 Referring to Figure 4.20, (a) can you provide any explanation for why the amount of chlorophyll *a* generally inversely correlated with zooplankton numbers at any locality? (b) Would you expect the numbers of zooplankton at these localities and depth to increase or decrease during daylight hours?

4.8 LONG-TERM CHANGES IN ZOOPLANKTON COMMUNITY STRUCTURE

Records showing long-term changes in plankton community structure are available for only a few marine areas, but they indicate that there can be considerable variation in the abundance and species composition of zooplankton on decadal time scales. Often these changes in plankton communities are correlated with changes in atmospheric and marine climate. Long-term climate changes that significantly alter marine ecosystems and biological production are known as **regime shifts**, and several examples are given below.

Some of the longest zooplankton records come from the north-east Atlantic Ocean, where continuous plankton recorders (CPRs) have been towed by commercial ships on regular routes for almost 50 years. Figure 4.23 shows a general decline in both phytoplankton and zooplankton abundance in this region over the past 40 years, except for small increases during the early 1980s. A similar pattern has been shown off southern California, where the macrozooplankton biomass was 70% lower in 1987–93 than in 1951–57. In this coastal area, ocean climate changes were correlated with the decrease in plankton. During the 40 years of declining plankton biomass, the sea surface

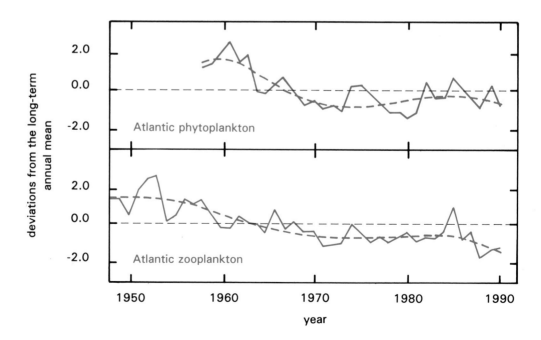

Figure 4.23 Long-term fluctuations in the abundance of phytoplankton and zooplankton in the north-east Atlantic Ocean. The blue solid line represents standard deviation units from the long-term annual mean; the blue broken line has been statistically smoothed to show the average trend.

temperature off California increased by about 1.5 C°, and the temperature difference across the thermocline increased. The increased stratification of the water column lessened wind-driven upwelling, and consequently the lower-nutrient regime depressed phytoplankton production and supported fewer zooplankton. It is not clear whether these changes are part of natural climatic cycles that will reverse in coming years, or whether the zooplankton decline is due to global warming. If temperatures continue to increase globally and stratification increases throughout the oceans, the biological impacts could be drastic in terms of lowering marine production in areas where it is presently enhanced by upwelling of nutrients.

Can changes in ocean climate increase biological production?

At about the same time that plankton biomass was declining in the California Current, plankton abundance was increasing in other parts of the Pacific. In the central North Pacific Ocean (ca. 26°–31° N to 150°–158° W), total chlorophyll a in the water column doubled from 3.3 mg m^{-2} during 1968–1973 to 6.5 mg m^{-2} in 1980–85. Farther north, in the subarctic Pacific Ocean, there was a doubling of zooplankton biomass and a similar increase in pelagic fish and squid abundance between the periods 1956–1962 and 1980–89. In both of these areas, the enhanced production has been correlated with an increase in the intensity of winter winds, which increase vertical mixing and bring more nutrients into the euphotic zone.

There may also be long-term changes in species composition of plankton communities. In the central North Sea, holoplanktonic calanoid copepods (see Section 4.2) dominated the mesozooplankton community from 1958 to the late 1970s. In the 1980s and early 1990s, meroplanktonic larvae of sea urchins and brittle stars (see Sections 4.3 and 7.2.1) became numerically dominant and more abundant than any single holoplanktonic species. This change in zooplankton composition was reflected in a 2- to 8-fold increase in the abundance of the macrobenthos of the area. The reason for a preferential increase in benthic echinoderms in the North Sea remains unclear.

4.9 SUMMARY OF CHAPTER 4

1   The marine zooplankton community includes many different species of animals, ranging in size from microscopic protozoans to animals of several metres in dimension. The holoplanktonic species spend their entire lives in the pelagic environment; meroplanktonic forms are temporary members of the plankton, and include the eggs and larval stages of many benthic invertebrates and fish.

2   Although zooplankton are routinely collected by towing fine-meshed nets through the water, not all species are representatively captured by this method. Some animals are too small to be retained in nets, others are capable of detecting and evading nets, and some species are too fragile to survive collection by nets and subsequent processing in chemical preservatives. Direct observations of zooplankton using scuba techniques, ROVs, or submersibles have greatly increased our knowledge of fragile and/or fast-swimming species.

3   The presence of meroplanktonic larvae in the water is linked to the reproductive patterns of the adults. In tropical regions, meroplankton are present throughout the year. In higher latitudes, the larvae of benthic

invertebrates and fish appear seasonally because reproduction in the adults is linked to higher temperatures and elevated phytoplankton production.

4   The vertical gradients of temperature, light, primary production, pressure, and salinity create distinctive environments at different depths in the water column. These vertical zones (epi-, meso-, bathy-, and abyssopelagic) are somewhat arbitrary in nature, but different species of zooplankton generally inhabit discrete depth zones within the ocean. The life styles, morphology, and behaviour of organisms living deeper in the water column differ from those exhibited by epipelagic species, and the biomass of zooplankton decreases exponentially with depth.

5   As light from the Sun diminishes with depth, bioluminescent light produced by organisms becomes increasingly important as a means of communication. Many different species display the ability to produce light, and the biological significance of bioluminescence varies with the species. Some use light displays to attract potential prey, others to deter predators; some may use bioluminescence to attract mates, or to form reproductive swarms.

6   Although most zooplankton have preferred depth ranges, many species move vertically in the water column with a diel periodicity. The most usual pattern is a nocturnal migration in which animals make a single ascent toward the surface at night, followed by a single descent to deeper water at sunrise. The adaptive significance of diel vertical migration may be different for different species. This behaviour may allow animals to conserve energy by remaining in colder waters except when feeding; it may reduce mortality from visual predators; or it may permit animals of limited swimming ability to sample new feeding areas with each ascent.

7   Diel vertical migration has several important biological and ecological consequences. It probably enhances genetic exchange by mixing the members of a given population; this results because vertical migrations are never precisely synchronized among all the members of a population. Some individuals begin migrations sooner or later than others, with the result that some members will eventually be lost from the original group and new members will be added. Secondly, diel vertical migrations increase the speed at which organic materials produced in the euphotic zone are transferred to deeper areas.

8   In high latitudes, extensive vertical migrations may be undertaken on a seasonal basis, and these are generally linked with reproductive cycles and development of larval stages. In such migrations, the adults are usually found in deeper waters during the winter when food is scarce; the developing young are present in surface waters during the spring and summer when phytoplankton is plentiful.

9   By moving vertically in the water column, zooplankton enter currents that are moving in different directions and at different speeds. Thus diel or seasonal vertical migrations that are attuned to particular current regimes can result in the retention of populations within favourable localities.

10   Present-day distributions of zooplankton have been established over geological time and reflect past dispersal patterns as well as the physiological and ecological requirements of the species.

11   Epipelagic zooplankton are often associated with specific water mass types, which are established by latitudinal gradients in temperature, salinity,

and other physico-chemical factors. Mesopelagic and bathypelagic species tend to have wider geographic distributions, reflecting increasing homogeneity in environmental conditions with increasing depth.

12   The numbers of species of epipelagic and mesopelagic zooplankton are higher in low latitudes, but the numbers of individuals tend to be relatively low. The reverse situation is found in high latitudes, where there are fewer species but with higher abundance.

13   Within the bounds of their geographic regions, zooplankton exhibit patchy distributions on a wide range of space- and time-scales. Patchiness may result from responses to physical turbulence or mixing, or to chemical gradients such as salinity changes. Patchiness may also result from interactions between prey and predators, or it may reflect other biological events such as reproduction.

14   Long-term records indicate that plankton abundance and species composition may change substantially over decadal time scales. Decreasing plankton biomass may be caused by climate changes that increase water stratification and depress upwelling; conversely, in other regions, increasing winds may enhance nutrient concentrations in the euphotic zone and lead to increased phytoplankton and zooplankton production.

Now try the following questions to consolidate your understanding of this Chapter.

QUESTION 4.7   Why do the planktonic Crustacea tend to have so many different growth stages?

QUESTION 4.8   Why do so many benthic species of animals produce meroplanktonic larvae?

QUESTION 4.9   Which planktonic organisms produce skeletal materials that contribute to sediments on the seafloor?

QUESTION 4.10   Of the major zooplankton groups listed in Table 4.2, which are predominantly carnivorous and which are predominantly herbivorous? (Refer to Section 4.2.)

QUESTION 4.11   What is the advantage of patchy distribution of plankton predators that actively seek out their food?

QUESTION 4.12   In Figure 4.22, only the copepodite V stage of *Neocalanus cristatus* is shown. Where and when would you expect to find adults of this species? (Refer to Section 4.6.)

5.1   FOOD CHAINS AND ENERGY TRANSFER

Food chains are linear arrangements showing the transfer of energy and organic materials through various trophic levels of marine organisms. Each **trophic level** is composed of organisms that obtain their energy in a similar manner. The pelagic food chain begins with the phytoplankton; these autotrophic **primary producers**, which build organic materials from inorganic elements, form the first trophic level. Herbivorous species of zooplankton that feed directly on the marine algae (e.g. Protozoa, many copepods, salps, larvaceans) make up the second trophic level, and they are referred to as **primary consumers**. Subsequent trophic levels are formed by the carnivorous species of zooplankton that feed on herbivorous species (**secondary consumers** like chaetognaths), and by the carnivores that feed on smaller carnivores (**tertiary consumers** including many jellyfish and fish). The total number of trophic levels will vary with locality and with the total number of species in the community. The highest trophic level is occupied by those adult animals that have no predators of their own other than humans; these top level predators may include sharks, fish, squid, and mammals. The total amount of *animal* biomass produced in all higher trophic levels, per unit area and per unit time, is called **secondary production** (as opposed to the primary production of plants). **Trophodynamic** studies examine the factors that affect transfers of energy and materials between trophic levels and that ultimately control secondary production.

Elements such as nitrogen, carbon and phosphorus, which become incorporated in organic components of plant and animal tissues, have a cyclical flow through food chains (Figure 5.1). Bacteria decompose waste materials and the tissues of dead organisms. **Decomposition** releases inorganic forms of essential elements, and these become available again for uptake by autotrophic organisms. Energy, however, has a unidirectional flow (Figure 5.1). Some energy is lost at each transfer to the next trophic level

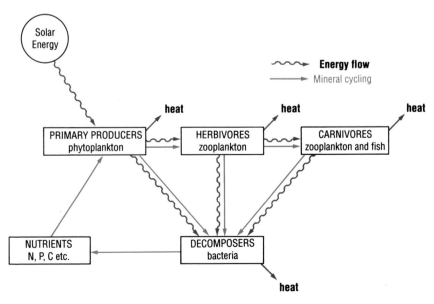

Figure 5.1   A schematic representation of mineral recycling and energy flow in marine ecosystems.

a. The Total number of Individuals decrease with each trophic level.

There is more primary production than secondary

because much of the chemical energy incorporated in organic compounds is converted to heat energy and is dissipated in respiration, when organic carbon is broken down to CO_2. As a consequence, the total energy will diminish at each trophic level, and this places a finite limit on the possible number of trophic levels in any community.

QUESTION 5.1 What are the consequences of energy loss due to respiration on (a) the relative total numbers of organisms in successive trophic levels, and on (b) the relative amounts of primary production and secondary production?

The size of individual organisms generally increases within each succeeding trophic level, but the generation time (or length of the life cycle) becomes progressively longer. The generation times of phytoplankton are measured in hours or days, those of zooplankton in weeks to months; those of fish in years; and those of marine mammals in many years. One might expect to observe considerable differences between the standing stocks of phytoplankton and fish or whales, but it is believed that because of the dissimilarities in generation times, the total biomass in each succeeding trophic level decreases only very slightly (Figure 5.2). The biomass of the exceedingly numerous, microscopic, and rapidly reproducing phytoplankton is probably never more than four times that of the small numbers of very large marine mammals which have long generation times.

We have seen that it is relatively easy to estimate primary production by marine phytoplankton (see Section 3.2.1). Because of the longer generation times of zooplankton and fish, and because it is difficult or impossible to follow populations of these animals in the sea for any length of time, it is much harder to obtain estimates of pelagic secondary production. There are some methods that can be applied to data collected in the field, and these are discussed in Section 5.3.1, along with their limitations. Secondary production can also be studied by growing marine zooplankton and fish under experimental conditions, and this approach is considered in Section 5.3.2. Another approach is to use estimates of primary production

Figure 5.2   The average biomass of organisms of different sizes in marine food chains. To eliminate differences in shape, the size of each type of organism has been converted to the diameter of a spherical particle having the same biomass as the organism. The upper line illustrates the average biomass of different organisms in the Antarctic Ocean, a region of high productivity. The lower line shows the biomass of typical organisms in the equatorial Pacific, an oceanic region of lower productivity.

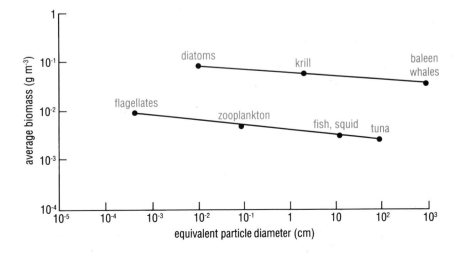

and our knowledge of marine food chains to predict secondary production and yields of fish. Using this indirect approach, it is necessary to know how much energy is transferred between each trophic level.

The efficiency with which energy is transferred between levels is called the **ecological efficiency** (E), and it is defined as the amount of energy extracted from a given trophic level divided by the energy supplied to that trophic level. Ecological efficiency is difficult to measure, and it can more easily be approximated from the **transfer efficiency** (E_T), defined as:

$$E_T = \frac{P_t}{P_{t-1}} \tag{5.1}$$

where P_t is the annual production at trophic level t, and P_{t-1} is the annual production in the preceding trophic level $t - 1$. In this equation, production can be defined either in terms of energy (e.g. measured conventionally in calories or, in the SI system, in joules) or in terms of biomass (e.g. carbon in grams). For the energy transfer between phytoplankton and zooplankton, E_T will equal the amount of herbivore production divided by the primary production. At the next step, the transfer efficiency will be the annual production of secondary consumers (i.e. carnivorous zooplankton) divided by the annual production of herbivores.

QUESTION 5.2 Calculate the transfer efficiency between phytoplankton and herbivorous zooplankton in a marine community where the net primary productivity is 150 g C m^{-2} yr^{-1} and the annual production of herbivorous copepods is 25 g C m^{-2}.

In marine ecosystems, values for transfer efficiencies have been estimated at about 20% for the transfer from plants to herbivores, and at 15–10% at higher levels. This means that there are corresponding energy losses between trophic levels of 80–90%, primarily through respiration (the heat losses shown in Figure 5.1).

QUESTION 5.3 Why would the respiration losses be greater, and transfer efficiencies lower, when considering higher trophic levels?

Note that equation 5.1 deals with the consumption of energy in succeeding trophic levels. A certain amount of production is not consumed directly. Not all living phytoplankton and zooplankton are eaten; some die naturally, and the energy contained in dead organic material becomes available for scavengers or microbial decomposers in a different pathway (see Section 5.2.1). This detritus may be cycled either in the water column or in the benthic community.

QUESTION 5.4 If a large amount of the primary production is not eaten by the herbivorous zooplankton, but dies and sinks out of the water column, what happens to the value of the transfer efficiency between these two trophic levels?

In addition to knowing how much energy is transferred between trophic levels, it is also necessary to know how many trophic levels there are in any particular locality for which secondary production will be estimated. Because of the energy losses incurred with each transfer between trophic levels, the number of links will partly determine the biomass of top-level predators (i.e.

fish, squid, or marine mammals). There is reasonable evidence to suggest that the number of links in the pelagic food chain varies with locality and may be determined by the individual size of the primary producers. The number of trophic levels varies from up to six in the open ocean, to about four over continental shelves, to only three in upwelling zones, as shown in Figure 5.3. Note that when the size of the dominant phytoplankton is small, the food chain is lengthened, as in open ocean areas. In such situations, marine protozoans (zooflagellates and ciliates) become important intermediary links; they may consume a major fraction of the primary production, and in turn they constitute an abundant dietary source for suspension-feeding copepods or other zooplankton that are incapable of feeding directly on very small

Figure 5.3   A comparison of food chains in three different marine habitats. The organisms representing each trophic level are only selected examples of the many marine species that could be present in that level. (Organisms not to scale.)

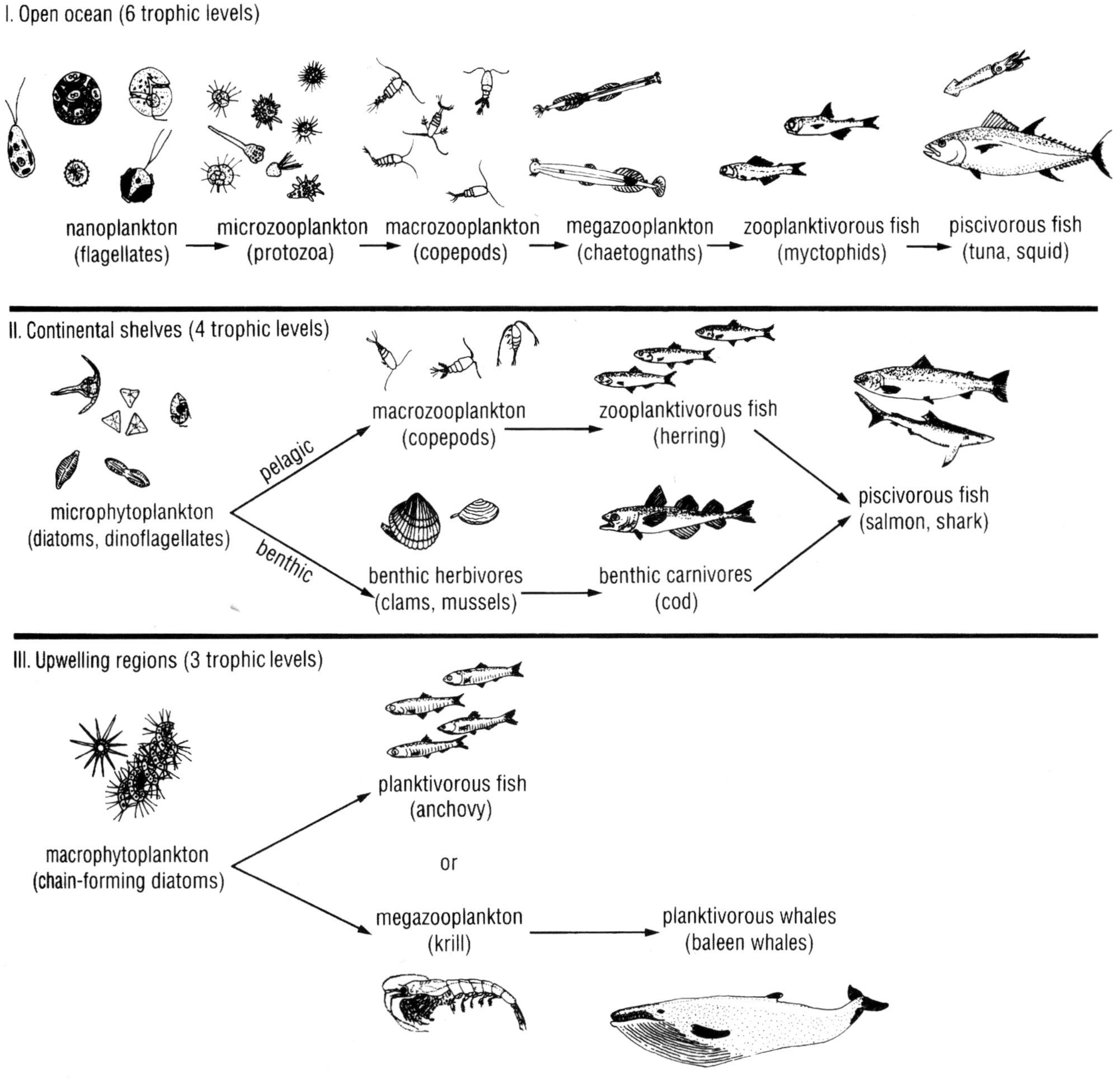

I. Open ocean (6 trophic levels)

nanoplankton (flagellates) → microzooplankton (protozoa) → macrozooplankton (copepods) → megazooplankton (chaetognaths) → zooplanktivorous fish (myctophids) → piscivorous fish (tuna, squid)

II. Continental shelves (4 trophic levels)

microphytoplankton (diatoms, dinoflagellates)

pelagic → macrozooplankton (copepods) → zooplanktivorous fish (herring)

benthic → benthic herbivores (clams, mussels) → benthic carnivores (cod)

→ piscivorous fish (salmon, shark)

III. Upwelling regions (3 trophic levels)

macrophytoplankton (chain-forming diatoms)

→ planktivorous fish (anchovy)

or

→ megazooplankton (krill) → planktivorous whales (baleen whales)

Table 5.1 The relation between recent estimates of primary production and fish production in three different marine habitats.

Habitat	Oceanic	Coastal	Upwelling
Percent ocean area	89	10	1.0
Mean primary productivity ($g\ C\ m^{-2}\ yr^{-1}$)	75	300	500
Total plant production (10^9 tonnes C yr^{-1})	24	11	1.8
Number of energy transfers between trophic levels	5	3	1.5*
Average ecological efficiency	10%	15%	20%
Mean fish production** ($mg\ C\ m^{-2}\ yr^{-1}$)	0.75	1000	44 700
Total fish production*** (10^6 tonnes C yr^{-1})	0.24	36.2	162

*The number of trophic levels in upwelling areas may be 2 (if fish are predominantly herbivores) or 3 as represented in Figure 5.3; 1.5 represents an average value for the number of energy transfers ($n = 1$ or 2).
**Calculated from equation 5.2.
***Corrected for percent ocean area occupied by each habitat (total area $= 362 \times 10^6$ km^2).

phytoplankton. In contrast, large chain-forming diatoms dominate in nutrient-rich upwelling regions, and short food chains result because large zooplankton or fish can feed directly on the large primary producers. Consequently, there is a high biomass of top-level predators in upwelling systems or other highly productive areas. Figure 5.2 compares biomass and length of food chains in regions of high and low productivity.

QUESTION 5.5 Refer to Figure 5.2. (a) How many orders of magnitude difference are there in the amounts of biomass produced in the Antarctic Ocean and in the equatorial Pacific? (b) What factors contribute to this difference? (Refer also to Figure 5.3 and Section 3.5.)

For a given locality, the number of trophic levels can be coupled with a quantitative estimate of primary productivity to predict yields of secondary production in any particular trophic level ($P_{(n+1)}$), according to the following equation:

$$P_{(n+1)} = P_1 E^n \tag{5.2}$$

In this equation, P_1 is annual primary production, E is the ecological efficiency, and n is the number of trophic transfers (which equals the number of trophic levels minus 1). A major difficulty in applying this equation lies in the accuracy of the values used for ecological efficiency and number of trophic levels. For example, the secondary production estimate can increase by an order of magnitude if the value for E is doubled. Until we can be more confident of the values for ecological efficiency and the

$n = 2$

$P_1 = 300$

$E = 0.1$

$300 \times (0.1)^2 = 3 g\ C\ m^{-2}\ yr^{-1}$

number of trophic levels in different locations, predictions of potential fish catches based on this method remain uncertain and unreliable. Nevertheless, equation 5.2 provides relative values that are useful for comparing production in different marine areas.

QUESTION 5.6 If the primary productivity of a coastal area is 300 g C m^{-2} yr^{-1} and herring (which feed on zooplankton) are the principal fishery, what would be the expected annual maximum yield of herring (in terms of g C m^{-2}), given an average ecological efficiency of 10%?

Table 5.1 couples general values for primary productivity (from Section 3.6) in the three major pelagic habitats (Figure 5.3) with numbers of trophic levels, and leads to the conclusion that upwelling regions should produce by far the highest numbers of fish (or whales) per unit area, and the open ocean the fewest. Even correcting for percent ocean area, the small upwelling areas should produce four times more fish than coastal areas. In fact, upwelling areas provide a significant fraction of the present world fish catch. Remember too that there is a major economic advantage in catching fish in upwelling and coastal areas because large numbers can be harvested within 30–80 km of the coastline. In contrast, the expense of commercial harvesting is much higher in the open ocean because the stocks are dispersed over a vast region.

5.2   FOOD WEBS

In reality, the concept of a food chain is a theoretical convenience and an attempt to reduce a complex natural system to simple dimensions. There are seldom simple linear food chains in the sea. Practically all species of organisms may be eaten by more than one predatory species, and most animals eat more than one species of food. The energy system is more accurately portrayed as a **food web** with multiple and shifting interactions between the organisms involved. Many species do not conveniently fit into the conventional trophic levels. Some species are omnivorous, feeding on both phytoplankton and zooplankton. Some also feed on **detritus**, the organic debris of faecal material, plant and animal fragments, crustacean molts, and abandoned larvacean houses and pteropod feeding webs. Some species change diets (and trophic levels) as they grow, or as the relative abundance of different food items changes. Still other species are parasites and obtain their energy from their hosts, and cannibalism is not uncommon within many marine species. Further, the benthic food chain is also linked to the pelagic production, as is illustrated in the continental shelf habitat in Figure 5.3. Some benthic species (e.g. barnacles, mussels) feed directly on phyto- or zooplankton, and other benthos are indirectly dependent on the pelagic production.

There are typically fewer species in high latitude communities, and polar marine food webs tend to be simpler than those of other localities. For this reason, we have chosen to present a schematic depiction of the food web of the Antarctic Ocean in Figure 5.4. Note that in this area, as well as in the Arctic Ocean, there are two basic types of primary producers: the pelagic phytoplankton and the algae which live within the ice. The latter, called **epontic** algae, are generally benthic species that are adapted for the low light intensities present at the undersurface of the ice. The abundant krill form the central point of the Antarctic food web as they are the dominant herbivores,

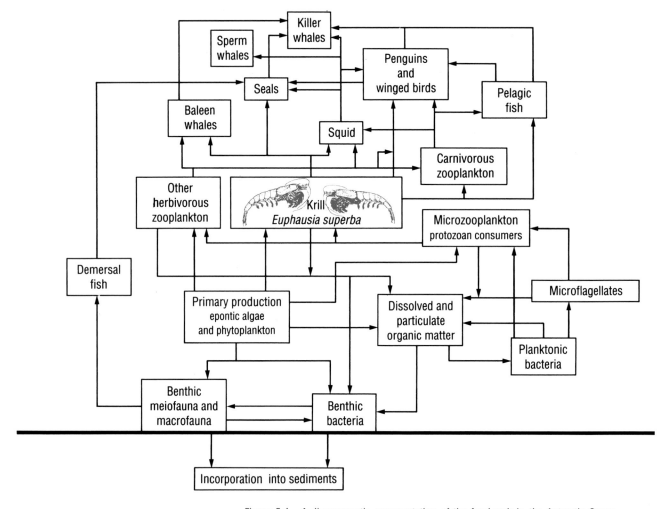

Figure 5.4   A diagrammatic representation of the food web in the Antarctic Ocean.

and they are an important food source for several species of carnivorous zooplankton, pelagic fish and squid, as well as for plankton-feeding baleen whales, seals, and seabirds. The abundance of krill may best be illustrated by pointing out that a single blue whale may consume up to eight tonnes (or more than 40 million individuals) of euphausiids daily during summer months. When several species rely heavily on a single food, as illustrated in Figure 5.4, competition for food may develop between the different predators if their shared food becomes a limiting resource. The decimation of Antarctic baleen whales by commercial whaling has demonstrated how the flow of energy through a food web may be altered when a competing species is eliminated or reduced. Table 5.2 presents the relative amounts of krill consumption by different predators before and after depletion of whale stocks. A large decline in baleen whale biomass made more of the euphausiid biomass available to competing species, and consequently increased the populations of Antarctic seals and birds by about a factor of three.

Competitive relationships among species may also diminish the biomass of top-level commercially-fished species. Larval fish, for example, may compete with chaetognaths, jellyfish, ctenophores, and other invertebrate carnivorous zooplankton for copepods or other prey. Consequently, only a

Table 5.2 Estimated changes in patterns of consumption of Antarctic krill by the major groups of predators, 1900–84.

Predator	Annual krill consumption (tonnes $\times 10^6$)	
	1900	1984
Baleen whales	190	40
Seals	50	130
Penguins	50	130
Fish	100	70
Squid	80	100
Total	470	470

certain fraction of the prey (and ultimately of the primary production) is converted to fish stocks. Competition for food and the resultant loss of energy to higher trophic levels is illustrated schematically in Figure 5.5. The same planktonic invertebrate predators also may eat the meroplanktonic larvae of the benthos, and thus they can decrease the production of shellfish, such as mussels and clams, as well as of noncommercial benthos.

Although food webs are more realistic than food chains, they are more difficult to quantify in ecological terms. Few marine systems have been well enough studied to attempt an **energy budget** analysis, in which the initial, measured, energy input of primary production is channelled into different trophic levels and pathways of the food web. The North Sea, however, has been intensively studied because of its long-time importance as a fishing region, its relatively small area, and its proximity to centres of marine research. Figure 5.6 outlines a quantitative analysis of energy flow through a food web of the North Sea based on major groups of organisms, rather than individual species.

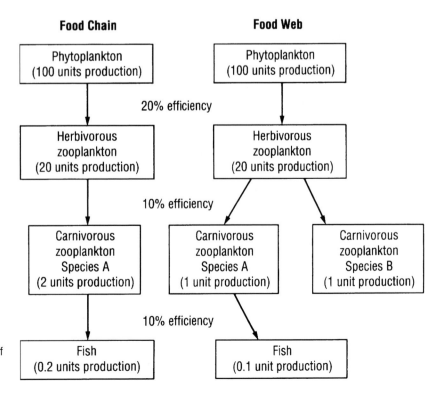

Figure 5.5 A comparison of a hypothetical marine food chain and food web. Both begin with 100 arbitrary units of phytoplankton primary production. The food chain produces 0.2 units of fish from this primary production, but the food web produces only one-half this amount of fish. In the food web, two carnivorous species (A and B) compete for the supply of herbivorous zooplankton. Half of the herbivores are consumed by Species A and half by Species B. Fish do not eat Species B, so their principal food supply is less by 50%.

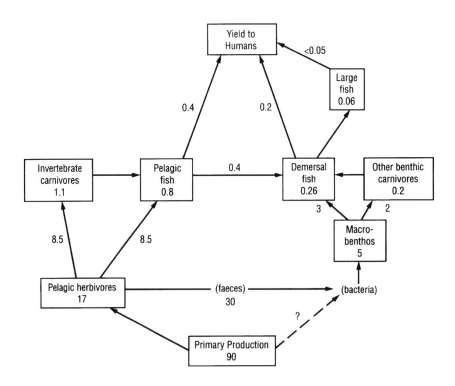

Figure 5.6   A North Sea food web. Numbers refer to annual production in g C m^{-2}.

The food web in Figure 5.6 is based on a primary productivity of 90 g C m^{-2} yr^{-1}, and it assumes (probably unrealistically) that the zooplankton consume all of the phytoplankton and excrete about 30% of their food as faeces. The pelagic herbivore production of 17 g C m^{-2} yr^{-1} has been estimated from field and experimental studies on the copepod *Calanus finmarchicus*, which is the dominant planktonic herbivore in the system. The production values for the next trophic level come from assuming that 50% of the herbivore production is eaten by planktonic invertebrate carnivores (such as chaetognaths and ctenophores) and 50% by fish. This equal allocation of energy is assumed, however, to result in lower production values for fish (even though they may additionally consume some invertebrate carnivores) because they have higher energy demands and burn a higher percentage of their food in respiration than do invertebrates. The partitioning of energy through the sinking of faeces, their decomposition by bacteria, and the incorporation of bacterial production into benthic invertebrates has been based on similar assumptions. The predicted yield to humans in terms of fish catch is less than 0.7 g C m^{-2} yr^{-1}, which is about 0.8% of the plant production.

Although the values of production in Figure 5.6 are extremely tentative and based on numerous assumptions, the model provides a scheme that links pelagic and benthic production and is an attempt to quantify a complex ecological system. Despite the difficulties, there are many potential rewards in studying food webs. Such studies allow us to examine the interactions of nutrient input, primary production, and secondary production, and to discover the determinants of production. They also can provide answers as to why there are particular patterns of species associations and of energy flow, and how these persist over time. It is also important to know how food webs will react to perturbations, such as pollution and commercial harvesting of high trophic levels.

5.2.1 THE MICROBIAL LOOP

The regeneration of nutrients in the sea is a vital part of the interaction between higher and lower trophic levels. This is accomplished by bacteria and planktonic protozoans interacting in a microbial loop that is coupled with the classic food chain formed by phytoplankton–zooplankton–fish (Figure 5.7). Particulate detritus formed through natural mortality of phyto- and zooplankton and nekton, or through the production of faecal pellets and structures such as crustacean molts, abandoned pteropod feeding webs or larvacean houses, is decomposed by bacteria which utilize the energy-rich detritus for growth. Bacteria also can utilize soluble organic materials released by the physiological processes of animal excretion and phytoplankton exudation, thereby efficiently converting dissolved nutrients into particulate biomass. Thus the microbial loop is of particular importance in increasing food chain efficiency through utilization of both the very smallest size fractions of particulate organic material (POM), as well as of the dissolved organic matter (DOM) which is usually measured as dissolved organic carbon (DOC).

The number of bacteria in the euphotic zone of the oceans is generally around 5×10^6 ml^{-1}. They may sometimes increase to 10^8 ml^{-1} in the presence of adequate nutritive materials and in the absence of bacterial grazers. In deep ocean waters, bacterial numbers may be less than 10^3 ml^{-1}. The number of bacteria in the sea is generally controlled through predation by nanoplankton, especially by various protozoans, but a few larger zooplankton (e.g. larvaceans) are also capable of capturing and consuming

Figure 5.7 A schematic illustration showing the coupling of the pelagic grazing food chain (phytoplankton to piscivorous fish) and the microbial loop (bacteria and protozoans). Dashed arrows indicate the release of dissolved organic material (DOC) as metabolic by-products. The DOC is utilized as a source of carbon by heterotrophic bacteria. The bacteria are consumed by protozoans, which in turn are eaten by larger zooplankton.

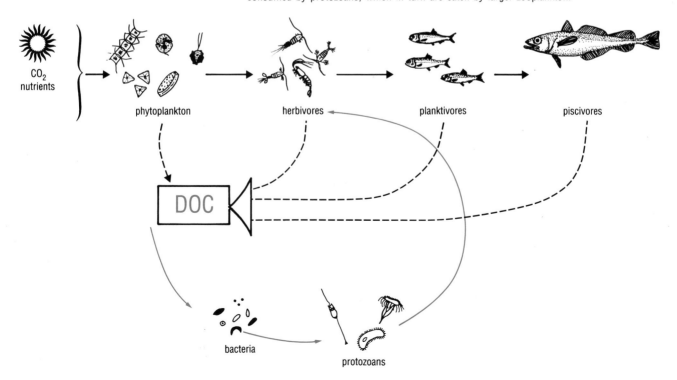

bacteria (see Section 4.2). Among the nanoplankton, the zooflagellates are particularly voracious in their consumption of bacteria. Heterotrophic zooflagellates are usually present in concentrations of about 10^3 ml^{-1}. However, when bacterial numbers start to increase, the zooflagellates quickly respond by consuming more bacteria and multiplying at a rate that tends to prevent very large increases in bacterial standing stock. Small bactivorous zooplankton are important links in transferring bacterial production to higher trophic levels as they, in turn, form a source of food for larger organisms, particularly for filter-feeding crustaceans. In general, filter-feeders like copepods and euphausiids are incapable of feeding directly on bacteria because their filtering appendages are too coarse to retain such small particles.

QUESTION 5.7 In addition to heterotrophic zooflagellates and larvaceans, what other zooplankton are capable of feeding on bacteria? (Refer to Section 4.2.)

The general pathways of the recycling processes described above are shown in Figure 5.7, where it is apparent that bacterial activity in the sea is closely linked with marine food webs. Changes in the phytoplankton standing stock, in particular, are often closely accompanied in time and distance by changes in bacterial biomass. This is illustrated in Figure 5.8 which demonstrates that as the chlorophyll concentration increases from about 0.5 μg to 100 μg l^{-1}, bacterial densities increase from about 10^6 ml^{-1} to 3×10^7 ml^{-1}. During exponential growth of the phytoplankton, bacteria can live off dissolved organic metabolites (exudates) that are released as part of the metabolic processes of phytoplankton growth. At the end of a phytoplankton bloom, when the algae enter a senescent stage, there is an accumulation of **phytodetritus** (i.e. nonliving particulate matter derived from phytoplankton) and an increased release of dissolved metabolites. It is particularly at this time that the bacteria can utilize these energy sources to multiply and produce a sharp pulse (or bloom) that follows the phytoplankton bloom. Thus the food web of temperate seas may shift seasonally from one that is based on high nutrients, diatoms, and filter-feeding copepods, to one that is dominated by the microbial loop and bactivorous zooplankton. A similar relationship between phytoplankton and bacteria influences the vertical distribution of bacterioplankton. Maximum numbers of bacteria generally occur at the pycnocline, where phytodetritus accumulates by sinking from the overlying euphotic zone. There, decomposition by bacteria contributes to the formation of oxygen minimum layers in stable waters (see Section 4.4). In general, it is estimated that bacteria, by using phytodetritus or dissolved organic exudates for their growth, may utilize up to 50% or more of the carbon fixed by photosynthesis.

Note in Figure 5.8, however, that the relationship between increasing phytoplankton and increasing bacterial numbers holds least well at very low chlorophyll concentrations (<0.5 μg l^{-1}), where bacteria are more numerous than expected. This indicates that, in very oligotrophic waters, bacteria constitute the dominant biomass of the microflora, and their numbers are independent of the very small amount of phytoplankton. In waters where nutrient concentrations are very low and limiting, there may be competition between bacteria and phytoplankton for essential elements. In this circumstance, predation on the bacteria by protozoans may influence the outcome of the competition.

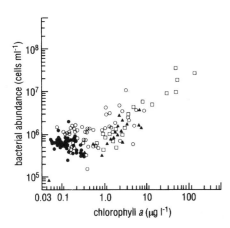

Figure 5.8 The relationship between chlorophyll a and bacterial abundance in the euphotic zone. Data are from ●, central North Pacific gyre; ○, Southern California Bight and Santa Monica Basin; and various additional marine (▲) and freshwater (□) locations.

The N Pacific gyre is an area of low Nutrient production.

QUESTION 5.8 In Figure 5.8, why are the values for chlorophyll *a* and bacterial abundance lowest in the central North Pacific gyre and highest in freshwater locations?

Marine viruses are the smallest and most abundant organisms in the sea, with concentrations ranging between about 10^3 and 10^9 ml^{-1}, yet their role in the microbial loop and generally in marine ecosystems remains highly speculative. There are significantly more viruses in near-surface waters compared with deeper layers, suggesting a coupling of viral particles with other upper-ocean biological processes. It is known that viral pathogens can infect marine bacteria and a variety of phytoplankton including diatoms and Cyanobacteria, and experimental work suggests that viral infection may affect the species composition of the phytoplankton community and significantly reduce primary productivity.

Studies of the microbial loop are relatively new in the science of biological oceanography. They have been hampered by the very small size of the microbes and protozoans and associated difficulties in collection, preservation, and identification. There is a need to understand the impact of this cycle on primary production in terms of nutrient competition and nutrient recycling, and on secondary production in terms of providing a link between bacterial production and its consumption by higher trophic levels.

5.3 MEASURING SECONDARY PRODUCTION

5.3.1 FIELD STUDIES

It is possible to obtain reasonable estimates of the amount of primary production in different marine areas (Section 3.2.1), and catch statistics of commercially fished species provide a minimum value of production at the other end of food chains. Information about secondary production in the intermediary trophic levels and for noncommercial top-level predators (i.e. jellyfish, ctenophores, 'trash fish') is much more difficult to obtain. There are, however, compelling reasons to tackle this problem. In some circumstances, primary production may not be a good indicator of production in higher trophic levels. For example, in highly eutrophic systems (see Section 3.4), the growth of phytoplankton may greatly exceed what can be consumed by herbivores, or the phytoplankton species which becomes dominant under these conditions may not be a suitable food source for herbivores; in either case, much of the primary production may enter the microbial–detritus circuit instead of the classic pelagic food chain. Relying entirely on fish catch statistics to provide values of secondary production in top trophic levels leads to underestimates, because it omits production of all the competing, unharvested species.

Secondary production can be estimated from field data. The production of a population of zooplankton is defined as the total amount of new biomass of zooplankton produced in a unit of time, regardless of whether or not it survives to the end of the time period. In this definition, biomass (*B*) is:

$$B = X \times \overline{w}$$

(5.3)

where X is the number of individuals in the population and $\overline{w}$ is the mean weight of an individual. It then follows that production (P_t), during a time interval from t_1 to t_2, can be expressed as:

$$P_t = (X_1 - X_2)\frac{\overline{w}_1 + \overline{w}_2}{2} + (B_2 - B_1) \qquad (5.4)$$

where the subscripts 1 and 2 indicate values obtained at time t_1 or time t_2, respectively. The expression $(B_2 - B_1)$ represents the increase in biomass observed during the time interval; the remainder of the equation (i.e. the decrease in population number times the average weight of an individual) represents an estimate of the biomass produced, but then lost through predation or water movements.

Ideally, one would wish to follow changes over time in numbers and growth under natural conditions in a single **cohort** of a population, a cohort being one identifiable generation of progeny of a species. However, these conditions can seldom be met in the sea and, in any event, it is usually impossible to follow and sample the same water mass for a period of time long enough to obtain meaningful measures of growth. The best that can usually be achieved is to attempt to follow changes in relative numbers and weights of distinctive life stages of abundant species. Because many copepod species are dominant members of the plankton and have easily identifiable age classes, this group of crustaceans is often selected for production measurements, and the example given below is for a copepod species producing one generation per year.

Figure 5.9 presents a hypothetical representation of the numbers of individuals in different, successive developmental stages of a copepod, ranging from newly hatched nauplii through copepodite stages I, III, and V (see Section 4.2). Note that the numbers (X) change with time due to a number of natural processes including mortality, aggregation, or water exchange. In order to calculate production, we also need to calculate changes in weight, or growth. This can be done if the time interval between the

Figure 5.9   A hypothetical representation of changes in the numbers of individuals in selected successive developmental stages of a copepod having one generation per year. Some stages are difficult to distinguish from each other, so for that reason all naupliar stages (NI-NVI) are lumped together and only copepodite stages CI, CIII and CV are considered here. (Refer to Section 4.2 for a discussion of copepod life cycles.) $\overline{w}_1$ and $\overline{w}_2$ indicate average weights of copepodite stages CI and CIII, respectively.

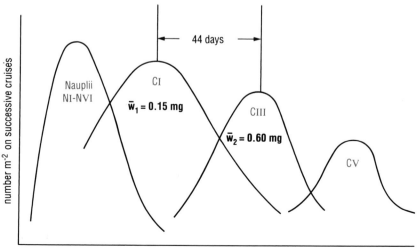

$x_1 = 80 \quad x_2 = 30$

$\overline{w_1} = 0.15 \quad \overline{w_2} = 0.60$

$P_t = (x_1 - x_2) \dfrac{\overline{w_1} + \overline{w_2}}{2} + (B_2 - B_1)$

$(80 - 30) \dfrac{0.15 + 0.6}{2} + (30 \times 0.6 + 80 \times 0.15)$

$50 \dfrac{0.75}{2} + (18 - 12)$

$= 50 \times 0.375 + 6$

$= \dfrac{24.75 \text{ mg m}^{-2}}{44 \text{ days}}$

$= 0.56 \text{ mg m}^{-2} \text{ d}^{-1}$

occurrence of maximum numbers of successive developmental stages is known, and the average weight of each stage is determined in the laboratory. In this example, there is a duration of 44 days between copepodite stage I and copepodite stage III, and a change in average weight from 0.15 mg to 0.60 mg between these respective stages. Assuming that there were, on average, 80 stage I copepodites m^{-2} and 30 stage III copepodites m^{-2}, we now have the information needed to calculate production.

QUESTION 5.9 Given the information above, and using equation 5.4, what was the average production per day of the copepod population during the 44-day period between peak numbers of stage I and III copepodites?

It is important to note that the value of P_t for any population will change with time because zooplankton grow at different rates during their life cycles. Production will generally be positive during periods of maximum growth; in temperate regions, P_t would be highest in spring when food is abundant and most of the zooplankton are young. The value for P_t may be negative in winter months when food concentrations are low and animals cease growing, or even lose weight. Because of these natural changes in P_t, growth increments need to be determined for a number of points in the life history of a species in order to calculate annual secondary production. Thus, annual secondary production (P_t) becomes a summation of production calculated for successive time intervals (P_{t1}, P_{t2}, etc.), so that:

$$P_t = P_{t1} + P_{t2} + P_{t3} \ldots P_{ti}. \tag{5.5}$$

The method described above may only be applied to species with distinctive life stages, and it is only practicable in regions where young are produced seasonally. Particularly in warm waters, many marine animals (including copepods and other Crustacea) reproduce more or less continuously, and thus there is a continuous input of young and a mixture of age and size classes. Further, many planktonic species do not have distinctly different life stages, and size or weight may not necessarily be good indicators of age. A good example of this is found in the thecosomatous pteropods (refer to Section 4.2); very young individuals quickly produce an adult-sized shell, but the body grows only gradually to occupy the entire inner space. Whereas crustaceans have a **determinate growth** pattern in which growth in each stage is limited by size of the exoskeleton, many of the other zooplankton have **indeterminate growth** and are capable of more or less continual growth during favourable conditions, or they shrink (lose biomass) during periods with low food concentrations. Often the attendant difficulties of working with population data collected from the field lead researchers to work with populations or individuals under experimental confinement, where conditions can be controlled and where individuals can be studied over prolonged periods of time. Various experimental approaches to secondary production studies are discussed in the following section.

5.3.2 EXPERIMENTAL BIOLOGICAL OCEANOGRAPHY

Because the oceans are so vast and the waters are in continual motion, and because of the reasons outlined at the end of the previous section, it is necessary to study some biological oceanographic processes under experimental conditions.

Many options are open to the experimentalist and some examples of different approaches are discussed below. In general, however, a choice can be made between:

(1) **laboratory-scale experiments**, which tend to study individual organisms in relatively small volumes of water;

(2) **enclosed ecosystem experiments**, which are carried out in very large containers of natural seawater in order to test the interactions of several trophic levels and their responses to perturbations; and

(3) **computer model simulations**, in which complex biological processes and ecological interactions can be studied on a computer, including the influences of physical and chemical environmental parameters on biological processes occurring in the sea.

No single method of study is likely to provide all the answers required, and all experimental and observational approaches, from laboratory test-tube studies to satellite data-gathering, have both advantages and limitations to the type of results that are generated. Figure 5.10 illustrates schematically how the three experimental approaches outlined above can interact with field studies to lead to a better understanding of natural events. Data gathered from the field and from experimental studies can be fed into mathematical computer models which attempt to integrate this information and simulate real conditions and events. These experimental options apply to studying problems in both pelagic and benthic marine ecology, but the examples discussed below concentrate on issues in planktonic ecology.

Laboratory experiments
Experiments in the laboratory can be conducted to determine food requirements of zooplankton and transfer efficiencies between trophic levels. The majority of such experiments have been carried out using crustaceans, principally copepods, because they can be easily captured in large numbers and with little or no damage, and they are also amenable to laboratory culture. The majority of experiments have also been done with herbivorous species, partly because their phytoplankton food can be cultured easily. A

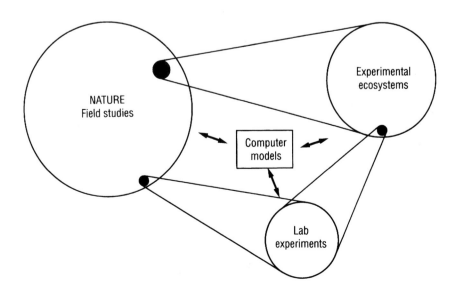

Figure 5.10   The interaction of field studies, experimental studies, and computer modelling in biological oceanographic research.

few carnivorous zooplankton, including certain chaetognaths and gymnosomes, have been utilized for experimental research but, in these cases, their zooplankton prey must also be amenable to culture or, alternatively, the prey must repeatedly be collected fresh and with minimal damage from the sea. The same principles and equations, using modified techniques, can be employed to study production in benthic animals.

Experimental studies of feeding and secondary production are based on the premise that only a certain fraction of the energy ingested in food can be utilized for production. The remaining fractions are expended in respiration or excretion, or are never utilized and pass through the animal in faecal material by the process known as egestion. The fractionation and utilization of energy can be expressed as:

$$G = R - E - U - T \tag{5.6}$$

where growth (G) is a measure of secondary production; R is the ration of ingested food; E is egested faecal material; U refers to excretory products (e.g. ammonia, urea); and T represents respiration. The units in this energy balance equation are given in joules or calories per unit weight. Excretory products are usually considered as a negligible fraction of the equation, and the equation can be simplified and rewritten as:

$$AR = T + G \tag{5.7}$$

where R, T and G are as defined above, and A is a constant relating the proportion of food assimilated (or actually utilized) to the amount consumed. AR is thus referred to as the assimilated ration. The rationale for establishing equations 5.6 and 5.7 is presented schematically in Figure 5.11.

Feeding experiments usually involve introducing from one to several tens of animals (depending on their size) into incubation beakers filled with a measured volume of seawater containing a known concentration of food particles. Controls are established in bottles containing food but no animals; these will show any changes in food concentration that occur independently of grazing or predation and will allow an assessment of error in counting food organisms. In the case of grazing experiments with filter-feeding herbivores, culture bottles are kept dark to minimize growth of the phytoplankton. Usually it is necessary to introduce a mechanical means of keeping cells in suspension and randomly distributed in the water. After containers have been incubated for a measured time interval, the food concentration is again determined. This can be done visually with a microscope, or electronic particle counters may be used if the particles are small and of appropriate shape.

There are a number of ways to calculate the **grazing rate** (number of algal cells eaten per herbivore per hour or day) or the **predation rate** (number of prey animals eaten per carnivore per hour or day). Both of these rates become more meaningful for comparative purposes if they are converted to **ingestion rate**, which is the weight or energy content of food ingested per animal per hour or day; in this case, weight can be expressed in terms of dry weight, organic matter, carbon, or nitrogen. Many factors affect feeding rates including temperature, type of food, and concentration of food. There is usually a direct correlation between the amount of food eaten and the food concentration, the relationship being expressed as:

$$R = R_{max}(1 - e^{-kp}) \tag{5.8}$$

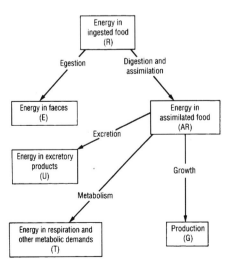

Figure 5.11 A schematic division of the losses and uses of the energy contained in consumed food.

where R is the ration of food ingested at a food or prey density p, R_{max} is the maximum ration taken at satiation, and k is a grazing constant linking food concentration and ingestion. Thus the ration increases with food concentration to some maximum level, as shown in Figure 5.12.

In order to establish the proportion of food that is actually digested in the gut and assimilated by an animal (AR in equation 5.7), an assimilation efficiency is calculated by comparing the ingested food with a quantitative measure of faeces produced. Thus, assimilation efficiency (A) can be calculated from:

$$A = \frac{R - E}{R} \times 100\% \qquad (5.9)$$

where R is the amount of food (ration) ingested and E the amount of faeces produced. (Note that in equation 5.7, A is the assimilation efficiency given in decimal form.) Although assimilation efficiency varies with type of food, age of an animal, and other factors, it is generally high (*ca.* $80->90\%$) for carnivorous animals, and somewhat lower (*ca.* 50–80%) for herbivores. Detritus feeders have the lowest assimilation efficiencies, usually being less than 40%.

Why are assimilation efficiencies different in carnivores, herbivores, and detritivores?

Assimilation is high in carnivores because of the similar biochemical composition of the prey and predator. It is lower in herbivores because of the greater difficulty in digesting plant food, particularly carbohydrates and especially cellulose. Assimilation efficiency is lowest in animals that feed on detritus because much of the food they consume, such as skeletal components, is indigestible.

Values for respiration losses (T in equations 5.6 and 5.7) can be determined in the laboratory. Bottles with and without experimental animals are prepared simultaneously with known volumes of seawater. Changes in dissolved oxygen concentration after an incubation period are detected by chemical methods or oxygen electrodes, and are attributed to respiration of the animals. Respiration rates are directly correlated with the environmental temperature, but they are inversely correlated with size of an animal, being higher per unit weight in smaller animals. In general, 40–85% of the assimilated food (AR) will be utilized for metabolic maintenance.

How would different respective respiration rates affect the daily food ration of small zooplankton compared with large zooplankton?

Because small zooplankton have higher respiration rates per unit body weight, they require much higher food intake relative to body weight. Very small zooplankton (e.g. crustacean nauplii) may require and ingest food equivalent to more than 100% of their body weight per day. In contrast, large zooplankton (e.g. adult euphausiids) may eat the equivalent of only about 20% of their body weight per day.

Referring to equation 5.7, growth or production (G) of zooplankton can be calculated indirectly from values obtained for ingestion rates (R), assimilation efficiency (A), and metabolic losses (T). In some circumstances, growth can be measured directly in laboratory experiments by measuring or weighing experimental animals at successive time intervals. There are two expressions that link the growth ($G = \Delta W / \Delta t$) of an animal with the

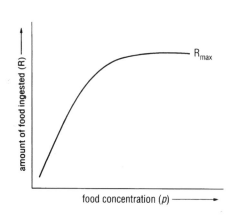

Figure 5.12 The relationship between the amount of food ingested (R) and the concentration of food available (p). Note that R increases up to a certain food concentration, then remains at the same level (R_{max}) with further increases in food.

amount of food ingested (R) by that animal. One is known as the **gross growth efficiency**, K_1, which is expressed as:

$$K_1 = G/R \times 100\%. \tag{5.10}$$

The other expression is the **net growth efficiency**, K_2, which is a ratio of growth (G) to assimilated ration (AR), so that:

$$K_2 = G/AR \times 100\%. \tag{5.11}$$

QUESTION 5.10   The gymnosome *Clione limacina*, feeding on the shelled pteropod *Limacina helicina*, increased in dry weight by an average of 5.0 mg per month, while consuming an average of 7.5 mg dry weight of prey. *Clione* assimilates its prey with 90% efficiency. (a) What is the gross growth efficiency for *Clione?* (b) What is the value of K_2 for this gymnosome?

Temperature and food concentration will influence both K_1 and K_2 values, and both values change with age of an animal. The number of eggs produced by experimental females may also be included as part of the total production in energy balance equations and, in general, fecundity increases with increased food ration. The efficiency at which assimilated food is converted into growth or progeny (K_2) usually ranges between 30% and 80% for zooplankton and fish. This is much higher than growth efficiency in terrestrial mammals, which tends to be on the order of 2–5%. Some of the difference arises because poikilothermic animals (planktonic invertebrates, fish) have lower metabolic costs than warm-blooded mammals. As well, terrestrial mammals expend energy in fighting the effects of gravity, whereas pelagic species are buoyed up in their seawater environment.

Experimental studies of energy requirements and energy partitioning as described above provide a more realistic basis for formulating ecological principles that govern food chains in the sea. These types of experiments permit us to judge the levels of food concentrations required to produce a certain number of animals at different trophic levels, at least under the conditions maintained in the laboratory. Conversely, knowing the amount of primary productivity in any particular marine region, and applying energy budget values obtained from the laboratory, permits a better prediction of the amount of secondary production that might be expected under those conditions.

The laboratory approach to establishing energy budgets in zooplankton has given us useful comparative values for different types of animals, and has allowed an assessment of transfer efficiencies based on ecological trophic theory and field surveys. Nevertheless, it is important to keep in mind that only some species are amenable to laboratory culture; the conditions are always artificial: and the results must be extrapolated to natural situations, including natural foods and changing environmental conditions.

Enclosed experimental ecosystems

An enclosed experimental ecosystem is a system in which a large volume of natural seawater is artificially enclosed in order to study populations of phytoplankton and zooplankton over time. This alleviates the problem of studying plankton populations at sea, where they are being continually displaced from one location to another by water movement. If the experimental system is large enough, it may be possible to study the

---

Handwritten notes (left margin):

a.

K_1 $\dfrac{5.0 \text{ mg}}{7.5 \text{ mg}} \times 100\%$

$= 67\%$

b.

$K_2 = \dfrac{5.0 \text{ mg}}{0.90 \times 7.5 \text{ mg}} \times 100.$

$= 74\%.$

fecundity = fertile / fertilizing.

interactions of several trophic levels, including such planktivorous species as ctenophores and fish.

Several conditions must be met in setting up an enclosure. In order to contain sufficient plankton and a reasonable semblance of the natural environment, the volume of seawater captured must usually be between 100 and 1000 m^3. It must be enclosed in such a way that natural sunlight can penetrate the water column. The depth profile of the water column in terms of temperature, nutrients, and salinity must also be preserved, and care must be taken that no toxic substances are accidentally introduced. These conditions can usually be achieved by using moored, transparent containers made of nontoxic material.

Figure 5.13 illustrates enclosures that were moored in a fjord in British Columbia, Canada, and used in studies known as the Controlled Ecosystem Pollution Experiments (CEPEX). Each 30-m deep container enclosed about 1300 m^3 of seawater. The purpose of the programme was to test the effects of small traces of pollutants on the ecology of the plankton community and on young fish. The employment of several enclosures permitted the addition of different types of pollutants in differing concentrations, as well as providing a control enclosure that could be compared with conditions in the surrounding, unenclosed water. Experiments were conducted with traces of copper (10–50 ppb), mercury (1–10 ppb), petroleum hydrocarbons (10–100 ppb), and several other potential pollutants that can be found in seawater, particularly in coastal areas.

Figure 5.14 provides an example of the results that were obtained in one experiment. At the start, diatoms (*Chaetoceros*) were the dominant primary producers in all enclosures, and they remained dominant in the control containers (J and K) throughout the experimental period. In the experimental ecosystems treated with copper (L and M), most of the diatoms died within three weeks (Figure 5.14a) and were replaced by small photosynthetic flagellates that began to increase after the first week following pollutant addition (Figure 5.14b). This change in the structure of the primary producers from large diatom chains (*ca.* 500 μm) to small flagellates of less than 20 μm is analogous, but on a microscale, to a change in size of terrestrial vegetation from trees to grass, and it had a profound influence on the total ecology of the enclosures. Changes in the phytoplankton community were accompanied by changes in the dominant species of the zooplankton community and, after about one month, the copper-treated and control containers had very different ecologies. This kind of result is almost impossible to obtain from either laboratory-scale experiments (which are too small) or from field experiments (where the water moves around too much).

QUESTION 5.11 Figure 5.14 shows that a small amount of copper in seawater caused a change in the primary producers, with long-chain diatoms being replaced by small flagellates. What difference might this make to the rest of the food chain in terms of the type of dominant species and number of trophic levels? (Refer to Section 5.1 for help if necessary.)

Enclosed ecosystem experiments have been performed at many different locations including in Loch Ewe in Scotland; at the Marine Ecosystem Research Laboratory in Rhode Island, U.S.A.; and in Xiamen in the People's Republic of China. Not all of the experiments have involved pollutants. Some have been designed to test the effects of natural perturbations, such as changes in light levels and in amount of vertical mixing, on plankton

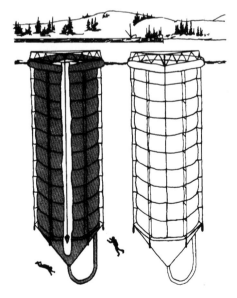

Figure 5.13 An artist's illustration of controlled experimental ecosystem enclosures floating in the sea. Each polyethylene container is 30 m deep and holds about 1300 m^3 of seawater. The scuba divers are drawn to approximate scale.

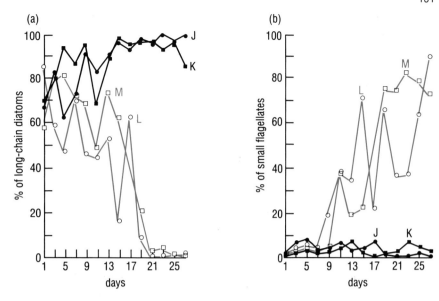

Figure 5.14 The results of experiments in enclosed experimental ecosystems. (a) The percentage of long-chain diatoms (*Chaetoceros*) in the total phytoplankton. (b) The percentage of small flagellates in the total phytoplankton. L and M, copper-treated containers; J and K, control containers.

ecology. Such studies provide valuable insights into the influence of physical variables on biological processes.

The disadvantages of using enclosed experimental systems of this type should also be mentioned. Perhaps the greatest drawback is the removal of small-scale physical turbulence from enclosed waters. Mixing is an important property of natural marine communities, and the damping out of turbulence within the containers can lead to spurious results if the experiments are continued for too long a time. Also, some caution has to be exercised in applying results obtained in experiments in one area to other locations. The waters of Loch Ewe, Scotland, for example, are not the same (physically, chemically, or ecologically) as those in the South China Sea, and similar experimental procedures may produce quite different results in the two localities.

Computer simulations of marine ecosystems
Computer models provide a third way of generating knowledge about the way marine ecosystems operate. Data gathered from many sources (e.g. field measurements, laboratory or controlled ecosystem enclosure experiments, satellite imagery) are entered into mathematical models designed either to simulate natural events or to provide predictions of future events. The type of model used depends on the questions being asked. A relatively simple model can be used to predict an oxygen balance in a particular body of water, but more complex models are required to examine trophic interactions between several levels in marine food webs.

The usual approach in setting up computer models that will examine ecological issues is to formulate a number of differential equations made up from non-linear empirical relationships which connect various forcing functions with ecotrophic structure (Figure 5.15). For example, the amount of available light and the concentrations of nutrients would be considered **forcing functions** in such equations; they set the constraints on production in the system being considered. A model would also consider non-linear

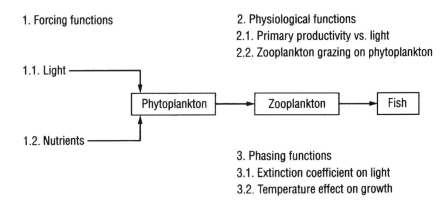

1. Forcing functions

1.1. Light

1.2. Nutrients

2. Physiological functions
2.1. Primary productivity vs. light
2.2. Zooplankton grazing on phytoplankton

Phytoplankton → Zooplankton → Fish

3. Phasing functions
3.1. Extinction coefficient on light
3.2. Temperature effect on growth

Figure 5.15 Ecosystem computer model construction showing examples of forcing, physiological, and phasing functions.

physiological functions, such as the reaction of light with phytoplankton (see Section 3.3). In addition to these two types of functions, there are a number of other environmental influences which modify both the forcing and physiological functions. For example, changes in temperature will modify most physiological functions, and the light extinction coefficient will modify the forcing function of light. These modifying influences form a third type of function; they are called **phasing functions** because they tend to speed up or slow down the interactions between forcing and physiological functions.

The layout of a simple computer model involving the interaction of different trophic levels is shown in Figure 5.16. In this system, the growth rate of the phytoplankton is governed by light and nutrients. The zooplankton graze the phytoplankton according to an equation in which the grazing rate is dependent on the concentration of phytoplankton; the one planktivorous species (a ctenophore) utilizes the zooplankton depending on the availability of the prey. The model also includes a recycling loop in which bacteria utilize dissolved organic material (DOC) lost from the phytoplankton and return part of it to the system via zooflagellates and microzooplankton. This model can be run on a desk-top computer, and an illustration of the kind of results that one can obtain are shown in Figure 5.17.

QUESTION 5.12 In the computer model layout shown in Figure 5.16 which are (a) the forcing functions and (b) the physiological functions?

Figure 5.17 shows the results of a computer simulation designed to test the effect of changing only one parameter, the extinction coefficient of light, on the output of the model in terms of the amount of phytoplankton and zooplankton produced. The amount of light was decreased by increasing the extinction coefficient (see Section 2.1.2) from 0.2 to 0.3 and to 0.7 m^{-1}. As one might expect, the amount of phytoplankton decreases during an 8-day period as the extinction coefficient increases. However, the standing stock of zooplankton <u>increases</u> with a decrease in light intensity from extinction values of 0.2 to 0.3 m^{-1}. This increase can be attributed to the fact that the phytoplankton grow more slowly at an extinction value of 0.3 m^{-1} than at 0.2 m^{-1}, and this in turn enables the slower-growing zooplankton to graze more of the phytoplankton, which would otherwise sink out of the water column as ungrazed phytodetritus. The very abrupt decline in the phytoplankton at day 4 (when $k = 0.2$) is due to nutrients becoming exhausted by the rapidly growing phytoplankton, and to the sinking of accumulating phytodetritus. The maximal zooplankton growth at a light intensity in between very bright and shady is a result of interactions within

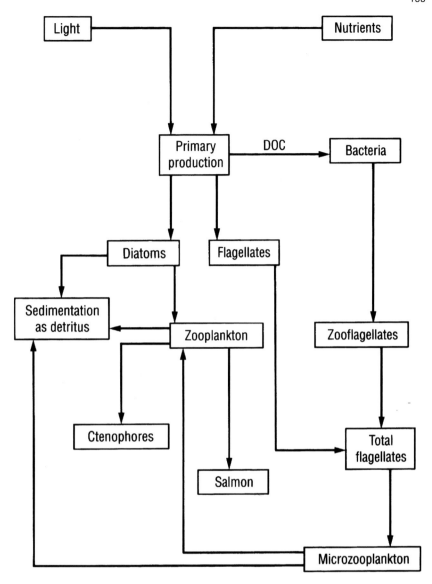

Figure 5.16 A simple trophodynamic model for use in computer simulations of trophic interactions. DOC, dissolved organic carbon.

the model, and could not be readily seen without the modelling approach. The use of ecosystem models has indicated that many other trophic relationships optimize at intermediate environmental values.

QUESTION 5.13 In the computer model results shown in Figure 5.17, why is the production of zooplankton lowest when $k = 0.7$?

low light ∴ low Primary production.

Computer simulation models have both advantages, as described in the example above, and disadvantages. They cannot replace the gathering of real data. Efforts can be made with models to simulate actual events. More often than not, however, data generated by actual events are the result of so many variables that it is statistically difficult to tune any one model to explain them. Thus a model may more often be used to project different scenarios of likely events. When compatibility with actual field data is required, a biological trophodynamic computer model must be coupled to a computer model of the physical environment. Such models have been produced, and an example is the General Ecosystem Model for the Bristol Channel and

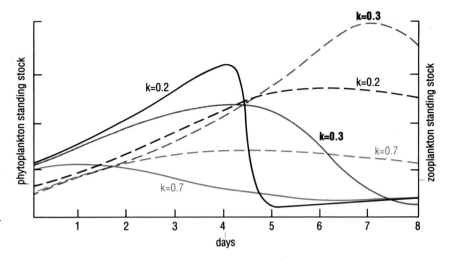

Figure 5.17 A computer model simulation predicting the production of phytoplankton (solid lines) and zooplankton (dashed lines) from different light extinction coefficients (*k*). (Standing stocks in relative units.)

Severn Estuary (known as GEMBASE) developed by the Institute for Marine Environmental Research in Plymouth, England. In this computer model, an ecological submodel is operated in seven different physical domains, and water exchanges between these subregions are determined by two physical models describing water flow. Integrated water exchanges produced from the physical models are then entered as input into the slower-scaled ecological models.

5.4   A COMPARISON OF MARINE AND TERRESTRIAL PRODUCTION OF ORGANIC MATERIAL

The preceding discussions of pelagic production and food webs have highlighted unique ecological features of the marine habitat. You may already have noted profound differences between marine and terrestrial ecology. The following points summarize differences in primary and secondary production in these two major environments. You may wish to compile your own list of other differences in physical features of the contrasting environments and of general anatomical and behavioural differences in marine and terrestrial organisms.

1   In the open ocean, the majority of the primary producers are microscopic phytoplankton (*Sargassum* weed being one of the few exceptions). Although macroscopic algae and sea grasses contribute to marine production in inshore shallow areas, by far the greatest amount of marine primary production is carried out by the small phytoplankton. In contrast, the great majority of terrestrial primary producers are large, highly visible forms like grass and trees.

2   The basic structure of phytoplankton and terrestrial vegetation is very different. The small size of phytoplankton enhances flotation and also, because of the surface to volume ratio, promotes uptake of nutrients directly through the cell wall from the surrounding water. Terrestrial plants, in contrast, require roots for anchoring and for nutrient uptake from the soil. They also tend to develop trunks and branches to fight gravity and maximize exposure to sunlight, and this requires the production of carbohydrates like cellulose and lignin which confer strength and rigidity. Phytoplankton, in

contrast, do not require large amounts of structural carbohydrate and are largely composed of protein.

3   Because of the small size and high protein content of the phytoplankton and their relatively low abundance compared to herbivore numbers, a major fraction of the marine primary production is usually consumed, digested, and readily assimilated by marine herbivores. Comparatively little of the pelagic primary production goes directly into the decomposer cycle. This is not the case in the terrestrial ecosystem, where much of the primary production is in the form of inedible or indigestible structural components, such as cellulose and lignin, that are contained in bark, tree trunks, and roots. Terrestrial animals rarely eat more than 5–15% of the total plant production, and they consume a lot of material that is largely indigestible. Much of the terrestrial photosynthetic production therefore enters indirectly into the food chain via the decomposition cycle.

4   In the pelagic environment, primary productivity ranges from about 50 to 600 g C m^{-2} yr^{-1}. In comparison, terrestrial primary productivity varies from virtually zero in arid deserts and regions that are too cold for plants (Antarctica) to maximum values of about 2400 g C m^{-2} yr^{-1} in grasslands, and about 3500 g C m^{-2} yr^{-1} in tropical rainforests. Although primary productivity of benthic marine plants in shallow areas may approach terrestrial values, in general the primary production per unit area in the sea is much lower than on our 'green' land.

5   A useful comparison can be made by examining the **production to biomass (*P/B*) ratio**; this is the relationship between the total annual production and the average biomass of living plant (or animal) material present throughout the year. Although the total biomass of the small marine phytoplankton may be relatively low, these algae are fast-growing. Consequently the *P/B* ratio for phytoplankton is roughly 100–300 in the marine pelagic ecosystem; this means that the phytoplankton biomass may turn over 100–300 times during a year. *P/B* ratios measured over one year are about an order of magnitude lower for marine zooplankton, and another order of magnitude lower for fish. In contrast, terrestrial plants have a very high total biomass (we can see this), but are generally slow-growing and long-lived; further, a large fraction of the primary production of these plants is used to maintain the respiration of that biomass. Therefore, terrestrial vegetation has much lower *P/B* ratios of about 0.5–2.0.

6   Most of the animals in the sea are cold-blooded (poikilothermic) invertebrates and fish and therefore have much lower energy requirements than terrestrial warm-blooded birds and mammals. As well, pelagic animals live in a buoyant fluid and use little energy in locomotion. In comparison, mammals and birds (and even the larger poikilothermic insects) expend large amounts of energy fighting gravity; walking, crawling, and flying are all more energetically costly than swimming. This means that a relatively larger fraction of ingested energy can be channelled toward growth and reproduction in marine pelagic animals. In fact, growth efficiency (the ratio of biomass production to quantity of food eaten) tends to be an order of magnitude higher in marine poikilothermic species. In contrast to land mammals and birds, they also produce larger numbers of young and usually expend no energy on parental care of the progeny. All of these features contribute to a much higher secondary production in the sea compared with that on land.

7   Whereas plants so evidently dominate the biomass on land, animals form the visually dominant group in the sea. Although the ocean contributes only about 50% of the world's plant production, it accounts for more than 50% of its animal production.

5.5   MINERAL CYCLES

All elements that become incorporated in organic materials are eventually recycled, but on different time scales. The process of transforming organic materials back to inorganic forms of elements is generally referred to as **mineralization**. It takes place throughout the water column as well as on the bottom of the sea, where much of the detrital material from overlying waters eventually accumulates. Recycling of minerals may take place relatively rapidly (within a season) in the euphotic zone, or much more slowly (over geological time) in the case of refractory materials which sink and accumulate on the seabed.

Figure 5.18 illustrates some of the ways in which elements are cycled by different groups of organisms. In the water column, where there is usually plenty of oxygen, decomposition of organic material takes place via oxidative degradation through the action of heterotrophic bacteria. Carbon dioxide and nutrients are returned for re-utilization by the phytoplankton. The cycle differs in **anoxic** areas where there is no free dissolved oxygen. Anoxic conditions are present in subsurface sediments on the seafloor and in a few special areas like the Black Sea, which is anoxic from about 200 m depth to the bottom because water exchange and mixing with the adjacent Mediterranean Sea are severely restricted by bottom topography. Under anoxic conditions, bacterial degradation takes place by anaerobic bacteria that utilize oxygen found in sulphate and nitrate radicals. This type of oxidation forms highly reduced compounds such as methane, hydrogen sulphide, and ammonia. Since these compounds are high in chemical energy, another group of bacteria (the **chemoautotrophs**) can utilize this energy to reduce carbon dioxide and make new organic material. The process of fixing carbon from CO_2 into organic compounds by using energy derived from oxidation of inorganic compounds (e.g. nitrite, ammonia, methane, sulphur compounds) is called **chemosynthesis**.

Ecologically, the most important aspect of recycling in the sea is the rate at which growth-limiting nutrients are recycled. Among the nutrients that can be in short supply in the sea, nitrate (NO_3^-), iron (bioavailable Fe), phosphate (PO_4^{3-}), and dissolved silicon ($Si(OH)_4$) are most often found in concentrations well below the half-saturation levels required for maximum phytoplankton growth (refer to Section 3.4). Silica limitation affects primarily those organisms that use this element to form skeletons; these include the diatoms and silicoflagellates among the phytoplankton (see Section 3.1), and the radiolarians among the zooplankton (refer to Section 4.2). The silicon cycle is relatively simple as it involves only inorganic forms; organisms utilize dissolved silicon to produce their skeletons, and this skeletal material dissolves following death of the organisms. The cycling of phosphorus is also relatively simple in a chemical perspective; at the usual alkaline pH of seawater, organic phosphate is relatively easily hydrolysed back to inorganic phosphate which is then available again for uptake by phytoplankton. Because phosphorus cycles rapidly through the food chain, it is seldom limiting in the marine

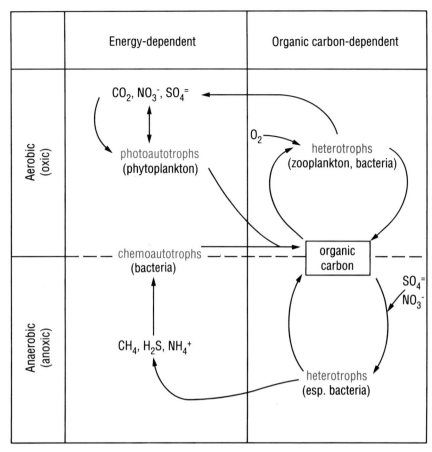

Figure 5.18   Marine organisms derive energy from light (photoautotrophs), from inorganic compounds high in chemical energy (chemoautotrophs), from organic carbon compounds (heterotrophs) or, in the case of a few bacteria, from a combination of these processes. Each of these sources of energy may become limited at some time or place in marine habitats. The supply of organic carbon compounds is largely limited by their rate of production from chemoautotrophs and photoautotrophs. Photoautotrophs are limited by the amount of light. Chemoautotrophs, which may be either aerobic or anaerobic species, are limited by the amount of highly reduced inorganic compounds derived from the metabolism of anaerobic heterotrophs. Thus each process is essentially dependent on the next process to recycle material through the entire system.

environment. Compared with silicon and phosphorus, the recycling of nitrogen, however, is a more complex process.

5.5.1 NITROGEN

The marine nitrogen cycle (Figure 5.19) is complex because nitrogen in the sea occurs in many forms that are not easily converted from one to another. These include dissolved molecular nitrogen (N_2) and the ionic forms of ammonia (NH_4^+), nitrite (NO_2^-) and nitrate (NO_3^-), as well as organic compounds such as urea ($CO(NH_2)_2$). The dominant form of nitrogen in the ocean is the nitrate ion, and it is often this form that is taken up by phytoplankton, although many species can also utilize nitrite or ammonia. A few phytoplankton species can also take up some small molecules of organic nitrogen, such as amino acids and urea. The rate at which nitrogen in a suitable state is made available for phytoplankton may limit primary production in oligotrophic waters throughout the year and in temperate waters during the summer. Remember too (from Section 3.4) that iron is necessary for the formation of reductase enzymes that are used in the conversion of nitrite and nitrate into ammonium, and this is used to make amino acids. If iron is present in limiting concentrations, then even an abundance of nitrate will not promote maximum phytoplankton production.

Regeneration of nitrogen in the water column results from bacterial activities and excretion by marine animals, especially the excretion of ammonia by zooplankton. As illustrated in Figure 5.19, the oxidation of ammonia to nitrite and then to nitrate is referred to as **nitrification**; the bacteria that mediate this change in chemical state are called **nitrifying bacteria**. The reverse process of forming reduced nitrogen compounds from nitrate occurs mostly in anoxic sediments and is called **denitrification**; these changes are carried out by **denitrifying bacteria**. The nitrogen cycle also involves **nitrogen-fixation**, in which dissolved nitrogen gas is converted to organic nitrogen compounds; this process can be carried out by only a few phytoplankton, notably some Cyanobacteria. Dissolved organic nitrogen (DON) and particulate organic nitrogen (PON) both serve as nutrients for bacterial growth. Bacteria break down proteins to amino acids and ammonia, and the latter is oxidized in the nitrification process. The eventual release of dissolved inorganic nitrogen (DIN) makes these forms available again for uptake by the phytoplankton. The various types of bacteria involved in this cycling can themselves serve as a direct source of food for some nano- and microzooplankton.

An important aspect of the marine nitrogen cycle concerns the source of the nitrogen used in primary production. Some fraction of the primary production is derived from nitrogen recycled from organic matter within the euphotic zone; another fraction is derived from **new nitrogen** which comes from sources outside the euphotic zone (see Figures 5.19 and 5.20). New nitrogen is primarily nitrate entering the euphotic zone from below the

Figure 5.19   The nitrogen cycle in the euphotic zone of the sea. The diagram illustrates the recycling of nitrogen that takes place within the euphotic zone and above the nutricline, as well as the input of 'new' nitrogen upwelled from deeper water. Note the interrelationships between DIN (dissolved inorganic nitrogen), PON (particulate organic nitrogen), and DON (dissolved organic nitrogen). The nutricline is where there is a rapid change in the concentration of a nutrient with depth.

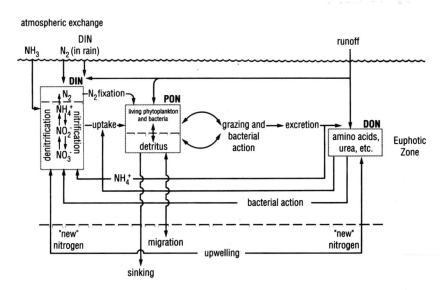

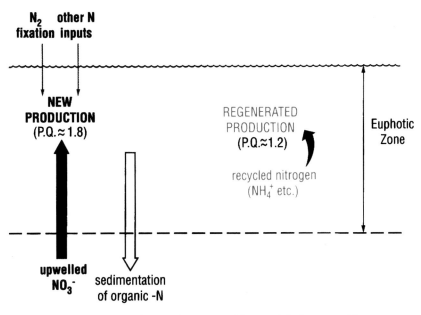

Figure 5.20   A comparison of production generated by recycled nitrogen and by new nitrogen. Regenerated production results only from the reduced nitrogen supplied within the euphotic zone by the excretion of organisms. New production depends on nitrogen supplied from outside the euphotic zone, of which the dominant source is the nitrate moving upward from below the nutricline. In a steady state, upwelled nitrogen is balanced by the downward sinking of nitrogen bound in sedimenting particles. On an annual time scale, new production in the open ocean is believed to be roughly one-third to one-half of regenerated production, but the actual value may vary considerably from these figures over short temporal and spatial scales. P.Q., photosynthetic quotient (defined in the text).

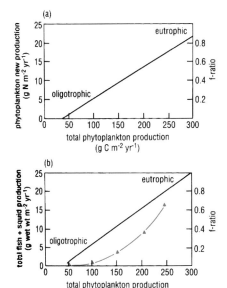

Figure 5.21   (a) Phytoplankton new production as a function of total (regenerated + new) annual phytoplankton production in different marine environments ranging from oligotrophic to eutrophic.

(b) Total carnivorous fish plus squid biomass production as a function of total phytoplankton production in marine environments ranging from oligotrophic to eutrophic. The blue line indicates the total amount of fish production that results from new production only; this is a sustainable yield because the nitrogen removed in the harvest will be replaced through upwelling of nitrate.

nutricline by vertical mixing, but it also includes smaller amounts of nitrogen entering by N_2-fixation and through river inflow and precipitation. Recycled nitrogen is primarily in the form of ammonia and urea. This comparison of regenerated and new nitrogen (and of **regenerated** and **new production**, see Figure 5.20) is important because only the continual input of new nitrogen can determine the total capacity of the ocean to produce a sustainable fish harvest (remember that removing fish from the ocean also removes nitrogen). It is also only the new nitrogen that can help to take up the excess CO_2 that is believed to be entering the ocean from human activities; in this case, increased production of phytoplankton removes more carbon dioxide.

In oligotrophic areas of the oceans (e.g. the large convergent gyres discussed in Section 3.5.1), there is little upward movement of water from below the euphotic zone and thus the amount of new nitrogen is very small. In upwelling regions, however, the amount of new nitrogen is very large. The ratio of new production to total (new + regenerated) production is referred to as the *f*-ratio. This is shown in Figure 5.21a where it is related to waters of different nutrient concentrations. The value of the *f*-ratio is probably 0.1 or less in oligotrophic waters, but may be as high as 0.8 in upwelling zones. An annual average for the whole ocean is estimated to be about 0.3 to 0.5. About one-third of the global pelagic primary production takes place in areas where new nitrogen is entering the euphotic zone; these coastal or upwelling areas represent only about 11% of the ocean surface (see Table 5.1). Elsewhere, primary production depends predominantly on nitrogen that is recycled within the euphotic zone.

a

$300 \times 0.3 = 90 \, g \, C \, m^{-2}$

regenerated production

$300 \times 0.7 = 210 \, g \, C \, m^{-2}$

New Production

f-ratio =

New Production / Total Production

$$\frac{210}{300} = 0.7$$

$\downarrow$ contraled buach.

from Table 3.5

QUESTION 5.14   (a) If the annual primary production of a marine area is 300 g C m^{-2} and 30% of that production is driven by regenerated nitrogen, what is the value of the f-ratio? (b) Where would you expect to find water with this amount of production and with this f-ratio? (Refer to Section 3.6.)

Figure 5.21b relates the production of carnivorous fish plus squid to total primary production and to new production. The total amount of fish and squid produced results from both regenerated and new production. Harvesting that fraction of fish produced from recycled nitrogen will lead to a loss of nitrogen from the system and thus a decrease in production, but removing fish produced from new nitrogen sources will result in a sustainable harvest because the nitrogen will be replenished. Thus a larger fraction of the total fish production can be removed from eutrophic waters than from oligotrophic waters without depleting nitrogen in the surface layers. Figure 5.21b shows that a total fish production of 2 g wet weight m^{-2} yr^{-1} in oligotrophic waters with an f-ratio of about 0.1 will give a sustainable annual harvest of 0.2 g wet weight m^{-2} yr^{-1} (from 0.1 of 2 g). In comparison, 20 g wet weight m^{-2} yr^{-1} of total fish production in eutrophic waters with an f-ratio of 0.8 provides a sustainable harvest of 16 g m^{-2} yr^{-1}, or an 80-fold increase in the sustainable harvest compared with only a 10-fold increase in total fish production.

QUESTION 5.15   Why does the amount of fish available for a sustainable harvest increase exponentially in going from oligotrophic to eutrophic waters in Figure 5.21? The answers should be apparent in Figures 5.21a and 5.21b.

In Figure 5.20, the **photosynthetic quotient** (PQ = moles of O$_2$ produced by the phytoplankton divided by the moles of CO$_2$ taken up) is used to diagnose the difference in the production processes involving the two types of nitrogen. Note that regenerated production based on recycled nitrogen has a lower PQ (≈ 1.2) than new production (PQ ≈ 1.8) that is based on nitrogen entering from the atmosphere and rivers and, especially, on upwelled nitrate. The reason for this is summarized in Figure 5.22, where different values of PQ are obtained depending on which type of nitrogen is being utilized and which chemical pathway is being followed in photosynthesis. If only carbohydrate material were being formed, there would be a stoichiometric yield of one mole of oxygen produced for each mole of carbon dioxide used. However, lipid material also is formed during photosynthesis, and because lipids are more highly reduced than carbohydrates, additional oxygen will be released and the PQ will approximate 1.2 instead of 1.0. If large amounts of nitrate (a new nitrogen source) are taken up and reduced in the process of forming proteins, even more oxygen is liberated relative to the CO$_2$ utilized, and thus the PQ will increase to about 1.8. If ammonia (a recycled nitrogen form) is used as a nitrogen source for protein manufacture by the phytoplankton, then oxygen is required in the process and the PQ would be 1.0 or less. Thus rapidly dividing phytoplankton populations, using nitrate, will have relatively high PQ values. Lower PQ values indicate that reduced states of nitrogen, like recycled ammonia, are being used in photosynthesis. Because it is difficult to make accurate measurements of the PQ or, alternatively, of the rate and amount of vertical nitrate transport, the f-ratio remains a matter of much discussion and debate among scientists.

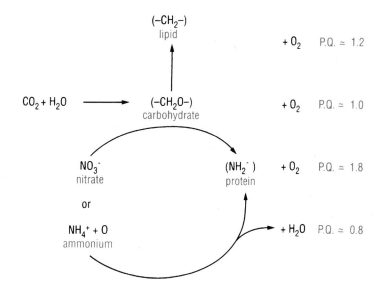

Figure 5.22 The relationship between photosynthetic pathways and the photosynthetic quotient (P.Q.). The basic equation for photosynthesis is shown in the middle; in this case, only carbohydrate is being produced and the P.Q. would be 1.0 (1 mole of O_2 is produced for each mole of CO_2 utilized). Note that the value of P.Q. varies according to the type of nitrogen source being utilized by the phytoplankton to form proteins. (See text for further details.)

5.5.2 CARBON

Carbon is another element that is essential for life, but unlike nitrogen, carbon is never present in the sea in limiting quantities. However, the carbon cycle (shown in Figure 5.23) has some special properties that involve both physical and biological processes.

Carbon dioxide enters the ocean from the atmosphere because it is highly soluble in water. If the concentration of CO_2 in seawater depended entirely on the partial pressure of CO_2 in the atmosphere (0.3 ml l^{-1}), on the relative concentrations of CO_2 in water and air, and on the temperature and salinity of the water, then the amount of CO_2 in seawater would be very low. In the sea, however, free dissolved CO_2 combines with water and ionizes to form bicarbonate and carbonate ions, as shown below.

CO$_2$ INPUTS
atmosphere;
 respiration;
mineralization; and
dissolution of CaCO$_3$

$$CO_2 \ + \ H_2O \ \rightleftharpoons \ H_2CO_3 \ \rightleftharpoons \ HCO_3^- \ + \ H^+$$

carbonic   bicarbonate
acid   ions

CO$_2$ UTILIZATION
photosynthesis; and
formation of CaCO$_3$

$$CO_3^= \ + \ H^+$$

carbonate
ions

These ions are **bound** forms of carbon dioxide, and they (especially bicarbonate) represent by far the greatest proportion of dissolved carbon dioxide in seawater. On average, there are about 45 ml total CO_2 l^{-1} of seawater, but because of the equilibrium chemical reactions shown above, nearly all of this occurs as bound bicarbonate and carbonate ions which thus act as a reservoir of free CO_2. The amount of dissolved CO_2 occurring as gas in seawater is about 0.23 ml l^{-1}. When free CO_2 is removed by photosynthesis, the reaction shifts to the left and the bound ionic forms release more free CO_2; so even when there is a lot of photosynthesis, carbon dioxide is never a limiting factor to plant production. Conversely, when CO_2 is released by the respiration of plants, bacteria and animals, more bicarbonate and carbonate ions are produced.

Note that hydrogen ions are liberated in the general chemical reactions shown above. This means that the pH of seawater is largely regulated by the concentrations of bicarbonate and carbonate, and the pH is usually 8 ± 0.5. When CO_2 is added to seawater due to mineralization processes and respiration, the number of hydrogen ions increases and the pH goes down (the solution becomes more acidic). If CO_2 is removed from the water by photosynthesis, the reverse happens and the pH is elevated. Thus seawater acts as a buffered solution.

Some marine organisms combine calcium with carbonate ions in the process of **calcification** to manufacture calcareous skeletal material. The calcium carbonate ($CaCO_3$) may either be in the form of calcite or aragonite, the latter being a more soluble form. After death, this skeletal material sinks and is either dissolved, in which case CO_2 is again released into the water, or it becomes buried in sediments, in which case the bound CO_2 is removed from the carbon cycle.

QUESTION 5.16 Which marine organisms (planktonic, nektonic and benthic species) incorporate carbon dioxide into carbonate skeletons and thus influence the carbon cycle? (Refer to Sections 3.1 and 4.2 for part of the answer.)

The simplified carbon cycle shown in Figure 5.23 summarizes the processes discussed above. In general, CO_2 is converted from inorganic to organic carbon by the photosynthesis of the phytoplankton. This is then consumed by the higher trophic levels, and some CO_2 is recycled as inorganic bicarbonate while some may be lost from the ocean surface in gaseous form. Carbon dioxide is absorbed at the ocean surface and is produced in the water column by respiration and mineralization processes. It is believed that more carbon dioxide is being absorbed by the oceans than is being lost to the atmosphere.

The total amount of soluble carbon dioxide (bicarbonate and carbonate ions plus dissolved CO_2) in the world's oceans is estimated to be about 38×10^{12} tonnes. This is about fifty times more than the total carbon dioxide in the atmosphere. The burning of fossil fuels is increasing the total carbon dioxide in the atmosphere at a rate of about 0.2% per year, and it is very important to know if this increase can be absorbed by the oceans, or if it will continue to accumulate in the atmosphere where it may contribute to global warming by the process often referred to as the 'greenhouse effect'.

The importance of the ocean's biology to the carbon cycle and to the balance of CO_2 in the atmosphere is threefold. Firstly, the amount of CO_2

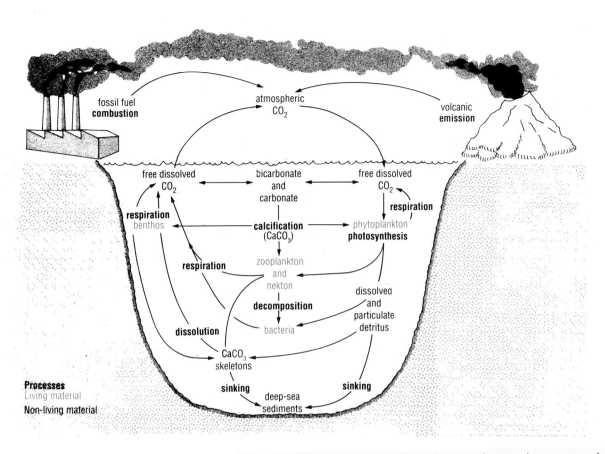

Figure 5.23 The basic scheme of the carbon cycle.

fixed by the food chain depends on how much new nitrate enters the euphotic zone to support photosynthesis (see Section 5.5.1). Secondly, the amount of carbon which can be permanently lost to the sediments depends on deep-water chemistry, ecology and sedimentation processes, and particularly on the bacterial loop (Section 5.2.1) which recycles dissolved and particulate organic carbon. Thirdly, the amount of carbon dioxide taken up in the carbonate skeletons of marine organisms has been, over geological time, the largest mechanism for absorbing CO_2. At present, it is estimated that about 50×10^{15} tonnes of CO_2 occurs as limestone, 12×10^{15} tonnes in organic sediments, and 38×10^{12} tonnes as dissolved inorganic carbonate. Determining the amount of carbon that is transferred along the various pathways in Figure 5.23 is a difficult problem in both geochemistry and biology, but a necessary exercise if we are to solve the global carbon budget.

QUESTION 5.17 Carbon dioxide is essential to the process of photosynthesis. Should it be considered as a nutrient, like nitrate for example, that can limit the rate of phytoplankton production in the oceans? *No .*

5.6 SUMMARY OF CHAPTER 5

1 Food chains are ways of describing the linear passage of energy and organic materials contained in food from the first trophic level of primary producers, through the consumer levels of herbivores and carnivores, to the top-level predators. There is an energy loss with each transfer between trophic levels because of metabolic demands and conversion of chemical energy to heat. However, chemical elements that are incorporated in food are

recycled through the decomposition of organic materials; this process releases dissolved inorganic compounds that can once again be taken up by phytoplankton and converted to organic compounds during photosynthesis.

2   Despite great differences in size between phytoplankton and consumers in higher trophic levels, the differences in generation times (hours to many years) among these organisms result in very similar biomass values in each trophic level of marine food chains.

3   Estimates of secondary production in different marine localities can be made by using the expression $P_{(n+1)} = P_1 E^n$. This equation combines quantitative values for primary productivity in an area with the number of trophic levels in the food chain, and with the ecological efficiency at which energy is transferred from one trophic level to another.

4   The number of trophic levels in a food chain is inversely correlated with the predominant size of the phytoplankton. Food chains in nutrient-rich upwelling areas are characterized by having large chain-forming diatoms, high primary productivity, few trophic levels, and a high biomass of fish or marine mammals. In the nutrient-poor open ocean, the primary producers are nanoplanktonic autotrophic flagellates with relatively low productivity; this leads to long food chains and, because of increased energy loss in longer food chains, there is a relatively lower biomass of top-level predators.

5   Food webs are more realistic, but more complex, depictions of energy flow through interacting species. They are a means of recognizing that many marine species compete for the same food items, that many animals change diets during life, that some organisms feed primarily on detritus, and that cannibalism is common in the sea. Such relationships may affect the amount of energy that is available for top-level predators; for example, competition for food between ctenophores and larval fish, or ctenophore predation on fish eggs and fish larvae, may significantly lower fish stocks that are harvested commercially.

6   Bacteria and planktonic protozoans interact in a microbial loop that is coupled with the classic phytoplankton–zooplankton–fish food chain. In this subsystem, bacteria decompose particulate and dissolved detritus; the resulting bacterial production is consumed by protozoans and by some larger zooplankton, such as invertebrate larvae and appendicularians. Thus bacteria regenerate dissolved nutrients for subsequent utilization by phytoplankton, they themselves form a source of food for planktonic bactivorous species, and the bacterial production is transferred to higher trophic levels by the intermediary links of protozoans which are fed on by larger plankton.

7   Major differences exist between marine and terrestrial food webs. The majority of marine primary production is carried out by fast-growing microscopic phytoplankton, most of which is consumed and assimilated by herbivores. In contrast, most terrestrial vegetation is large, slower growing, and contains much indigestible structural material. Only 15% or less of the total terrestrial plant production is eaten, and only a fraction of this is digestible and assimilated into herbivore production. Further, the dominant marine animals are poikilothermic, with lower metabolic energy demands than the homoiothermic birds and mammals that live in terrestrial habitats. This difference in energy utilization, coupled with the fact that most marine primary production is eaten, means that energy is transferred with greater efficiency through marine food chains, and that there is a much higher secondary production in the sea compared with that on land.

8   Whereas it is possible to obtain fairly accurate estimates of primary production through various techniques, and fish catch statistics provide minimum values for energy output from marine food webs, it is much more difficult to quantify secondary production in the intermediate trophic levels occupied by zooplankton and smaller nekton. Although some techniques have been applied to measure secondary production from field data, these are often impracticable because of the vast geographic areas under consideration and the continual movement of the water and resident organisms. Many researchers have therefore resorted to different experimental options. These include laboratory-scale experiments, which attempt to quantify each aspect of energy partitioning in a species; the use of controlled ecosystem experiments, which are carried out on a larger scale and attempt to study several interacting trophic levels at one time; and computer model simulations, in which data from various sources are entered into mathematical models that attempt to simulate natural processes.

9   The term 'mineralization' describes the process whereby elements that have passed through food webs are recycled. Ecological studies are particularly concerned with the recycling rates of essential nutrients that may be present in limiting concentrations in the sea; these include nitrate, iron, phosphate and, occasionally, dissolved silicon. Of these nutrients, nitrate is the one that is most often present in sufficiently low concentrations to limit plant growth. Nitrogen has a complex cycle in the sea because it occurs in many forms; nitrate is the dominant form most often utilized by phytoplankton, but ammonia, nitrite, and dissolved molecular nitrogen can also be used by some species. The physiological activities of organisms produce particulate and dissolved organic nitrogen in various chemical species, and different types of bacteria mediate the conversions from one type of nitrogen compound to another.

10   An important distinction is made between regenerated nitrogen (primarily ammonia and urea) that is recycled in the euphotic zone by pelagic organisms, and new nitrogen (primarily nitrate) that enters the euphotic zone from upward movement of deep water or, in smaller amounts, from river inflow and precipitation. The amount of new nitrogen relative to regenerated nitrogen (the f-ratio) is high in upwelling regions and low in oligotrophic areas. Where production is based primarily on new nitrogen (nitrate), the photosynthetic quotient (PQ) is high, indicating the formation of proteins from nitrate with release of oxygen. If photosynthetic production is based on regenerated nitrogen forms (i.e. ammonia), the PQ is low as oxygen is required in the reaction. It is the continual input of new nitrogen that can elevate primary production levels and ultimately sustainable fish harvests.

11   Carbon is essential for life, and is never present in limiting quantities in the sea. This is because dissolved carbon dioxide enters into equilibrium reactions with bicarbonate and carbonate ions. As more CO_2 enters the sea from the atmosphere, or as the result of physiological activities (primarily respiration), more bicarbonate and carbonate ions are formed, thus increasing the amount of CO_2 which can continue to enter the sea. Conversely, when there is a biological demand for dissolved CO_2, the chemical reactions are reversed and CO_2 is released from its bound ionic states. As the quantities of CO_2 entering the atmosphere from human activities increase, it becomes increasingly important to determine how much can be absorbed by the seas and how much will accumulate in the atmosphere where it may contribute to global warming.

$$\frac{1000 \times 20}{100} = 200$$

$$\frac{200 \times 10}{100} = 20$$

$$\frac{20 \times 10}{100} = 2.$$

$$\frac{2 \times 10}{100} = 0.2$$

$$\frac{0.2 \times 10}{100} = 0.02 \text{ g wet weight T } m^{-2} yr^{-1}.$$

5.23

$$A = \frac{5 - 0.75}{5} \times 100\% = 85\%$$

Now try the following questions to consolidate your understanding of this Chapter.

QUESTION 5.18 Assume that in an open ocean region the primary production is 1000 g wet weight m^{-2} yr^{-1}, and there is a 20% transfer efficiency between the primary producers and primary consumers, and 10% efficiency between all successive trophic levels. What is the maximum amount of the primary production that can be converted to fish in the highest trophic level?

QUESTION 5.19 Considering the relative age when different types of phytoplankton and different top predators first appeared in the geologic record (see Appendix 1), what can you deduce about the evolution of the types of food chains shown in Figure 5.3?

QUESTION 5.20 What might happen to a regional fishery if most of the piscivorous fish (e.g. tuna) were removed, leaving mostly planktivorous fish (e.g. sardines) to harvest? Consider the effects on the regional food chain in terms of energy transfer and relative numbers of organisms in different trophic levels.

QUESTION 5.21 In one region of the North Atlantic, the production of sand eels, which feed on zooplankton, has been determined from fish catch data to be 0.5 tonnes wet weight per hectare per year, and the phytoplankton production in the same area is 200 g C m^{-2} yr^{-1}. Assuming that carbon makes up 50% of the dry weight of fish and that the dry weight is 20% of the wet weight, what is the average ecological efficiency of this system? (Note: 1 hectare $= 10\,000$ m^2.)

QUESTION 5.22 In each of the following situations, would it be best to use a laboratory experiment, an enclosed experimental ecosystem, or a computer model simulation?

(a) to study the effect of pesticide runoff from agricultural land on neritic waters; *enclosed.*

(b) to examine the potential environmental impacts of damming an estuary; *computer*

(c) to investigate physiological properties of plants or animals. *Lab.*

QUESTION 5.23 The chaetognath *Sagitta elegans* consumes 5 mg of copepods per day and produces 0.75 mg of faecal material per day. What is the assimilation efficiency of this carnivore?

QUESTION 5.24 Would it be practical to add nutrients to a small bay in order to enhance oyster cultures? *NO*

QUESTION 5.25 Assuming that the amount of new nitrogen remains constant, what would happen to the productivity of an ocean area if many of the resident fish were removed by a commercial fishery?

Fish make up the largest fraction of the nekton, but large crustaceans, squid and related cephalopods, sea snakes, marine turtles, and marine mammals can be important nektonic species in certain areas. Large nektonic animals and seabirds can have profound influences on marine communities in terms of predation. As well, many of these animals figure importantly in commercial harvests as sources of food, fur, or other commodities, or they have done so in the past. Fish dominate the present marine catch, and squid are being taken in increasing numbers; the catch of marine mammals and marine turtles is declining through public pressure for conservation of many of their species.

6.1 NEKTONIC CRUSTACEA

Although there are pelagic swimming crabs and shrimp that fall in the category of nekton, relatively little is known about the biology of the different species and few of them are abundant enough to be of commercial interest. Ninety-five per cent of commercially harvested crustaceans are demersal species that are caught in benthic trawls. However, considerable attention has been given to euphausiids (see Section 4.2) as an exploitable resource, as one species in particular is exceedingly abundant.

krill etc.

With the decline in whale numbers and the consequent cessation of most commercial whaling, *Euphausia superba*, the Antarctic krill and predominant food of baleen whales (see Figure 5.4), became an alternative fisheries target. Although rarely used for human food, these large (5–6 cm long) euphausiids can be dried and processed into feed for livestock, poultry, and farmed fish. Russian and Japanese fleets began harvesting krill in the 1960s, and the peak catch was 446 000 tonnes in 1986. Economic considerations closed the Russian harvest, and in 1994 the commercial harvest was only about 100 000 tonnes, taken by fleets from Japan and Chile. This is an insignificant fraction of the amount consumed by natural predators, estimated at about 470 million tonnes annually (see Table 5.2). The potential krill harvest has been estimated to be at least 25–30 million tonnes a year, or about one-third of the present world fish catch. However, the economic costs of fishing in the remote Antarctic are relatively high and, although krill form vast swarms, the congregations are widely scattered and sometimes located at depths of 150–200 m. Nevertheless, once a swarm has been located by echo-sounding, a single net haul from a large fishing vessel may commonly catch 10 tonnes of krill. The ecological consequences of removing vast numbers of krill on the balance of populations in the Antarctic ecosystem (including recovery of whale numbers) are not clear but, because krill are central to the Antarctic food web, there is reason for caution in expanding this harvest.

There is also a commercial harvest of a smaller euphausiid, *Euphausia pacifica*, along the northeastern Japanese coast. This particular fishery depends upon the unique fact that, in this area during the spring, the euphausiids form surface schools and are therefore easily accessible. The harvest of about 60 000 tonnes per year is mainly processed for use as feed in fish farms. The euphausiids provide a rich source of both protein and vitamin A; the latter is believed to enhance the texture and pigment of the flesh of the farmed fish.

6.2 NEKTONIC CEPHALOPODS

Squid (Figure 6.5g), cuttlefish, and octopods (e.g. *Octopus*) are the molluscan members that make up the Class Cephalopoda. Squid constitute approximately 70% of the present catch of cephalopods, and estimates indicate that the harvest could be increased appreciably. The potential world catch of squid is estimated conservatively to be 10 million tonnes annually. Despite their abundance, surprisingly little is known of the biology and ecology of many species.

Squid range in size from a few centimetres up to the legendary deep-sea giant squid (*Architeuthis*) that exceeds 20 m in length (with outstretched tentacles) and 270 kg in weight, and thus attains the status of being the largest of all living invertebrates. All squid swim by propulsion, ejecting jets of water from their siphon, and these streamlined cephalopods rival fish in swimming ability and manoeuverability. Some of the larger squid species are capable of speeds of about 10 m s^{-1}. They are also rivals with some fish for food, as squid typically eat 15–20% of their body weight per day, taking a variety of zooplankton as well as smaller fish and other squid as prey.

Many of the very abundant squid species are extensively fished and form a major source of human food in some countries. In 1981, the Japanese began using driftnet fishing to harvest squid in the Pacific Ocean. Driftnets are panels of monofilament webbing measuring 8–10 m in width and up to 50 km in length; mesh size is usually 90–120 mm. At night, the nets were placed vertically in the open ocean and allowed to drift with the winds and currents for about 8 hours in order to snare squid and fish. By 1989, Japan, Korea, and Taiwan were deploying about 800 driftnet vessels in the Pacific to harvest 300 000 tonnes of squid annually, and an estimated 200 other vessels were operating in the Atlantic and Indian oceans. In addition to squid, these almost invisible nets indiscriminately captured a large number of other species. In the North Pacific, salmon were the most common (and illegal) by-catch; in the South Pacific, albacore tuna were a valuable catch totalling 60 000 tonnes in 1988. By 1989, there was growing concern about the numbers of animals being captured incidentally by this method. In addition to nonselective capture of other fish species (including sharks) and turtles, it was estimated that between 750 000 and 1 000 000 seabirds and 20 000 to 40 000 marine mammals were being killed annually in the nets used in the Pacific alone. There were no estimates of the numbers of large zooplankton, such as salp chains and jellyfish, that may also have been destroyed by the nets. The seriousness of removing such vast numbers of animals from the oceans led the United Nations General Assembly to accept a resolution calling for an international moratorium on all large-scale, high seas, driftnet fisheries in 1993. Although some driftnets continue to be used, the problem has abated.

QUESTION 6.1 Assuming a nightly deployment of 800 driftnets during the 1980s, how many kilometres of net were set out each night by the squid fleets in the Pacific Ocean? *40,000 km.*

Squid can be harvested by more selective fishing methods. In Japan, about 500 000 tonnes of *Todarodes pacificus* are caught each year using a technique that captures only squid. This particular squid undertakes an extensive annual migration of about 4000 km, moving from the spawning grounds in the northern part of the East China Sea (about 32°N) to the vicinity of the Kurile Islands (45°N) before returning.

6.3   MARINE REPTILES

There are comparatively few reptiles that have adapted to a marine life. The best known are the eight species of marine turtles, but there are more than six times as many species of sea snakes, and there is one marine lizard, a large seaweed-eating iguana of the Galapagos Islands. Several crocodiles live in coastal waters, the largest being the infamous Australian species *Crocodylus porosus*.

Marine turtles usually are found in tropical waters, but some migrate or are carried by currents to temperate shores. Some turtles feed on jellyfish or fish in the open ocean, others (the green turtle) on shallow-water seagrasses, but all undertake long migrations to return to land in order to lay their eggs at specific nesting sites on sandy shores. High mortality is inflicted on the eggs, which are eaten by natural predators and are also prized by humans. Newly hatched young also have high mortality rates as they are preyed upon by birds and crabs during their scramble for the sea, and by predatory fish during their early life in the water. Adults have been hunted nearly to extinction for their meat and decorative shells. All sea turtles are now considered to be threatened or endangered species and conservation methods are in force in many countries, including bans on capture and importation of turtle products. Efforts are being made in several areas of the Indo-Pacific and Caribbean to gather eggs and keep these until hatching, at which time the young are released directly into the sea. It remains to be seen whether these protective measures will restore population numbers.

Sea snakes breathe air by means of nostrils and lungs, but they are truly marine animals that inhabit coastal estuaries, coral reefs, or open tropical water. Most of the approximately 60 species remain at sea to bear their live young. They school in large numbers and feed on small fish or squid which they kill with venom injected by fangs. Sea snakes are extremely poisonous and, although not all are aggressive, they have caused human deaths. They themselves have few predators except for sea eagles, sharks, and saltwater crocodiles. Sea snakes are presently restricted to warm waters of the Indian and Pacific oceans, but there has been concern that a new sea-level canal through the Isthmus of Panama would allow their passage into the warm waters of the Caribbean and Atlantic. They are presently excluded by a freshwater barrier in the canal.

6.4   MARINE MAMMALS

There are three orders of mammals that have evolved from different terrestrial ancestors and independently adapted to life in the sea. These three orders include respectively: the whales, dolphins and porpoises: the seals, sea lions, and walruses; and the dugongs, manatees, and sea cows. All share the mammalian characteristics of being warm-blooded (homoiothermic) and nursing their young, and they all rely on breathing air.

The order **Cetacea** comprises the 76 or so species of marine mammals known as whales, porpoises, and dolphins. The ancestors of this group were land animals that entered the sea about 55 million years ago. The largest of these marine mammals are the baleen whales (Figure 6.1); these include the biggest animals that have ever lived, the blue whales, which can attain a length of 31 m.

Baleen whales form a separate suborder (**Mysticeti**) of about ten species. Like the largest of the sharks, most of these immense whales feed primarily on zooplankton that they strain through specialized horny plates called **baleen** or whalebone. The brush-like baleen hangs down from the roof of the mouth on both sides, and food that collects on the baleen is periodically removed by the tongue of the whale. The humpback and finback whales also

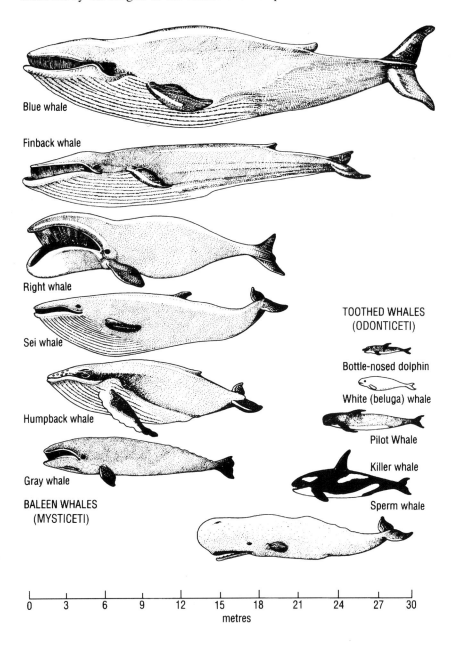

Blue whale

Finback whale

Right whale

Sei whale

TOOTHED WHALES
(ODONTICETI)

Bottle-nosed dolphin

White (beluga) whale

Humpback whale

Pilot Whale

Killer whale

Gray whale

Sperm whale

BALEEN WHALES
(MYSTICETI)

0	3	6	9	12	15	18	21	24	27	30

metres

Figure 6.1   Relative sizes of baleen and toothed whales.

are capable of capturing schools of relatively large fish, such as mackerel and herring, and the grey whale suction-feeds on benthic animals.

Some of the large baleen whales (e.g. greys, humpbacks) make extensive seasonal migrations, usually breeding in winter in tropical waters and moving poleward to feed in summer. Smaller cetaceans do not undertake long migrations, but move in response to changing food supplies or physical changes.

QUESTION 6.2   Of what advantage is it to migrating whales to have warm-water breeding nursery areas and high-latitude summer feeding sites?

The suborder **Odonticeti** includes the other 66 species of cetaceans, all of them equipped with teeth and characterized by having a single blowhole instead of the two of baleen whales. The odontocetes include the remaining whales, dolphins, and porpoises (Figure 6.1). The toothed whales are formidable predators in the sea, taking squid or fish as prey or, in the case of some killer whales, even other whales or seals and sea lions. Unlike some of the baleen whales, these animals are not reliant on surface-living prey, and they may undertake dives to depths of several hundred metres. The sperm whale holds the record among marine mammals for deepest dives; it is believed to descend to over 2200 m in search of giant squid. Some odontocetes hunt prey by **echolocation**, in which they emit pulses of sound and monitor the returning echoes, and at least some species show cooperative behaviour in herding and capturing their prey.

Some scientists have suggested that the cetaceans as a whole, and possibly the sperm whales alone, consume a greater quantity of prey than is taken by the entire world commercial fishery. For example, between 1979 and 1982, 18 species of cetaceans consumed between 46 000 and 460 000 tonnes of prey annually on Georges Bank, off the northeastern United States, compared with a commercial harvest in the same area of 112 000–250 000 tonnes. In the Mediterranean, where squid figure importantly in the human diet, cetaceans are estimated to eat about 2.3 times more squid than are taken by humans. And before commercial whaling fleets decimated whale populations in the Antarctic, baleen whales may have taken about 190×10^6 tonnes of krill annually, a figure that represents more than twice the total world catch of all marine species (see Section 6.7.1). Given these figures, it is not surprising that fishermen often regard cetaceans as competitors for fish and squid stocks.

The Inuit have been hunting marine mammals since time immemorial, but the earliest recorded whaling began off northern Europe between A.D. 800 and 1000 and whaling became a major commercial enterprise in the 1700s and 1800s. Whales were exploited primarily for their oil, which was used in lamps, and for whalebone (baleen) used to stiffen women's apparel; whale meat was of secondary importance, except in Japan. The advent of increased mechanization, motorized high-speed ships, and explosive harpoons in the early 1900s resulted in rapid declines in whale populations and threatened extinction of some species. Even after the establishment of the **International Whaling Commission** (IWC) in 1946, the annual catches of whales continued to climb to about 65 000 during the 1960s. Pre-exploitation and present population estimates of 13 species are given in Table 6.1; of these species, nine have been listed as endangered or vulnerable species since 1970. Given the low fecundity and slow development times of the great

Table 6.1 Past and present population estimates of whales. All estimates are taken from the International Whaling Commission and most are highly speculative.

Common name	Scientific name	Population estimates Pre-exploitation	Present
Baleen whales:			
Blue	*Balaenoptera musculus*	228 000	<10 000*
Fin	*Balaenoptera physalus*	548 000	150 000*
Sei	*Balaenoptera borealis*	256 000	54 000*
Bryde's	*Balaenoptera edeni*	100 000	90 000
Minke	*Balaenoptera acutorostrata*	140 000	725 000
Bowhead	*Balaena mysticetus*	30 000	7800*
Northern right	*Eubalaena glacialis*	No estimate	<1000*
Southern right	*Eubalaena australis*	100 000	3000*
Humpback	*Megaptera novaeangliae*	115 000	10 000*
Grey	*Eschrichtius robustus*	>20 000	21 000*
Toothed whales:			
Sperm	*Physeter catodon*	2 400 000	1 950 000*
Narwhal	*Monodon monoceros*	No estimate	35 000
Beluga	*Delphinapterus leucas*	No estimate	50 000

*Listed as an endangered or vulnerable species by the International Union for the Conservation of Nature or by the United States Government.

whales, severely depleted populations may take decades to recover. Only one species, the minke whale, is known to have greatly increased in abundance. The relatively small minkes were never heavily exploited, and the southern population might have increased due to the decline of larger baleen species with which it may compete for krill. In 1986, the IWC voted to establish an indefinite ban on commercial whaling in the hope of re-establishing endangered stocks. In 1994, the IWC designated 28×10^6 km^2 around Antarctica as a whale sanctuary, thus providing permanent protection for about 90% of the world's whales. These measures do not, however, slow cetacean mortality due to other factors. The smaller dolphins and porpoises are captured incidentally by fishing gear; and cetaceans that reside in, or temporarily enter, coastal waters are subject to increasing habitat destruction and pollution. There is no doubt that pollution threatens beluga whale populations in the St. Lawrence Estuary of eastern Canada, and river dolphins in many other areas.

A second order of marine mammals includes the seals, sea lions, and walruses. These familiar animals are known taxonomically as **pinnipeds** (order Pinnipedia), meaning 'feather-footed' to describe their four swimming flippers. In contrast to whales, these animals spend part of their time on land or on ice floes, where they congregate for breeding and resting. The 32 species of pinnipeds are found in all the seas of the world, and there is one freshwater species in Lake Baikal, but the majority of species and the largest populations are found in the cold waters of the Arctic and Antarctic. Most feed primarily on fish or squid, but walruses also use their tusks to dig molluscs and other benthic animals from the sea bottom. Pinnipeds typically live and travel in herds, and some may undertake long migrations at sea.

Although seals and sea lions have been heavily exploited in the past for their fur and oil, and walruses for their ivory, hunting pressure has lessened for most of the species. However, the Caribbean monk seal is now believed extinct, and the Hawaiian and Mediterranean monk seals remain endangered.

Manatees and dugongs belong to the mammalian order **Sirenia**. They are the only herbivorous aquatic mammals, and they rely on larger plants, not algae, for nourishment. Their food requirements restrict them to living in shallow coastal waters, estuaries, and rivers. All four species of this order reside in warm waters and do not come on to land. Manatees and dugongs are thought to have been highly social animals, as old records report huge congregations of these animals before hunting decimated their numbers. The few remaining individuals have tended to become solitary or form only small family groups. The sirenians have been particularly vulnerable to hunting pressure because of their inshore habitats and their slow and placid behaviour, and they are prized for their meat, oil and hides in many cultures. At one time, dugongs had a widespread distribution which included Atlantic waters; today, they are restricted to the Indian and Pacific oceans. All three species of manatees are found only in tropical Atlantic waters.

A fifth species of sirenian, Steller's sea cow (*Hydrodamalis gigas*), became extinct within historical times. The existence of these huge animals was documented only once, by Georg Wilhelm Steller who was acting as a physician and naturalist on an expedition sailing in the far North Pacific under Commodore Vitus Bering. The *St. Peter* was shipwrecked in 1741 on a small island near the western end of the Aleutian chain. Although Captain Bering and many of his crew died from sickness in the first days after landing, the remaining survivors found adequate food in the resident otters and seals. They also discovered sea cows in nearby waters and were eventually able to capture these for a source of meat and oil. The animals were described as attaining a length of up to 10 m and a weight of 10 tons. They fed on kelp and other large seaweeds, chewing the plant blades with horny plates covering the palate and jaws. They formed small herds, and were slow-moving and of a passive temperament. The shipwreck survivors brought word of their discoveries when they returned to Russia in 1742, and future whaling and fur-hunting expeditions to the Bering Sea began to rely on sea cows as a source of food during the winter. Although 2000 Steller sea cows were estimated to live in the region, the last was killed in 1768, only 27 years after the discovery of this population. Fossil evidence indicates that, as recently as 20 000 years ago, this sea cow inhabited coastal areas as far south as California; although the cause of its extinction in these areas is a matter of conjecture, the animal was probably easy prey for pre-historic as well as modern humans.

6.5 SEABIRDS

Like the marine reptiles and mammals, seabirds have evolved from land species that readapted to life in the sea. There are now approximately 260–285 species of seabirds, depending on how the term is defined; these species represent roughly 3% of the world's birds. Those birds that are most highly adapted to the marine environment include the auks, albatrosses, petrels, penguins, and gannets, all of which have few representatives on land or freshwater and spend 50–90% of their lives at sea. At the other extreme are the shorebirds, like sandpipers and plovers, that depend upon marine food sources but are incapable of swimming.

The many species of oceanic birds have developed diverse methods of feeding and take different types of prey (Figure 6.2); this is reflected in species differences in structure of the bill and wings. Some species

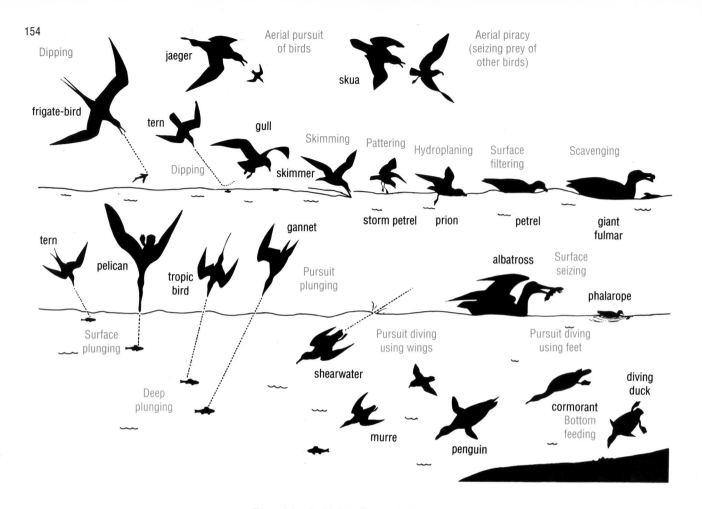

Dipping

jaeger

Aerial pursuit
of birds

skua

Aerial piracy
(seizing prey of
other birds)

frigate-bird

tern

Dipping

gull

skimmer

Skimming

Pattering

Hydroplaning

Surface
filtering

Scavenging

storm petrel

prion

petrel

giant
fulmar

tern

pelican

tropic
bird

gannet

Pursuit
plunging

albatross

Surface
seizing

phalarope

Surface
plunging

Deep
plunging

Pursuit diving
using wings

shearwater

murre

penguin

Pursuit diving
using feet

cormorant

diving
duck

Bottom
feeding

Figure 6.2 Seabird feeding methods. In most cases, the examples given represent only one
of several types of birds that feed in the depicted manner.

(skimmers, gulls, petrels) skim neuston from the immediate surface layers of
the ocean, others (pelicans, terns, gannets) plunge deeper into the water to
seize zooplankton, squid, or fish. Penguins, cormorants, murres, and puffins
actively pursue their prey underwater, using their wings or feet for
swimming. Although emperor penguins may dive to depths of more than
250 m, the majority of seabirds are essentially dependent on the uppermost
layers of the sea for their food. The impact of seabird predation on oceanic
surface life has often been neglected, but it may be considerable.

Although seabirds are found world-wide, the largest colonies are located
adjacent to highly productive ocean areas where food is plentiful and
concentrated. Millions of penguins (six species) are present in the Antarctic,
where they depend for food on the vast numbers of krill or the abundant fish
and squid that occur in this productive ocean. Equally large numbers of birds
form island colonies in the upwelling coastal regions off western South
America (see Figure 6.10). At sea, birds frequently form feeding
aggregations along oceanic fronts which, like upwelling regions, have
relatively high biological productivity (see Section 3.5). Far fewer birds are
present in low-productivity tropical regions. Seasonal changes in the marine
environment can be reflected in the distribution of birds, and some species
undertake long annual migrations in response to seasonal food availability
and suitable weather for breeding.

Shorebirds known as red knots undertake one of the longest migrations that is linked to exploiting seasonally available marine food resources. The American subspecies (*Calidris canutus rufa*) spends the austral summer at the southern tip of Argentina, feeding primarily on young mussels in intertidal areas. It migrates northward in March, stopping to feed on clams, mussels, and worms along the South American coast. By late May, over 100 000 birds flock in Delaware Bay, along the eastern seaboard of the United States. Their arrival is timed to coincide with the breeding cycle of horseshoe crabs (Figure 7.7g) that come ashore in thousands to lay millions of eggs. Their eggs provide a high energy food source for the birds, and this fuels their remaining flight to islands in the Canadian Arctic. Each female lays four eggs shortly after arrival at the breeding grounds; these weigh about 75 g, or more than 50% of the female's weight. The period spent in the Arctic coincides with peak abundances of insect and aquatic life, and the return southward migration in July and August occurs when local marine invertebrate populations are highest along the Atlantic coast of America. The birds increase their weight by about 40% within the few weeks spent feeding intertidally, then resume their migration to South America. (See Section 8.5 for additional information on the impacts of shorebirds feeding on benthic animals.)

Seabirds exhibit natural fluctuations in population densities which can be caused by climate change and subsequent fluctuations in prey availability. This has been documented, for example, in the bird colonies found on rocky uninhabited islands off the coast of Peru (see Section 6.7.2). Over evolutionary time, birds evolve adaptations that permit them to exist within a range of natural climatic variability. Unfortunately, seabirds are increasingly faced with new sources of mortality inflicted by human activities, and these occur with a tempo that precludes slow evolutionary adaptation.

All seabirds depend on nesting sites on land for breeding, and it is here that they encounter their greatest risks. Seabirds on land are exceptionally vulnerable to predation as it is difficult for them to defend themselves, and their eggs and young, against land mammals and snakes. Many species nest on inaccessible rock islands that are naturally free of mammalian predators. However, there are numerous examples of the deliberate or accidental introduction of predators, such as cats, rats, and pigs, that have resulted in disturbance or destruction of bird colonies.

QUESTION 6.3 What are the predators of Antarctic species of penguins? (Refer to Figure 5.4.) *leopard Seals, Orca's Shua's.*

Some seabirds have been exploited for their feathers, meat, eggs, or body oil. Only one, however, has been exterminated within historical time. This was the great auk (*Pinguinus impennis*) (Figure 6.3), which lived only on isolated islands in the North Atlantic. This Northern Hemisphere bird was the ecological equivalent of the southern penguins; like penguins, it was large (to 1 m tall), flightless, and dived to pursue its marine prey. The giant auk was discovered in 1534, when there were several hundred thousand birds, but it was hunted to extinction within 300 years. At first hunted for meat by local fishermen, these seabirds were later harvested commercially for their feathers and oil. The last great auk was killed on Funk Island on June 3, 1844. Thousands of giant auk carcasses discarded on the rocky island provided fertilizer to nourish grasses, and today the island is a sanctuary for puffins and murres.

Figure 6.3   The giant auk as painted by John James Audubon in about 1835, nine years before this flightless bird was hunted to extinction.

Increasingly, seabirds are encountering greater mortality from coastal pollution. In all oil spills, seabirds are usually the most obvious victims, and the effects of oil pollution are too well known to document here. Although most spills are localized, the mortality may be high; for example, an estimated 500 000 birds perished from the effects of the oil spill of the *Exxon Valdez* in Alaska. Pesticide residues, working their way up the food web and accumulating in the bodies of seabirds, have caused thinning of egg shells and reduced hatching success in pelicans, ospreys, and other species. Other toxic chemical pollutants which enter the sea and may affect birds include organochlorines, PCBs (polychlorinated biphenyls), and heavy metals like mercury. Also of concern is the continuing loss of feeding and reproductive habitats through development of coastlines (see also Chapter 9).

Increased fishing efforts have also affected seabird numbers. Large numbers of birds have been incidentally captured and drowned in driftnets in the North Atlantic and Pacific (see Section 6.2). Perhaps larger numbers are affected by reduction in food, through harvests of their prey. In Norway, for example, puffins have declined because of overfishing of the immature herring that are their main food.

As is the case with many other animals, seabirds are faced with new sources of mortality and an accelerated pace of change. Their past evolutionary experience, developed over 60 million years, will be of little value in establishing defences against oil spills and net-capture. This is particularly true for those seabirds that have low fecundity and require a long period before attaining breeding age. For some species, continued survival may depend on legislated and enforced conservation and on protection of fish stocks and nesting and breeding habitats.

6.6   MARINE FISH

Fish constitute the largest and most diverse group of marine vertebrates. They are taxonomically separated into the following three classes:

Class Agnatha. This class encompasses the most primitive of living fish, the jawless lampreys and hagfish. This group evolved about 550 million years ago in the Cambrian, but presently has only about 50 species.

Class Chondrichthyes. The sharks, skates, and rays which belong to this class are also known as **elasmobranch** fish; they are characterized by having a cartilaginous skeleton and lacking scales. This is also an ancient group, first appearing about 450 million years ago, and there are presently about 300 species.

Class Osteichthyes. This class includes the **teleost** fish which have a bony skeleton. This is the most successful group and comprises the vast majority of living fish, with somewhat over 20 000 marine species. The teleosts evolved about 300 million years ago.

Agnatha

The bodies of hagfish (Figure 6.4a) and lampreys are elongate and eel-like, and the animals lack scales. The mouth is surrounded by a sucking disk, and most of the species are predators on other fish. The scavenging hagfish burrow into the bodies of dead or dying prey to feed on internal parts; lampreys are parasites which attach themselves to fish by their sucker-like disk and cut into the flesh to feed on soft parts and body fluids. All hagfish species are marine, whereas the different species of lampreys may be either marine or freshwater. Even the marine species of lampreys spend part of their lives in freshwater; the young live in rivers where they feed on small invertebrates and possibly fish fry, and after metamorphosing, they move to the sea to complete their development.

Chondrichthyes

Sharks (Figure 6.4c) are typically thought of as fast-swimming voracious predators that consume large prey, but many also act as scavengers in the sea. Paradoxically, the largest members of this group are docile feeders on plankton; these include the basking shark (*Cetorhinus maximus*) and whale shark (*Rhincodon typhus*) which attain lengths of 14 m and 20 m, respectively. Both species have small teeth and strain plankton from water using specially modified gills. Skates and rays have flattened bodies, and

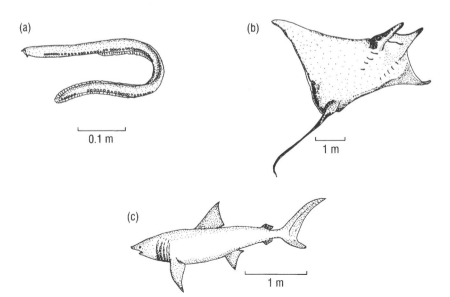

(a)

0.1 m

(b)

1 m

(c)

1 m

Figure 6.4   Primitive fishes belonging to the Class Agnatha (a) and Class Chondrichthyes (b and c). (a) hagfish; (b) manta ray; (c) shark.

most are adapted to a bottom-dwelling habitat. The majority are predators on benthic organisms (especially crustaceans, molluscs, and echinoderms), but some feed on fish, and the large manta rays (Figure 6.4b) are plankton feeders.

Sharks and rays typically have internal fertilization and low fecundity, producing only small numbers of relatively large eggs. Most sharks and all rays give birth to live young. Skates lay their eggs in protective cases that are attached to a substrate, and the young hatch from these within a few weeks or months.

There is an increasing demand for shark meat and shark fins, the latter being considered a delicacy in Asia. Numbers of sharks have sharply and rapidly declined in regions where fishing has intensified, and many sharks are also captured incidentally in commercial fishing operations for other species.

QUESTION 6.4   Why are shark populations likely to be slow to recover from over-exploitation? *low fecundity,*

Osteichthyes

The many species of teleost fish inhabit diverse types of marine environments, and consequently they are a heterogeneous group in terms of anatomy, behaviour, and ecology.

The most familiar teleosts are those that are harvested commercially (Figure 6.5a–f; Figure 6.9) and, because of their economic importance, more is known about the biology of these particular species. Such fish feed on many different prey items depending on their size, location, and the availability of prey at different times. Some are strictly planktivores (plankton-feeders), other fish are piscivores (fish-eaters), or a combination of both. The most numerous of these fish occupy lower trophic levels; these include herring, pilchards or sardines, and anchovies, all of which eat chiefly zooplankton, although adult anchovies can feed directly on large chain-forming diatoms as well. Larger fish, such as cod, hake and pollock, may begin their lives by feeding on small zooplankton; as juveniles, they switch to preying on larger zooplankton (e.g. euphausiids), and then become piscivores as adults. The largest of the pelagic teleosts are piscivorous species such as tunas, jackfish, and barracuda. Some fish, such as cod, haddock and hake, feed both in mid-water and on the sea bottom and are

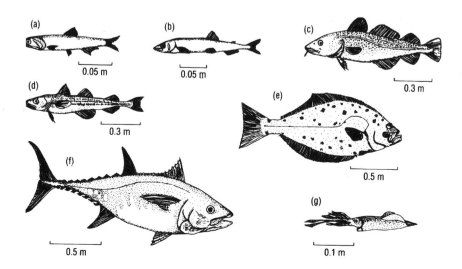

Figure 6.5   Commercially harvested types of teleost fish (a–f) and squid. (a) anchovy; (b) capelin; (c) cod; (d) pollock; (e) halibut; (f) tuna; (g) squid.

capable of catching fish or benthic invertebrates. True demersal fish spend all their lives on or near the sea bottom, where some (e.g. sole) feed only on the benthos (clams, worms, and crustaceans being favoured foods) and others (e.g. halibut, turbot) eat smaller fish.

Food supplies for oceanic fish vary in abundance due to physical factors. Some species respond to predictable seasonal variability in food concentration by migrating to certain feeding sites when prey becomes particularly abundant. For example, the migrations of tuna in the Pacific put them in areas where swarms of pelagic crabs are seasonally available. However, for many fish, variability in food concentration may cause significant change in their growth rate and survival, and this is reflected in the variability in fish catch from year to year.

Fish associated with special benthic habitats, such as coral reefs, are themselves specialized to feed on corals or resident plants and animals; these species are considered in more detail in Section 8.6. As with most other animal groups, the largest fish populations are found in temperate waters, but species diversity is much higher in tropical and subtropical waters.

How do mesopelagic and bathypelagic teleosts differ from epipelagic species?

Fish residing in deeper waters (>300 m) are not as numerous as epipelagic species, and they are not exploited commercially. The most diverse of the roughly 1000 mesopelagic fish species, both in numbers of species and individuals, are the 300 + species of **stomiatoids** (Figure 6.6a–c, f, g) and the 200–250 species of lantern-fish (also called **myctophids**) (Figure 6.6d, e). Bioluminescence (see Section 4.4) is common in both groups, and the name 'lantern-fish' refers to light production from numerous photophores that are arranged in specific patterns in the different species of myctophids. Stomiatoids are distinguished by having photophores arranged in definite rows, as well as by having the dorsal fin located far back on the body. The photophores in both groups of fish contain symbiotic bacteria that produce the light which may be used to lure or locate prey, or to find mates in these dark depths.

The majority of mesopelagic fish are small, ranging from about 25–70 mm in length at maturity; the largest mesopelagic species are about 2 m long. Many of the stomiatoids have elongate, relatively streamlined bodies, but the hatchet-fish with their large, upwardly-directed eyes are named for their laterally-flattened and squared shapes (Figure 6.6c, g). Stomiatoid fish typically have large jaws with numerous sharp teeth, and they feed on zooplankton, squid, and other fish. Some species have the capacity to unhinge their jaws in order to ingest large prey (Figure 6.7), and many species have extensible digestive organs to accommodate large food items (Figure 6.8). The best known stomiatoid genus, *Cyclothone* (Figure 6.6b), contains many species, and these fish live between 200 m and 2000 m depth in large schools. The shallower-living species are silvery or partly transparent; the deeper residents are typically black. The lantern-fish perform diel vertical migrations, some rising to the very surface to feed on planktonic crustaceans and chaetognaths, and this group comprises a major food source for tuna, squid, and porpoises. Lantern-fish range in size from 25 mm to 250 mm in length. Many display sexual dimorphism in the arrangement of photophores, suggesting that they recognize sex by male–female differences in light patterns.

160

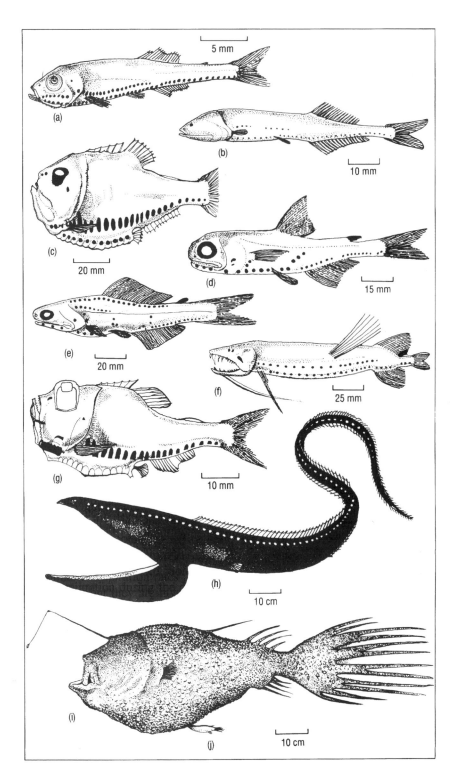

Figure 6.6 Mesopelagic (a–g) and bathypelagic (h–j) fishes. (a) *Vinciguerra attenuata*, a stomiatoid. (b) *Cyclothone microdon*, a stomiatoid. (c) *Argyropelecus gigas*, a hatchet-fish. (d) *Myctophum punctatum*, a lantern-fish. (e) *Lampanyctus elongatus*, a lantern-fish. (f) *Bathophilus longipinnis*, a stomiatoid. (g) *Argyropelecus affinis*, a hatchet-fish. (h) *Eurypharynx pelecanoides*, a gulper-eel. (i) A female *Ceratias holboelli* with an attached parasitic male (j), deep-sea angler-fish.

161

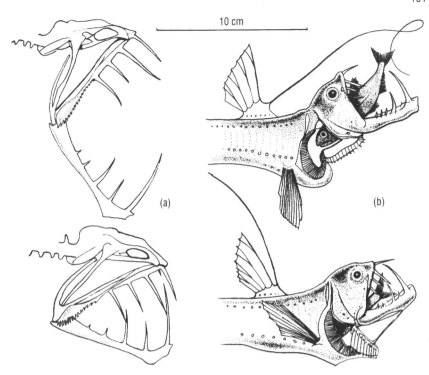

Figure 6.7 Swallowing mechanism of the stomiatoid *Chauliodus sloani*, a deep-sea viper fish. (a) The positions of the skull and jaw bones when the mouth is open and closed. (b) The attitudes of the fish when ingesting prey.

In bathypelagic waters (below 1000 m), there are about six times fewer fish species. The greatest diversity is found in the 100 or so species of ceratioid angler-fish, so-named because of the characteristic bioluminescent lures which the females dangle in front of their mouths. Population sizes also diminish in deeper water and, as potential mates become more difficult to find, some fish (and also some invertebrates) exhibit reproductive and development patterns that differ considerably from those of shallower-living species. One extreme strategy has developed in some of the angler-fish in which young males live freely, but later attach themselves to females

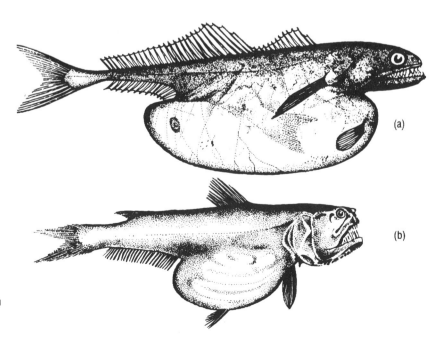

Figure 6.8 The ingestion capability of deep-sea fish. (a) *Chiasmodon niger* with a curled-up fish in its stomach that is longer than itself. (b) *Evermannella atrata* containing a squid. Sizes range up to 150 mm long.

(Figure 6.6i, j). The males undergo a morphological transformation and remain small (about 15 cm long); they live as an external parasite on the much larger (*ca.* 1 m long), free-living female and serve only to fertilize her eggs. Gulper-eels (*Eurypharynx*) (Figure 6.6h) are also residents of the bathypelagic zone. These dark-coloured, elongate fish, with a funnel-like throat, attain lengths of 1–2 m and are capable of swallowing large fish as prey.

In contrast to elasmobranchs, almost all teleosts have external fertilization and high fecundity (see Section 4.3). Whereas some species attach their eggs to a substrate, most lay large numbers of small floating eggs, and the hatching larvae form part of the meroplankton. Teleosts typically spawn many times, and growth is continuous through life. These characteristics make them less vulnerable to commercial harvesting than the cartilaginous sharks and rays.

6.6.1 FISH MIGRATIONS

The swimming abilities of most epipelagic fish make them independent of ocean currents, and they are able to migrate from one area to another, selecting favourable conditions in terms of food availability or reproductive sites and associated physical parameters. Whereas many species may undertake oceanic migrations ranging from several hundred to several thousand kilometres between, for example, feeding and spawning areas (see Figure 6.14), other fish may undertake migrations between the sea and freshwater.

Anadromous fish, such as salmon, sturgeon, shad, smelt and sea lampreys, breed in freshwater. The young then migrate to sea, where they spend most of their adult life. The length of time spent at sea is species-specific, but the adults eventually return to their specific freshwater sites to breed and spawn. Some species, like the Pacific salmon, die after mating; but others, such as the Atlantic salmon, do not and may return several times to their breeding site.

Catadromous fish are those that breed in the sea, but spend the majority of their adult life in freshwater. Some of the longest migrations are undertaken by the catadromous American (*Anguilla rostrata*) and European (*A. anguilla*) eels. Adults migrate from rivers in Europe and eastern North America to breeding sites in the Sargasso Sea, where they spawn many small floating eggs in deep water and then die. The larvae remain at sea for one or two years before arriving at the coasts of America and Europe, respectively; there they metamorphose into elvers which enter the estuaries and freshwater rivers. They remain for 8–12 years in their freshwater habitats before returning to the sea as mature adults.

6.7 FISHERIES AND FISHERIES OCEANOGRAPHY

Marine fisheries constitute a multibillion-dollar industry supplying about 20% of the animal protein consumed by humans, and also producing animal feeds for domestic livestock and poultry, fish oils for paints and drugs, pet foods, and some food additives. As the human population continues to expand, the increasing demand for high-quality protein and other marine resources has focused attention on the present stocks of commercial marine species and on the feasibility of increasing, or at least maintaining, the present harvest. It has become apparent that fisheries management has not always been successful in maintaining fish yields and conserving stocks (see Section 9.1), and that our information concerning the biology and ecology of many species may be insufficient to establish reliable estimates of yields. One branch of marine science, **fisheries oceanography**, addresses these problems.

6.7.1 WORLD FISH CATCH AND FISHERIES MANAGEMENT

In an evolutionary sense, the most successful of the larger marine animals are the extremely abundant species of fish that are hunted commercially. These include herring, anchovies, sardines, cod, and mackerel, some of which are illustrated in Figures 6.4 and 6.9. These fish, and others, are among the top ten species of fish that make up the world fish catch (Table 6.2); Table 6.3 lists the major fishing nations of the world. The total world reported catch of marine fish (including shellfish and squid) in 1993 was 84 million tonnes, down from a high of 86 million tonnes in 1989. At least a further 27 million tonnes of **by-catch** – unwanted marine species caught incidentally – is thrown back into the sea every year, most of it dead or dying. The largest fraction (64%) of the global marine catch comes from the Pacific Ocean, with 28% from the Atlantic and 8% from the Indian Ocean. Humans directly consume about 70% of the total world catch (in live weight units); the remaining 30% is used as poultry or livestock feed, and this comes mostly from the smaller species of fish such as anchovies, herring, and sardines (or pilchards).

Table 6.2 Principal species of fish comprising the total world fish catch (FAO statistics, 1993). Relative dominance of species may change from year to year due to climate change and/or exploitation pressure.

Species	1993 catch ($\times 10^6$ tonnes)
Peruvian anchoveta (*Engraulis ringens*)	8.3
Alaska pollock (*Theragra chalcogramma*)	4.6
Chilean jack mackerel (*Trachurus murphyi*)	3.4
*Silver carp (*Hypophthalmichthys molitrix*)	1.9
Japanese pilchard (*Sardinops melanostictus*)	1.8
Capelin (*Mallotus villosus*)	1.7
South American pilchard (*Sardinops sagax*)	1.6
Atlantic herring (*Clupea harengus*)	1.6
Skipjack tuna (*Katsuwonus pelamis*)	1.5
* Grass carp (*Ctenopharyngodon idella*)	1.5

* Freshwater and cultured species

QUESTION 6.5 In Table 6.3 how can you explain the high fish catch of two of the world's smaller countries. Chile and Peru? (Refer to Section 3.5 if necessary.) *Near to upwelling area, large export.*

The management of the world's fishing industry is very complicated because it involves not only biological and econological knowledge of many species, but it also must take into account economic considerations, competition between nations, labour unions, and public marketing strategies. It is beyond the scope of this text to deal with economic and political problems in fisheries, but it is germane to consider oceanographic topics that may supply explanations for comparative abundance of fish species and for fluctuations in fish populations (as shown, for example, in Figure 6.9 and 6.10).

Table 6.3 Principal fishing nations of the world and 1993 FAO catch statistics (including aquaculture). Relative dominance of nations may change from year to year due to ecological or economic changes.

Country	1993 catch ($\times 10^6$ tonnes)
China	*17.6
Peru	8.5
Japan	8.1
Chile	6.0
U.S.A.	5.9
Russian Federation	4.5
India	4.3
Indonesia	3.6
Thailand	3.3
Korean Republic	2.6
Norway	2.6
Philippines	2.3

*Approximately 50% of the fish catch from China is from aquaculture

The history of fisheries science has been briefly reviewed in Section 1.4. Early studies on the population dynamics of fish stocks led to the development of what are generally referred to as **stock/recruitment theories**, 'stock' referring to population numbers of adult fish, and 'recruitment' to the numbers of juvenile fish entering the adult population. These fisheries theories were based upon a central premise that reproduction, survival, and productivity of fish populations were largely independent of changes in the physical environment of the fish, or of changes in biological components (i.e. interacting species) within the community under consideration. The basic argument put forward was that the recruitment of new fish stock was a function of the numbers of eggs produced and subsequent survival of young. Because total egg production is a function of the size of the adult population and survival was considered constant, it was maintained that the size of the adult stock could be controlled by manipulating fishing pressure through regulating the number of boats, the size of nets, and the total allowable catch. This basic premise, with later variations, became the basis for the management of fisheries for nearly 100 years.

In cases when reliable assessments can be made of the numbers of juvenile fish that will enter a fishery within the next one or two years, the stock/recruitment approach has been of some value. On the whole, however, fisheries scientists have not been successful in managing fish stocks, or in making long-term predictions. It is now clear that the abundance of fish is determined by a variety of factors, and data are being compiled from a number of fields to answer fundamental questions about fish ecology. Plankton ecology may provide some answers, as most fish eggs and fish larvae are meroplanktonic, and many adult fish depend on the plankton community for food. Changes in the physical environment may also have strong influences on fish populations. Fisheries oceanography concerns the search for knowledge about the natural regulation of fish populations and seeks to apply this information to fisheries management.

6.7.2 FLUCTUATIONS IN THE ABUNDANCE OF FISH STOCKS

Changes in the abundance of some species of fish are associated with long-term changes in the oceanic climate (see Section 4.8). Figure 6.9 shows a

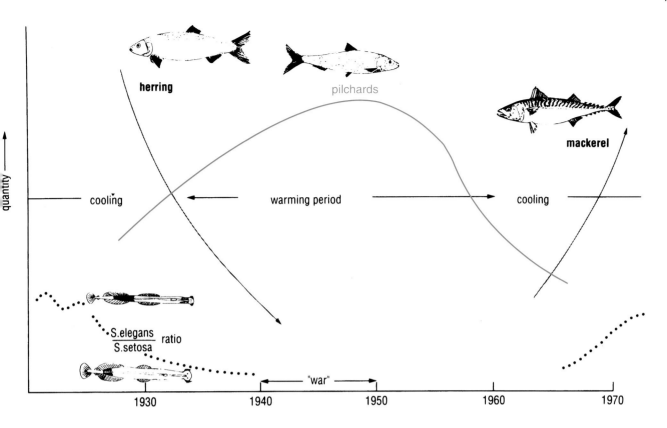

Figure 6.9   The Russell cycle in the western English Channel: Long-term fluctuations in the abundance of herring (*Clupea harengus*), pilchards (*Sardina pilchardus*), and mackerel (*Scomber scombrus*) as related to changes in oceanographic climate and the ratio of numbers of the chaetognaths *Sagitta elegans* and *Sagitta setosa*. (Arbitrary units)

regime shift, known generally as the Russell cycle, that has taken place in the western English Channel. Herring decreased in abundance from about 1930 as water temperatures warmed. At the same time, pilchards started to become more abundant, reaching their highest populations during the 1940s and 1950s when herring were scarce. As water temperatures cooled in the 1960s, pilchards disappeared from the area, and mackerel became the dominant pelagic fish. During the same periods, there was a change in the dominant chaetognath species, from *Sagitta elegans* during cool intervals to *S. setosa* in warmer water. It is generally agreed that this is a natural cycle caused by climate change, and that the changes in fish species are independent of fishing activity.

Figure 6.10 presents the history of the Peruvian anchovy catch as an example of a managed fishery that failed to take natural environmental change into account. Waters off Peru are normally enriched by coastal upwelling that leads to very high productivity (see Section 3.5.2) and a short food chain (Figure 5.3). Until about 1970, enormous numbers of plankton-eating anchovies were produced, and these were the major food for millions of seabirds (boobies, brown pelicans, gannets, and cormorants being dominant). The birds are collectively referred to as **guano** birds, the Spanish name referring to the faecal droppings of the birds. Guano from the birds built up over time on their nesting islands to depths of up to 50 m and, because it was rich in nitrates and phosphates, it was collected and sold for fertilizer.

The Peruvian anchovy fishery was developed in the late 1950s and was to be a model of fisheries management. It was predicted, based on stock size, that between 9 and 10 million tonnes of anchovies could be harvested annually without decreasing the stock. Figure 6.10 shows the increase in annual fish catch from less than one million tonnes in 1958 to 13 million tonnes in

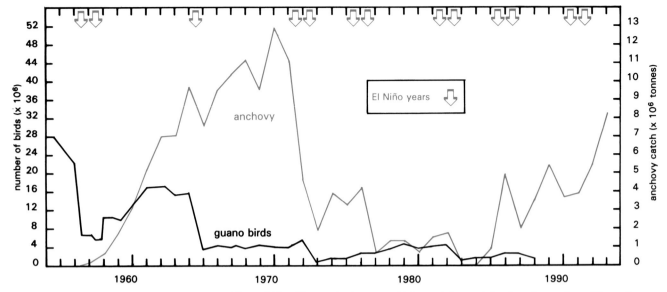

Figure 6.10   Changes in the guano bird population and the Peruvian anchovy (*Engraulis ringens*) catch along the west coast of South America. Arrows indicate El Niño years and the relative intensity of these climatic changes. Anchoveta numbers represent the total catch from Peru, Chile, and Ecuador.

1970, making it the largest fishery ever based on a single fish species. However, the fishery collapsed in the 1970s and it has taken 20 years to recover to an annual catch of about 8 million tonnes of fish (see Table 6.2). One of the factors that was not fully understood in the 1960s and 1970s when this fishery was being 'managed' was the effect of El Niño events on the anchovy stock.

El Niño is the name given to a warm, nutrient-poor surface current that flows over the cold coastal upwelling off the coast of Peru. The phenomenon has been well documented for about 50 years, and was recorded for hundreds of years before that. El Niño events occurred in 1957–58, 1965, and 1972–73 (Figure 6.10). Each time they occurred, there was a decrease in the guano bird populations which fed on the anchovy. With the intrusion of warm oligotrophic surface waters, the anchovy migrated deeper in the water — too deep to be reached by these diving birds, and many birds starved. In 1957, up to 20 million birds died during the El Niño, but recovery of bird populations following natural El Niño events was usually relatively rapid. However, the fishery continued during these El Niño episodes in spite of the fact that the anchovy were being displaced from their natural, near-surface habitat. A lack of understanding the effects of El Niño on the anchovy, coupled with overestimates of maximum sustainable yield as calculated from stock/recruitment theory, led to the eventual depletion of the stock. The fishery harvested less than 2 million tonnes of anchovy from 1977 to 1985. Bird populations dropped to about one-tenth of their original numbers and, although fish are now increasing in number, the seabirds have not recovered from levels of about 3 million.

The examples of the Russell cycle and the Peruvian anchovy fishery have been given to demonstrate that environmental changes may significantly affect both the yield of fish and the type of fish present in any area. When overfishing is coupled with a natural environmental change that is also decreasing the stock, the consequences may be severe and the recovery of the stock may take decades, if it occurs at all. It is now clear that the role of the environment cannot be excluded from fisheries theories, and that oceanographic data may contribute useful and necessary information to create new management theories. Because marine fish stocks occupy large expanses of ocean, satellite remote sensing is now being applied to examine those variations in ocean conditions that cause natural fluctuations in the distribution, abundance, and availability of commercial stocks.

6.7.3 REGULATION OF RECRUITMENT AND GROWTH IN FISH

Most teleost fish species are highly fecund, each female usually spawning between 10^3 and 10^6 eggs per year. An extremely small variation in the mortality of the progeny (e.g. from 99.90 to 99.95%, see Question 6.11) may cause a very large change of several hundred per cent in the adult population size. Thus it is important that fisheries management theories include an ecological understanding of the factors controlling the survival and recruitment of larval and juvenile fish into adult stocks. Factors determining growth rates of fish are also important because they determine size and partly control survival in young fish.

A number of hypotheses have been formulated to explain fluctuations in adult abundance due to differences in the recruitment and growth of young fish, and none are mutually exclusive. Some of these hypotheses are listed as follows:

  1. **Starvation hypothesis**. If there is not enough planktonic food in the sea, larval fish mortality will increase and few, if any, will survive to become adults.

  2. **Predation hypothesis**. Predators, including larger fish and some carnivorous zooplankton, may consume large numbers of both larval and juvenile fish. Heavy predation results in few young surviving to become adults.

  3. **Advection hypothesis**. Physical oceanographic processes may transport the young fish away from their nursery areas to unfavourable environments where they will not survive.

  4. **Growth hypothesis**. The maximum size attained by fish at the time of harvest, multiplied by the number of fish captured, gives the biomass yield to the fishery. Size and numbers of fish are also used to establish fish quotas in terms of allowable tonnage. Numbers are determined by survival (see hypotheses 1–3 above), and size is determined by growth. The growth hypothesis is based on the consequences of fish growth being inhibited by either biotic (e.g. food) or abiotic (e.g. temperature) factors.

In order to see how these hypotheses might be integrated with present biological oceanographic knowledge, several mechanisms of fish recruitment and growth are discussed below.

The tendency of teleost fish to produce very large numbers of eggs is quite unlike the reproductive patterns of most terrestrial animals of similar size. A single female cod, for example, may lay more than a million eggs per year. Obviously only a very small percentage of these eggs are needed to replace the adult stock. The mortality of eggs and fish larvae is very high, as illustrated in Figure 6.11, and may result from predation, advection, or, in the case of larvae only, from starvation.

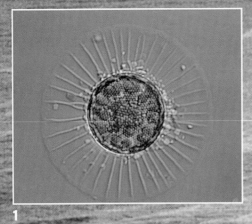

1

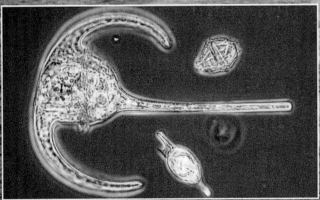

2

Plate 1   The diatom *Planktoniella sol*. The species is named for its resemblance to the sun, and its disklike shape retards sinking. Diameter, *ca.* 150 μm.

Plate 2   A chain-forming diatom, *Thalassiosira sp.*, one of the common members of the phytoplankton community in temperate waters. Length of chain *ca.* 180 μm.

Plate 3   Dinoflagellates: the largest specimen (*ca.* 200 μm long) is a species of the common photosynthetic genus *Ceratium*.

3

Plate 4   *MAIN PHOTOGRAPH*   A red tide off the coast of Japan caused by a bloom of dinoflagellates.

Plate 5   A coccolithophorid, *Scyphosphaera apsteini*, with calcareous plates known as coccoliths. Diameter, < 20 μm.

Plate 6   A photosynthetic silicoflagellate, *Dictyocha fibula*, with a skeleton formed of silica. Width, including spines, *ca.* 50 μm.

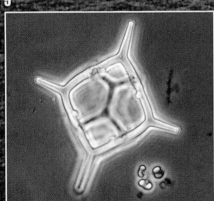

5

6

9

Plate 9   A radiolarian, *Astromma aristotelis*, with a skeleton formed of siliceous spicules. Diameter including spines, *ca.* 100 μm.

Plate 8 The global distribution of ocean chlorophyll concentration obtained from remote sensing by satellite. Note the difference in colour between the central oceanic gyres with relatively low productivity and the coastal and upwelling zones with high productivity.

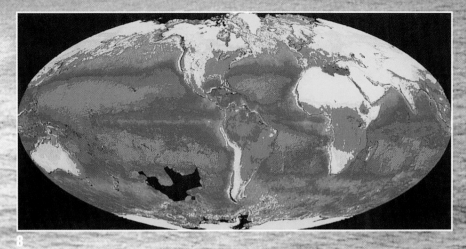

8

Plate 7 A colour-enhanced satellite photograph of warm (3) and cold (4) core rings formed from the Gulf Stream. (1) indicates the warmest region of the Gulf Stream, which cools as it moves northward to region (2). (5) indicates cold Labrador Current water moving sothward and warming as it flows to region (6). Different colours indicate different surface temperatures: red, warmest; pink, coldest.

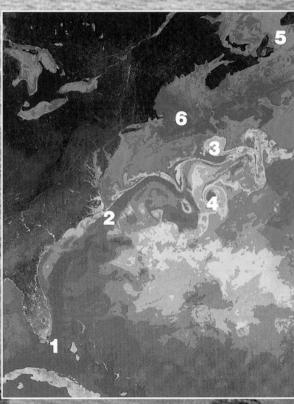

7

Plate 10 A ciliate protozoan, *Strombidium viride*, with a pigmented eyespot located near the cilia. Length, *ca.* 60 μm.

Plate 11 A tintinnid ciliate (*Favella sp.*). Note the external test and the brown food spots within the protozoan, and the cilia projecting on the right side of the photo. Length, *ca.* 200 μm.

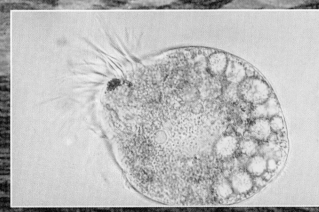

10

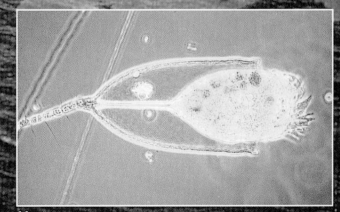

11

12

Plate 12   The medusa *Pelagia noctiluca*. Note the slender trailing tentacles and larger central, yellow oral lobes. Bell diameter about 6 cm.

Plate 13   A colonial siphonophore (*Agalma*). Note the multiple swimming bells and trailing tentacles (left).

Plate 14   The common ctenophore *Pleurobrachia* with its two tentacles spread in the fishing position. Body about 2 cm.

Plate 15   The ctenophore *Beroe*, a major predator of *Pleurobrachia*. About 5 cm long.

Plate 16   A primitive heteropod, *Atlanta peroni*. Note the coiled, transparent, calcareous shell and large eyes at the base of the tentacles. About 5 mm in diameter.

16

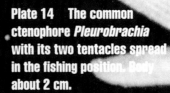

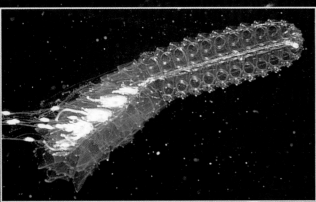

13

14

15

Plate 17 A heteropod (*Carinaria*) with expanded transparent body and very reduced shell located below the single swimming fin. The mouth is directed upward at the end of a short snout. About 5 cm long.

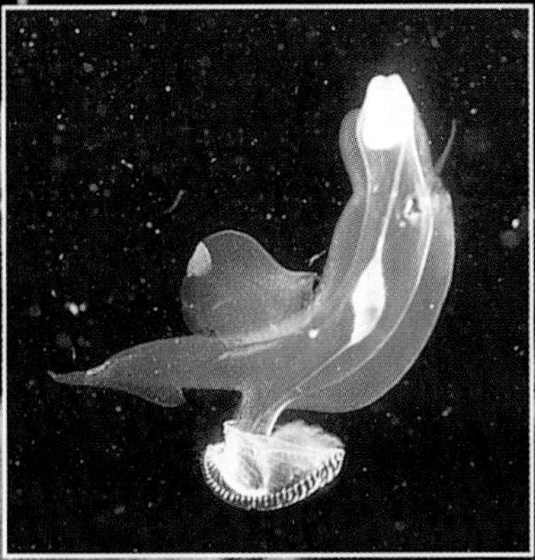

17

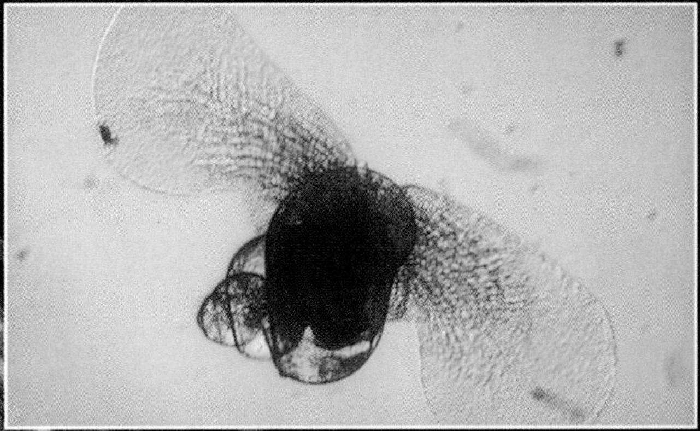

18

Plate 18 The thecosomatous pteropod *Limacina*, with paired swimming wings and a coiled shell of a only a few millimetres in diameter.

19

Plate 19 A theocosomatous pteropod (*Cuvierina*) with a cigar-shaped shell (about 10 mm long) and paired swimming wings.

Plate 20 A pseudothecosome mollusc (*Corolla*) with an expanded wingplate lying between the lower rounded body and upper proboscis. Wingspan about 5 cm.

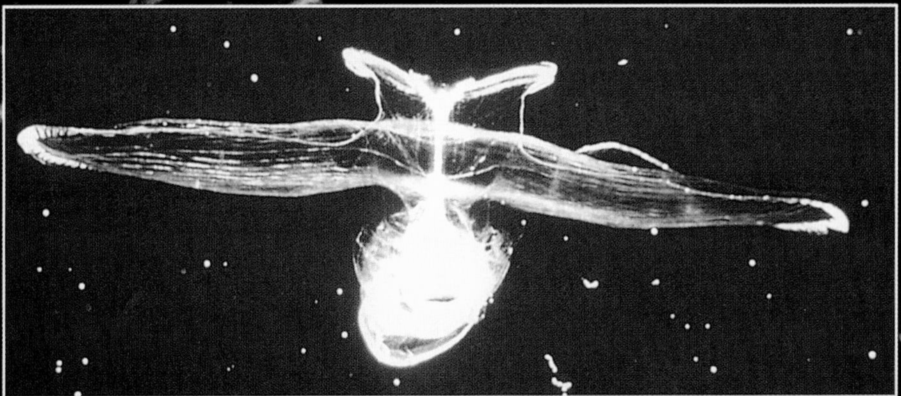

20

Plate 21 The gymnosome *Clione limacina*, a molluscan predator of species of *Limacina* (thecosomes). Note the shell-less streamlined body and paired swimming wings. Length 3–7 cm.

Plate 22 The siphonophore *Forskalia* with attached amphipods that appear as white round spots on the upper swimming bells. Multiple tentacles project below.

Plate 23 A deep-sea shrimp (*Pasiphaea*) showing the red coloration that is typical of many mesopelagic zooplankton. Scale in mm.

Plate 24 An appendicularian (larvacean), *Stegasoma magnum*, which has been dyed with carmine particles to show the (red) filters of its transparent mucous house. The animal itself is located centrally, with a small white body and slender tail. Size about 3 cm.

Plate 25 The solitary stage of the salp *Pegea socia* (length, ca. 25 mm). It will bud asexually to produce the chain of individuals shown in Plate 26.

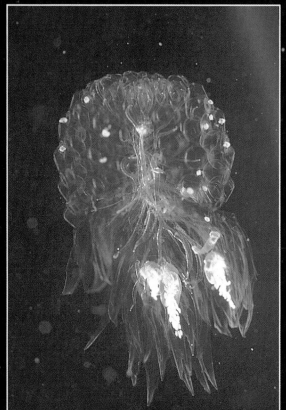

22

23

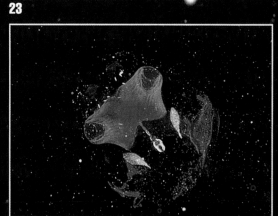

24

21

25

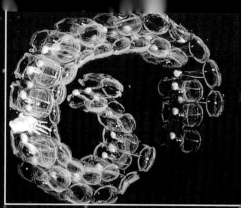

26

Plate 26 The aggregate stage of the salp *Pegea socia* in which the individuals are grouped in chains. Chain length, ca. 13 cm.

Plate 27 A common neustonic cnidarian (coelenterate), *Velella velella*, also known as the by-the-wind sailor. A transparent sail projects above the water surface, multiple tentacles trail below. About 3 cm wide.

Plate 28 The neustonic snail, *Janthina*, suspended from the sea surface by a raft of bubbles. The opaque structures between the animal's shell and its raft are egg capsules. Shell size about 15 mm.

27

29

Plate 29 The neustonic nudibranch mollusc *Glaucus atlanticus*. Its fingerlike cerata contain stinging nematocysts which the animal obtains by feeding on *Velella* and *Physalia*. To 30 mm

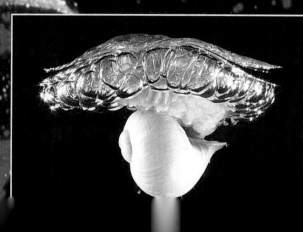

Plate 30   These stromalites in Western Australia are formed from calcium carbonate deposited by Cyanobacteria. Such formations are known to have been in existence two billion years ago, and thus these structures represent one of the longest continuous biological lineages.

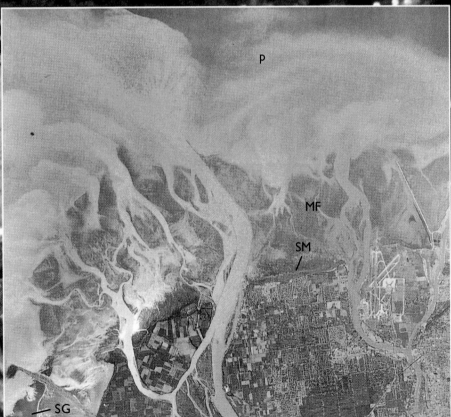

P

MF

SM

SG

Plate 33   An aerial view of the Fraser River estuary in British Columbia, Canada. Several branches of the Fraser River run through the city of Vancouver and discharge large volumes of silt-laden freshwater into the Pacific Ocean. The locations of several of the estuarine communities can be seen including the narrow saltmarsh zone (SM), a seagrass bed (SG), the extensive mudflats (MF), and the pelagic community (P).

Plate 31 A rocky shore community in which acorn barnacles, stalked barnacles and mussels are competing for attachment space. Note that the acorn barnacles in the centre have grown very tall and narrow because they are being pressed from the sides by organisms competing for space.

Plate 32 *MAIN PHOTOGRAPH* Sea otters feeding on sea urchins. These mammals are keystone species regulating the ecology of some North Pacific kelp forests.

Plate 34 Corals and coral reef fish from the Great Barrier Reef off Australia.

Plate 35 Corals and associated fish from reefs off Tahiti.

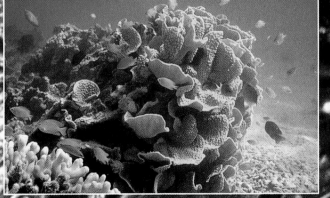

34

35

Plate 40   MAIN PICTURE
Calyptogena, a large clam (to 40 cm) that forms a major portion of the biomass around certain hydrothermal vents and cold seeps.

Plate 36   Expanded coral polyps and calcareous coral cups. The tentacles of the polyps contain batteries of nematocysts used in food capture.

Plate 37   One of the many types of cnidarians (a hydrozoan) found in association with stony corals on tropical reefs.

36

Plate 41   An aerial view of the oil spilled from the tanker *Amoco Cadiz,* coming ashore at Roscoff, France, in 1978. Note that although oil covers most of the foreshore, beach areas in the lee of the islands and bays are relatively free of oil.

41

38

Plate 38   Tropical mongrove trees with extensive root systems.

Plate 39   *Riftia pachyptila,* a vestimentiferan tubeworm that is a dominant animal at the Galapagos hydrothermal vents. A plume of red tentacles projects from the leathery tube of each individual.

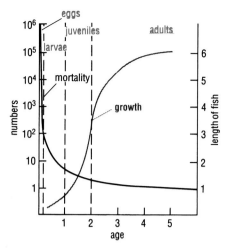

Figure 6.11   Idealized population mortality and individual growth curves for a species of teleost fish during its life cycle from egg to adult. Arbitrary units of age and length, and variable time intervals (indicated by dashed lines) for life stages.

1   The hypothesis that starvation may regulate larval fish survival is concerned with what is called the **critical phase** in the life of fish. This phase begins immediately after hatching when the young larvae still have the remnants of the yolk sac on which to subsist. In order to survive, the larvae must begin to eat sufficient planktonic food before the yolk is exhausted. This means that the larvae must hatch at a time in phase with abundant plankton concentrations. If a larval fish hatches too early, or too late, relative to its food supply, it will die (Figure 6.12).

2   The predation hypothesis assumes that larger organisms have fewer predators and are also better able to escape from predatory attacks. Thus if larval or juvenile fish can grow fast enough, mortality will be lessened and more fish will survive. This is illustrated in Figure 6.13 which also shows that this hypothesis is partly dependent on the planktonic food supply, although here food concentrations are not restricted to the critical phase as in Figure 6.12.

QUESTION 6.6   (a) Referring to Figure 6.13b, is there a growth rate for young fish below which there would theoretically be 100% mortality due to predation? (b) Would you expect to find 100% mortality from predation in nature? a Yes b No

3   The advection hypothesis can be illustrated in a number of different ways depending on the type of fish species. For example, plaice tend to spawn on specific sites that are associated with favourable nursery areas to which the larvae are carried by currents. This is shown in Figure 6.14 for a population of plaice that spawns in the southern North Sea. In some years, however, strong storm activity may disrupt the current system that carries larval fish to their nursery area, and the plaice larvae may then be transported to areas that are unfavourable for survival.

4   The growth hypothesis derives from the growth curve shown in Figure 6.11. Growth is dependent on a variety of parameters affecting the

Figure 6.12   The critical phase of larval fish survival requires that planktonic food (often copepod nauplii) must be present in the water at the time of hatching (e.g. at time A). If the food organisms occur later (e.g. at time B), all the fish larvae from one particular spawning will die from starvation. (Arbitrary units.)

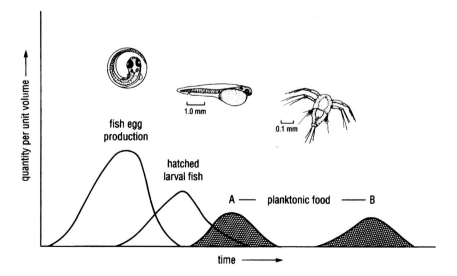

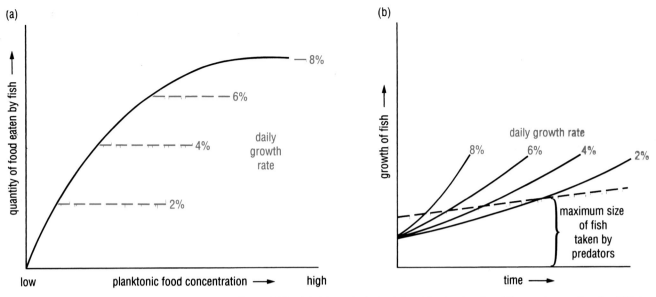

Figure 6.13 (a) A hypothetical relationship showing the difference in growth rates (2% to 8% per day) of juvenile fish depending on the concentration of planktonic food (usually, copepods).

(b) The growth rates from (a) are shown over time, together with the size at which maximal predation of the fish takes place. Note that this size also increases slightly with time as the predators themselves grow. Juvenile fish that grow slowly (e.g. at 2% day^{-1}) are exposed to predation over a longer time period and therefore have lower survival rates than faster-growing (e.g. 8% day^{-1}) fish. (All units are arbitrary values.)

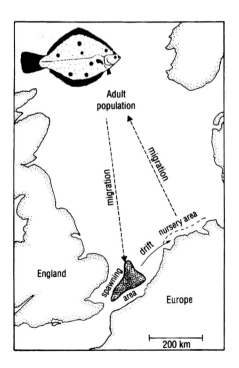

Figure 6.14 The larval drift and migrations of plaice (*Pleuronectes platessa*) in the North Sea. Larvae hatch in the spawning area, then drift northeast to a food-rich nursery area. They remain in the nursery area for their first year of life, growing from 1.5 cm to 20 cm in length before migrating to the northwest.

rate of growth and the length (size) at maturity. In general, the growth rate is directly proportional to temperature, but size at maturity is inversely proportional to temperature. Thus an increase in temperature, which is governed by the physical ocean climate, can have the dual effect of producing more rapidly growing fish, but ones that are smaller at maturity. In addition, the growth efficiency (K_1 or K_2, see equations 5.10 and 5.11) of fish varies with the type of food consumed. Prey with high protein content (e.g. copepods) produce faster growth than foods (e.g. small ctenophores) with very high water content and low protein. Growth is also influenced by the metabolic costs associated with particular types of prey; for example, a predator that has to chase its prey would have a higher metabolic cost than one that filters its food. Thus, as prey type changes (due to changes in the ecosystem), the growth efficiency of fish will also change, and this will affect the growth curve (Figure 6.11).

In summary, each of the four hypotheses discussed above can be shown to have some experimental support. However, none can be shown at present to be the only mechanism determining the fluctuations in the abundance of fish, and it is likely that more than one mechanism is operable. It is also possible that some other factors, such as fish disease, may at times be important in regulating recruitment of young to adult stocks. In order to improve the management of fisheries, these mechanisms need to be further researched through experimentation, field data, and ecosystem computer models.

QUESTION 6.7 Does Figure 6.11 suggest that, in order to increase the biomass yield of fish, it is better for a fishery to allow larger numbers of small fish to survive, or smaller numbers of large fish? What other factors might affect the decision? *large numbers of small fish*

6.7.4 FISHING AND THE USE OF NEAR REAL-TIME OCEANOGRAPHIC DATA

The previous section has concentrated on mechanisms that might be relevant in producing seasonal and long-term variations in the abundance of fish. Another problem faced by fishermen is where to find the main concentrations of fish. In general, fishermen have been almost too successful in finding fish, and some stocks have been seriously depleted due to the massive investment in the mechanized harvest of fish. However, it still remains a problem to forecast the exact location of fish schools on time scales that relate to the time which a fishing boat can economically spend at sea.

Certain types of oceanographic data, such as surface temperatures and depth of the thermocline, can greatly assist fishermen in rapidly locating fish and in reducing the cost of remaining at sea, but only if these data are collected during a time that is near to the harvest period. This type of information is referred to as 'near real-time data'; the data are collected very close to the time of an event for which the data are required.

Fish become easy targets for fishermen when they are concentrated into schools. They may form schools when feeding, or during periods of reproduction and migration. Feeding schools are often associated with very productive waters, such as might be present on a particular bank (i.e. a shelf-break frontal zone as defined in Section 3.5.4), or at a boundary between two water masses (e.g. a planetary frontal zone, Section 3.5.3). Reproduction may take place in other localities, with the fish migrating together between feeding and spawning sites.

The location of particular types of fish is often an important factor in the regulation of fisheries. For example, in the North Pacific Ocean, salmon generally stay in subarctic waters of $<14°C$, whereas the squid and tuna fisheries are in subtropical waters of $>14°C$. The boundary between these two water masses is not a fixed line, but varies geographically by several hundred kilometres depending on the physical oceanography. Near real-time oceanographic information on the location of this frontal zone is important to the high seas fisheries. This is particularly true for the squid fishery which is carried out in the open ocean and, by international agreement, is not allowed to stray into areas where salmon may be inadvertently collected as well. Ocean surface temperatures can be detected by satellites, and the location of the 14° boundary is relayed to the Japanese squid fishing vessels via a land-based station which also monitors the location of the fleet.

It is also important for inshore fixed-depth fisheries to be able to predict the location of fish schools from near real-time oceanographic data. Figure 6.15 illustrates the location of cod stocks relative to a trap fishery that is maintained at a fixed depth throughout the year. (This illustration serves as a good example because the location of the fishing effort is fixed every year; however, it could also apply to a net fishery operating at a fixed depth.) A trap, consisting of a large net that directs swimming fish into a central area from which they cannot escape, is set from the surface to the seafloor. Various physical disruptions in the water temperature can result in the location of the cod at very different depths and regions relative to the trap locations. Under some circumstances (conditions b and d in Figure 6.15), the position of the traps is such that few, if any, fish would be captured. At such times, fishermen might profitably employ a different type of fishing method, using nets towed from boats or, in the case of condition 'd', the traps might be relocated in deeper water if feasible. The application of near real-time

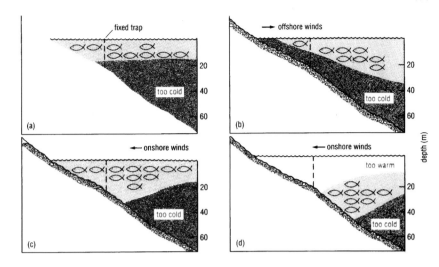

Figure 6.15   The trap fishery illustrated here (and other inshore fisheries) can be severely affected by local water temperature conditions, which in turn are determined by prevailing wind and air temperature conditions. (a) shows a June situation in an area having a late, cold spring with light variable prevailing winds. The surface water has warmed enough to a depth of about 18 m to allow cod to come close inshore at shallow trap depths. In (b), prevailing offshore winds have driven the suitable water layer off the shore, and cold water coming to the surface severely restricts the area in which trap fishing might be successful. In (c), prevailing onshore winds early in the season have had the opposite effect, pushing the suitable water layer onto shore, and deepening and expanding the potential for good trap fishing. Finally, in (d), a late-season situation is shown in which prolonged warm weather and onshore winds have combined to produce a nearshore layer of water too warm for cod; the fish stay below it out of reach of the trap.

data on the thermal structure and movement of the water masses near the traps could greatly assist this fishery, and such information should become an integral part of fisheries management.

As mentioned earlier, fish schools also form during reproductive periods, and the congregation of herring on certain banks in the North Sea is a well-known phenomenon associated with the spawning of these fish. Unfortunately, it has also become the time and place where the herring are most easily caught — just when they are about to create a new population. There is obviously a division of opinion in such a case as to whether one catches the fish when it is most economical to do so, or whether it would be better to wait for a time when less biological damage would be done.

The migration routes of fish can also be greatly influenced by local oceanographic conditions, which in turn affect the fishery. An example from North America is given in Figure 6.16, illustrating migration routes of Fraser River sockeye salmon off the west coast of Canada. During 1955–77, most of the fish returned to the river through the Strait of Juan de Fuca, where they were exposed to both Canadian and U.S. fishermen. However, during 1978–83, up to 80% of the fish returned through routes north of Vancouver Island, where they were available only to Canadian fishermen. This resulted in a change of considerable economic impact to the fisheries of the two adjoining nations. A computer model that used the temperature and salinity of coastal water (the latter indicating amount of freshwater inflow from rivers) was developed to predict the diversion in 1978–83 as indicated in the inset graph.

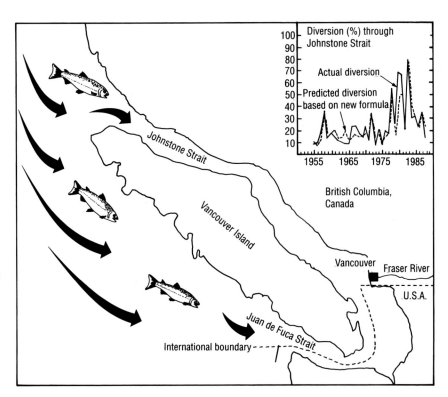

Figure 6.16 The migration routes of sockeye salmon returning to the Fraser River (British Columbia, Canada) for spawning. The inset graph shows the percentage of fish that were diverted through Johnstone Strait during return to the Fraser River from 1955 to 1988 (solid line). From 1955 to 1977, most fish returned to the river through the Strait of Juan de Fuca; from 1978 to 1983, the majority of the salmon returned through Johnstone Strait. The dashed line in the inset graph indicates predicted diversions through Johnstone Strait based on a computer model simulating changes in freshwater flow and water temperature.

6.8 MARICULTURE

One way of managing fishery resources and increasing yield is to attempt to control many environmental variables by growing marine species in enclosures or impoundments. Such systems, because they concentrate the target species and are located in coastal areas, also have the advantage of easy and economical harvesting. The cultivation of marine species is called **mariculture**. It can be regarded as the marine counterpart of agriculture, as many of the principles and problems associated with increased production through artificial culture are similar. Interest in mariculture continues to increase as the human population expands and demands more protein, and as open-sea fishing reaches the limits of exploitation of wild stocks.

A variety of pelagic and benthic organisms are cultured including some seaweeds, crustaceans (shrimp or prawns), molluscs (mussels, oysters, scallops, clams, and abalone), and fish (e.g. salmon, mullet, sole, turbot, eels). Most of these species are raised for human consumption, but some are cultured for other commodities, such as pearls, food additives (e.g. alginates from seaweeds), and domestic animal feed.

One of the simplest forms of mariculture involves the transplantation of wild stocks (e.g. fish, oysters) into new areas where natural conditions are favourable for increased production. Successful transplants have involved the introduction of a Pacific salmon into Lake Michigan in the U.S. and into New Zealand and Chile. Another very simple technique is the trapping of a wild population in enclosures, where the animals are held without artificial food until harvesting; this is done with shrimp in Singapore and some fish (e.g. mullet) at various localities. More intensive culture requires more manipulation in terms of providing feed and fertilizers, controlling the physical environment, and eliminating predators and disease.

The most economical species to culture are those that occupy lower trophic levels, feed on naturally available food, and can produce high biomass in crowded conditions. Mussels, for example, feed by filtering plankton, and they are routinely cultured in many coastal areas. In nature, mussels attach to hard substrates and their numbers are often controlled by the amount of space available and by the degree of predation caused by starfish and boring snails. In culture, mussels are provided with ropes suspended from rafts for attachment, thus increasing the space available and also eliminating benthic predators; the mussels feed on plankton in the surrounding water. Such intensive culture can yield up to 600 tonnes of mussels per hectare annually; about 50% of this is drained body weight, available for consumption.

Animals from higher trophic levels require the provision of either artificial feed or of cultivated or captured prey and, because they are generally larger in size and mobile, they also occupy more space. However, the relatively high costs of culture may be justified by higher market value. For example, salmon can be grown in enclosures (Figure 6.17) and fed on synthetic feed (which has the advantage of chemically known constituents) or on processed euphausiids captured at sea (see Section 6.1); the high expenses of this culture are recovered in the high market price of the salmon.

Another form of salmon enhancement is to rear salmon eggs in hatcheries and then release the young salmon into the sea. This is known as **ocean ranching**, from analogy with raising some cattle by releasing them into ranchlands. Depending on the species, released salmon spend between 2 and 5 years in the open ocean before returning to the same stream in which the hatchery is located. Ocean ranching is less expensive than pen culture, but the returns of adult fish depend upon survival at sea.

The examples given above all utilize only one species; this type of system is known as **monoculture**, and it represents an extreme simplification of a natural community. In order to increase profitability, some systems attempt to culture more than one species in the same enclosure; this is known as **polyculture**. In a monoculture system, uneaten food and faecal material fall to the bottom where they are either flushed by natural water movement, or are otherwise cleaned from tanks to prevent excessive bacterial growth and

Figure 6.17 An aerial view of a salmon farm in British Columbia, Canada. The salmon pens are located in a protected cove that shelters the enclosures from storm damage.

the formation of anoxic water. A polyculture system introduces species that feed on this detritus and can also be harvested for commercial profit. Herbivorous fish (e.g. some carp, mullet) are commonly grown in ponds together with prawns that feed on detritus, fishmeal, and/or filamentous algae. It is also possible to combine fish that feed on zooplankton with those that eat benthic plants. Polyculture is presently more highly advanced in freshwater systems, but it can be expected to expand as more mariculture systems are developed.

The leading countries in mariculture tend to be those with high human populations and high protein demands. Fish (carp) culture began about 4000 years ago in China, and this country continues to lead the world in terms of quantity of production from combined marine and freshwater culture (see Table 6.3). The second leading country is Japan, which has developed some of the most advanced mariculture facilities and techniques. In North America, only about 2% of fishery products come from culture, and that figure includes cultivated freshwater species as well as marine organisms. In Europe, Spain is the leading producer with mussel culture. On a global basis, roughly 5×10^6 tonnes of marine species are grown in culture annually.

QUESTION 6.8 What is the percentage of marine species produced in culture compared with the total marine fish catch harvested from the oceans?

$$84 \times 10^6 \text{ t.}$$

$$\frac{84}{100} \times 5 = 4.2\%$$

There is no doubt that mariculture will continue to expand throughout the world in response to increasing demands for more protein for human consumption and, to a lesser extent, to supply luxury foods such as salmon and lobster, or other commodities. There are several constraints to this expansion, however. Some problems are technological, like the selection of suitable sites or disease control, and some of the constraints are economic, balancing high costs of feed, fertilizer, and manpower against market prices. A much more serious problem looms with increasing coastal pollution throughout the world. Mariculture depends on coastal sites, whether cultured species are grown in enclosures in the natural environment or in land-based facilities that depend on water pumped from a nearby marine source. Eggs and young stages of marine animals are particularly sensitive to pollutants, and adults may accumulate chemical or biological substances from polluted water that make them dangerous to humans. Shellfish, for example, filter the cholera bacterium from sewage-polluted water, and consumption of infected animals spreads and compounds this disease.

6.9 SUMMARY OF CHAPTER 6

1   The nekton comprises the larger, pelagic, marine animals whose swimming abilities are such that their movements are independent of ocean currents. Included in this category are larger crustaceans (some euphausiids, shrimp, and swimming crabs), squid, sea snakes, marine turtles, and marine mammals, with adult fish making up the dominant fraction. Seabirds are also considered here because they are dependent on the sea for food and may have considerable influence on the neuston and epiplanktonic communities.

2   Few commercially harvested crustaceans are pelagic, but some of the larger, very abundant euphausiid species are presently fished in the Antarctic and off Japan. The superabundance of the Antarctic krill (*Euphausia*

superba) makes it an attractive fishery target, and it is likely that this harvest will increase, despite high economic costs.

3   Squid form another abundant invertebrate group targeted by fisheries. More needs to be known about the biology and abundance of these animals before fisheries management can be effective in protecting these stocks. It has become evident that driftnet fishing for squid is very unselective, and that vast numbers of seabirds, fish, turtles, and marine mammals have been inadvertently captured and killed by this method. More selective, alternative squid-fishing techniques are available, although their use increases the costs of fishing.

4   Eight species of turtles, one lizard, and about 60 species of snakes are the only reptiles to have become marine. The turtles have become endangered species from hunting of the adults and their eggs.

5   There are about 110 species of marine mammals. The largest are the baleen whales that feed by filtering zooplankton or fish (or benthic invertebrates, in the case of the grey whale) through their plates of baleen. The toothed whales (including dolphins and porpoises) are predators in the sea. These two groups of cetaceans together consume a much greater quantity of marine biomass than is removed from the oceans by the entire commercial fishery.

6   Many species of whales, pinnipeds (seals, sea lions and walruses), and sirenians (manatees and dugongs) have been extensively hunted. Their low fecundity and long development times from birth to maturity make them especially vulnerable to rapid depletion of population numbers through commercial harvests. Many of the species are now endangered, and recovery of populations is slow.

7   The most highly adapted of the seabirds spend 50–90% of their lives at sea, but all remain dependent upon land for nesting sites. The highest numbers are found in association with very productive waters, where zooplankton and fish are concentrated. Non-migrating species are subject to natural mortalities caused by climate change and subsequent declines in their prey (e.g. the effects of El Niño on the guano birds off Peru). Human-induced mortalities such as overfishing, habitat destruction, introduction of predators, and coastal pollution increasingly threaten seabirds.

8   The great majority of marine fishes are teleosts with a bony skeleton, and the 20 000 or so species show considerable diversity in terms of anatomy, behaviour, and ecology. The most abundant species are epipelagic plankton-feeders with very high fecundity, and many of them (e.g. herring, sardines, anchovies) form the basis for some of the most profitable marine fisheries.

9   There are fewer deep-water species of fish, and they are not as numerous as their shallower-living relatives. Many of the meso- and bathypelagic fishes tend to be relatively small. Many of the species have photophores and use bioluminescence to locate or lure either prey or mates, or to evade predators.

10   In 1993, the total world catch of all marine species of fish (including squid and shellfish) was about 84 million tonnes per year. About 64% of this catch was taken from the largest of the oceans, the Pacific Ocean, with China, Peru, and Japan being the leading fishing nations.

11　Although various attempts have been made in the last 100 years to manage fisheries, these have met with little success. Many fish stocks or fishing regions have been depleted or are in danger of being overfished.

12　Fisheries management has traditionally been based largely on stock/recruitment theories that have ignored the role of the environment in causing natural fluctuations in the numbers of fish. It is becoming increasingly apparent that the recruitment of larval fish into adult stocks can vary greatly depending on whether there is sufficient food for young fish, whether predation is high or low, whether larval fish are transported by currents into unfavourable habitats, whether disease affects the population, and whether growth is slowed or hastened by temperature, food availability, or other factors.

13　Biological oceanography can assist fisheries in two ways. The first way is by increasing our understanding of what factors cause natural fluctuations in the abundance of fish. The second way is in providing near real-time oceanographic data in connection with the actual process of fishing; better information on the location of fish schools can reduce the cost of the fisheries.

14　Mariculture is another way of increasing the yield of fisheries resources. At present, only a few of the many marine species are under culture, but cultivation is expected to expand from its present production of about 5×10^6 tonnes per year.

Now try the following questions to consolidate your understanding of this Chapter.

QUESTION 6.9　From a knowledge of primary productivity in relation to the physical movement of water (Section 3.5), what other locations in the oceans in addition to those discussed in this Chapter might support good fishing?

QUESTION 6.10　Can the total tonnage of 'fish' be increased by catching smaller and smaller fish, or even by harvesting zooplankton?

QUESTION 6.11　Suppose each female fish in a particular species lays 10^3 eggs each year. In one year, 99.90% of the eggs or progeny die; in another year, the mortality is 99.95%. Assuming that in each case the remainder became mature adults, what would be the difference in the numbers of the adult populations resulting from these two year classes?

QUESTION 6.12　What might be some ecological consequences of fisheries continuing to remove increasingly large numbers of fish and squid from the oceans?

QUESTION 6.13　Other than selecting species for culture on the basis of trophic level position, what additional physiological and biological features might make certain organisms more amenable and attractive as possible culture organisms than others? Consider the requirements of all life stages in formulating your answer.

QUESTION 6.14　What percentage of the world's ocean is now a designated whale sanctuary? (Consult Table 5.1.)

Relative to the pelagic zone, the seafloor presents a greater variety of physically diverse habitats that differ from each other in terms of depth, temperature, light availability, degree of immersion (tidal vs. subtidal), and type of substrate. Hard, rocky substrates provide sites of attachment for sessile species like barnacles and mussels which remain in one place throughout their adult life, and they provide crevices and depressions that can be used by mobile animals as refuges from predators. Soft-bottom substrates (e.g. mud, clay, sand) offer both food and protection for burrowing animals. At least partly owing to the greater variety of benthic habitats, the number of species of benthic animals (estimated at >1 million) is much greater than the combined number of pelagic species of larger zooplankton (about 5000), fish (>20 000) and marine mammals (ca. 110).

As in the pelagic environment, vertical gradients of temperature, light, and salinity are especially important in establishing distinctly different living regimes for benthic organisms. Figure 1.1 shows the ecological divisions of the seafloor based on depth and topography. Some of the ecological-depth divisions have well-defined boundaries, others are more arbitrary zones, but each of these benthic habitats presents distinctly different living conditions. The animals that inhabit different zones will generally be of different species, each uniquely adapted to the particular environment in which it is found.

The smallest benthic zone (Figure 1.1) is the **supralittoral** or **supratidal** zone, an area just above high water mark and immersed only during storms. On steep shores, this zone will receive spray from breaking waves, and it is sometimes referred to as the 'splash' zone. On flat beaches, the area may be marked by heaps of seaweeds cast ashore. Few species are adapted to live in this transitional region between the sea and land.

The **littoral** or **intertidal** zone lies between tide marks and is thus immersed at high tides and exposed at low tides. The extent of this zone depends upon local topography and tidal range. This area lies within the euphotic zone, and benthic algae as well as phytoplankton are available for grazing and filter-feeding benthic herbivores. These in turn support a diverse and abundant carnivore community.

The **sublittoral (subtidal)** zone extends from the low tide mark to the outer edge of the continental shelf, at a depth of about 200 m. Part of the sublittoral area also lies within the euphotic zone, but benthic plants decline from low numbers to zero in the deeper regions. Rocky substrates become scarce and are replaced by soft substrates. The sublittoral zone occupies about 8% of the submerged seafloor.

The remaining benthic habitats are located below the euphotic zone. The **bathyal** zone extends down the continental slope from 200 m to 2000 or 3000 m (the lower boundary is indefinite), and it occupies approximately 16% of the submerged seafloor. The **abyssal** zone, extending from 2000 or 3000 m to 6000 m, is by far the largest ecological region, encompassing almost 75% of submerged benthic habitats. This zone is also characterized by having a temperature of 4°C or less. The deepest areas of the sea are the trenches, extending downward from 6000 m to somewhat over 11 000 m depth; ecologically, this benthic habitat is referred to as the **hadal** zone (from 'Hades', the Greek mythological underworld). This last zone is the least well known because of its inaccessibility, and relatively few species have so far been described from it.

The vast majority of larger benthic species live in depths less than 200 m, and there are many more species in shallow tropical waters than in shallow cold seas. Some of the different types of shallow-water and deep-sea benthic communities are considered in more detail in Chapter 8.

7.1   BENTHIC PLANTS

A variety of marine plants attach to the seabed or live within sediments in shallow depths. All are restricted to the euphotic zone; that is, they are confined to intertidal and shallow subtidal regions.

Certain intertidal marine communities are dominated by large **angiosperms** (flowering plants) that only flourish in sheltered regions, where accumulated sediments allow rooted plants to develop. These communities include tropical mangrove swamps, with a variety of salt-tolerant trees and shrubs (see Section 8.7); estuarine saltmarshes which are dominated by marshgrasses (see Section 8.5); and meadows of seagrasses which occur low in the interidal zone. All of these benthic **macrophytes** (large, visible plants) are highly productive, but they contain a large proportion of materials that are indigestible to most marine animals. They therefore form large amounts of detritus, which may be exported by tidal currents into other marine areas, and this decomposing detritus contributes to the high productivity of coastal waters.

Marine macrophytes also include the conspicuous algae that are most abundant on rocky shores in temperate zones. The algae (which are rootless) have developed anchoring structures called holdfasts. Among the important members of this group in terms of production are the long-stemmed kelps (brown algae) (Figure 7.1c and Figure 8.3) that anchor to rocky substrates in the subtidal zone (see Section 8.3). Some kelp (e.g. *Macrocystis*) have extremely fast growth rates and form large underwater forests. Other types of seaweeds (Figure 7.1) include the common macrophytic algae (e.g. *Fucus*) that cover rocks in the intertidal zones; they can be very abundant, but their rate of production is usually only about half that of the kelps. Kelp and seaweeds may be a direct source of food for some herbivores, and they also form abundant detritus that is ultimately consumed by detritivores. As much as 30% of their production may be lost in exudates that contribute to the pool of DOM (see Section 5.2.1).

Dissolved organic matter

Some green algae (e.g. *Halimeda*) and red algae (e.g. *Lithothamnion*) have the ability to incorporate calcium carbonate in their tissues, which is a most effective defence against being eaten by herbivores. These hard **coralline algae** grow as encrustations over rocks or shells or coral reefs, and they can contribute materially to the formation of carbonate deposits.

Epiphytic algae (e.g. *Ectocarpus*) grow on the surfaces of other larger plants (seaweeds, kelp or seagrasses). The epiphytes are generally thin-walled and filamentous, and therefore can be easily consumed by marine herbivores.

The least obvious benthic producers are the unicellular algae that live on sand grains (**epipsammic** species), or that form mats on the surface of mud.

Figure 7.1 Examples of seaweeds.
(a) *Enteromorpha* (up to 500 mm); (b) *Ulva* (up to 250 mm); (c) *Alaria* (up to 2 m);
(d) *Chondrus* (up to 150 mm); (e) *Gigartina* (up to 200 mm); (f) *Delesseria* (up to 250 mm);
(g) *Fucus vesiculosus* (up to 1 m). (a) and (b) are green algae; (c) and (g) are brown algae; and (e) and (f) are red algae.

These **microphytes** (microscopic plants) include motile pennate diatoms, blue-green algae, and dinoflagellates (see Section 3.1 for general descriptions). These organisms are often extremely abundant and despite their small size, they are an important source of primary production in shallow waters. Certain dinoflagellates have even taken up residence within the tissues of benthic animals; the best known are the symbiotic algae of corals that are described in Section 8.6.

The rock-like reefs called **stromatolites** (Colour Plate 30) that are located in shallow waters off western Australia and around the Bahamas are of great interest from an evolutionary perspective. They are formed by mats of microphytic photosynthetic Cyanobacteria (see Section 3.1.3) that deposit calcium carbonate which builds up in successive layers at a rate of about 0.5 mm per year. Although the present reefs began to form relatively recently in geologic time, the constituent Cyanobacteria are similar to microbes that flourished two billion years ago, and thus they represent one of the longest continuous biological lineages known.

Different benthic plants characteristically occupy different tidal levels, and this zonation is partly determined by their different abilities to absorb particular wavelengths of light. The complete spectrum of visible light is available at the sea surface, but different wavelengths are quickly absorbed and scattered within the water column (Figure 2.4). Green algae (e.g. *Ulva*) typically grow in shallow water, and their pigments absorb both long and short wavelengths (Figure 3.4a). Brown and red algae also contain green chlorophyll, but they have particular accessory pigments that mask its colour. Compared with green algae, brown algae (e.g. kelp, *Fucus*) are most abundant in somewhat deeper water; their main pigment, fucoxanthin, is more efficient at capturing blue-green light (Figure 3.4b). Some red algae (e.g. *Gigartina*) are characteristically subtidal; their red pigments (phycoerythrin and phycocyanin) are also efficient at absorbing subsurface light that cannot be absorbed by chlorophyll *a* (Figure 3.4b). There are numerous exceptions to this depth-distribution pattern, however. For example, certain red algae (e.g. *Porphyra*) may be found in the high intertidal zone, and some green algae (e.g. *Ulva*) may occupy lower regions. This is because other factors, such as resistance to wave action, tolerance to drying during tidal exposure, and selective grazing by herbivores also determine the position of plants in intertidal areas (see Section 8.2.1). In the sediment surface, blue light is absorbed first and red light penetrates the farthest; thus small algae growing within sand or mud may also show a different distribution from that described above.

QUESTION 7.1 Why are algae commonly grouped in colour categories, such as 'red' algae, 'brown' algae, and 'green' algae?

[handwritten margin note: Because they use different wave-lengths of light for photosynthesis]

7.1.1 MEASUREMENTS OF BENTHIC PRIMARY PRODUCTION

Production of small benthic plants can be measured by carbon or oxygen exchange, or estimated from chlorophyll concentration as described in Section 3.2.1. Production of benthic macrophytes, however, is commonly measured by harvesting and weighing the plants, with production being reported in terms of carbon per unit area. In temperate areas where seasonal growth must be taken into account, harvesting is usually done at the period of maximum biomass; some allowance is made for estimating the biomass that has grown, but died and disappeared before and after the measurement. Growth rates of fronds of macrophytes can sometimes be obtained by punching holes in the plants; the hole will move away from the meristerm (growing region) of the plant, and growth can be expressed in terms of distance moved and incrase in the size of the hole.

Attached plants have the advantage of being washed with turbulent water that brings a continual renewal of dissolved nutrients. Nutrient concentrations are often elevated in coastal waters, and the rates of nutrient uptake are generally high in benthic plants. Usually the productivity per unit area of large attached algae is an order of magnitude greater than that of phytoplankton. Values for benthic productivity are given throughout Chapter 8, where specific types of communities are considered in more detail. Although benthic photosynthetic production exceeds that of phytoplankton in the water column in many coastal areas, only a small fraction of the seafloor receives sufficient light to support attached plants. On a global scale, production by benthic plants accounts for less than 10% of the total primary production in the sea.

7.2 BENTHIC ANIMALS

Benthic animals (or **zoobenthos**) are separated into two ecological categories based on where they live relative to the substrate. The **infauna** are those species that live wholly or partly within the substrate; this category includes many clams and worms (polychaetes) as well as other invertebrates (Figure 7.2). Infaunal species usually dominate communities in soft

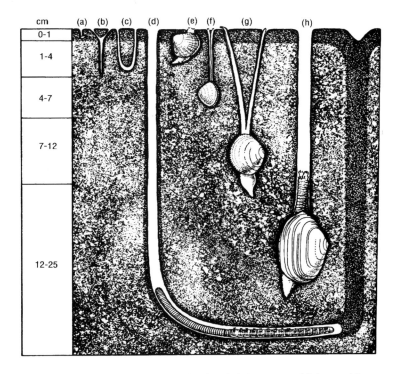

Figure 7.2   Representative infauna, showing their burrows and living positions.
(a) *Hydrobia*, a snail; (b) burrow of *Pygospio*, a polychaete; (c) burrow of *Corophium*, an amphipod; (d) *Arenicola*, a polychaete; and the clams (e) *Cardium*, (f) *Macoma*, (g) *Scrobicularia*, and (h) *Mya*.

substrates, and they are most diverse and abundant in subtidal regions. There are a few infaunal species in hard substrate communities as well, rock-boring clams being one example. The **epifauna** (Figure 7.3) are those animals living on or attached to the seafloor; about 80% of the larger zoobenthos belong to this category. A few common examples of epifauna include corals, barnacles, mussels, many starfish, and sponges. Epifauna are present on all substrate types, but they are particularly richly developed on hard substrates, and they are most abundant and diverse in rocky intertidal areas and coral reefs. A third category can be added to include those animals that live in association with the seafloor but also swim temporarily above it; animals such as prawns and crabs, or flatfish such as sole, form the **epibenthos**.

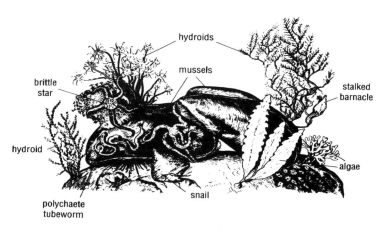

Figure 7.3   Representative epifauna and epiflora.

It is also convenient to classify benthic animals into size categories. In this case, size is relative to the mesh size of sieves used to separate animals from sediments. The following categories encompass all sizes of benthos:

Macrofauna (or macrobenthos): those animals retained by a 1.0-mm-mesh sieve. These are the largest benthic animals, including starfish, mussels, most clams, corals, etc.

Meiofauna (or meiobenthos): those animals retained by a 0.1-1.0-mm-mesh sieve. These are small animals commonly found in sand or mud. The group includes very small molluscs, tiny worms, several small crustacean groups (including benthic copepods), as well as less familiar invertebrates (see Section 8.4.2 and Figure 8.5).

Microfauna (or microbenthos): those animals that are smaller than 0.1 mm in dimension. This smallest size category is largely made up of protozoans, especially ciliates (Figure 7.4).

7.2.1 SYSTEMATICS AND BIOLOGY

Benthic communities contain an extremely diverse assemblage of zoobenthos. Many of these marine species have no terrestrial or freshwater counterparts and are unfamiliar animals. Some of the dominant types of

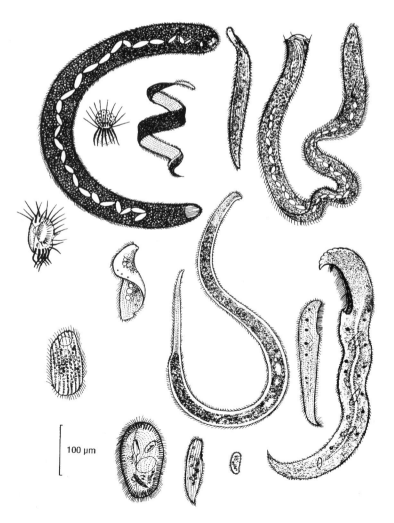

Figure 7.4 Examples of microfauna: ciliate protozoans showing diversity of form.

Table 7.1   Major taxonomic groups and representatives in marine benthic communities.

Phylum	Subgroups	Common names/representatives
Protozoa	Foraminifera	forams
	Xenophyophoria	—
	Ciliophora	ciliates
Porifera		sponges
Cnidaria	Hydrozoa	hydroid polyps
(formerly	Anthozoa	sea anemones; corals
Coelenterata)		
Platyhelminthes	Turbellaria	flatworms
Nematoda		roundworms
Nemertea		ribbon worms
Annelida	Polychaeta	polychaete worms
	Pogonophora	beard worms
	Vestimentifera	vestimentiferan worms
Sipuncula		sipunculids (peanut worms)
Echiura		echiurids (spoon worms)
Hemichordata	Enteropneusta	acorn worms
Mollusca	Gastropoda	snails; nudibranchs
	Bivalvia	clams; mussels
	Polyplacophora	chitons
	Aplacophora	aplacophorans
	Scaphopoda	tusk-shells
	Cephalopoda	octopuses
Echinodermata	Asteroidea	starfish
	Ophiuroida	brittle stars
	Echinoida	sea urchins; sand dollars
	Holothuroidea	sea cucumbers
	Crinoidea	feather stars; sea lilies
Ectoprocta		bryozoans (moss animals)
Brachiopoda		lamp shells
Arthropoda		
(Class Crustacea)	Ostracoda	ostracods
	Copepoda	cyclopoids; harpacticoids
	Tanaidacea	tanaids
	Isopoda	isopods
	Amphipoda	amphipods
	Cirripedia	barnacles
	Decapoda	crabs; lobsters; shrimp
Chordata	Ascidiacea	tunicates (sea squirts)

animals in benthic communities are listed in Table 7.1 and illustrated in Figures 7.5–7.7. They are described below, with particular attention being directed to their positions in benthic food webs.

The best known of the benthic Protozoa are the **foraminifera**, whose planktonic relatives were described in Section 4.2. Several thousand benthic species are known, and they form a dominant element of the micro- and meiobenthos, particularly in deep-sea sediments. Although they are unicellular organisms, benthic forams are not necessarily small in size; some attain lengths of 25 mm. There are both epifaunal and infaunal species, and in general the various species feed on benthic diatoms and algal spores in shallow water, and on other protozoa, detritus, and bacteria in all depth zones. Newly discovered relatives, the **Xenophyophoria** (Figure 7.5a), are especially abundant in hadal zones. They are the largest of all protozoans, with diameters of up to 25 cm but only 1 mm thick. Their extended pseudopodia (to 12 cm long) form tangled masses on the seafloor, and these sticky structures probably collect organic matter from surface sediments. **Ciliates** (Figure 7.4) are important members of the microbenthic community; many are adapted to attach to sand grains or to live freely within the interstitial spaces of sediments. The fragility of ciliates has hampered sampling and ecological studies, but these protozoans are no doubt an important link in shallow water between the microflora (e.g. benthic diatoms) and larger animals and, at all depths, between bacteria and deposit-feeding invertebrates.

The most primitive multicellular animals are the **sponges** (Figure 7.5b), which may constitute a large fraction of the macrobenthos in some marine regions. Known to exist from late Precambrian times (>600 million years ago), this ancient group now has roughly 10 000 species, almost all of them marine. They are named for their porous nature (Phylum Porifera), and the many cavities of sponges provide protective refuges for myriads of small animals such as worms and crustaceans. All sponges are **sessile**, that is, they are attached and immobile. Most filter-feed by producing currents that draw

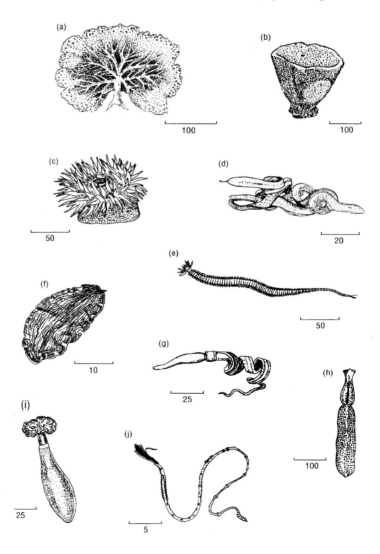

Figure 7.5   Representative benthic animals:
(a) unicellular xenophyophore; (b) sponge;
(c) sea anemone; (d) nemertean; (e) polychaete;
(f) flatworm; (g) enteropneust hemichordate;
(h) echiurid; (i) sipunculid; and
(j) pogonophoran. (All scales in mm.)

suspended particles through the sponge. The pores of the sponge act like a sieve, allowing only the smallest particles to pass and be captured by special flagellated cells. Food consists largely of bacteria, nanoplankton, and small detrital particles. The sponge skeleton is composed of calcium carbonate or siliceous spicules embedded in the body wall, or of spongin fibers. Because of their hard spicules, and reputedly because of their bad taste, sponges have few predators, the exceptions being some coral reef fish and some snails and nudibranchs. They have both asexual and sexual reproduction, and are capable of regenerating from only fragments of a whole organism.

The Phylum Cnidaria has many benthic representatives in addition to the pelagic species described in Chapter 4. This group too has had a long evolutionary history, and there are now species living in most marine environments. Most of the bottom-dwelling species are epifaunal but a few exceptional species have adapted to live within sand or mud. Although there is considerable diversity within the group, all benthic cnidarians are characterized by having a radial symmetry, and all are suspension feeders that capture prey using nematocyst-laden tentacles. Some species also trap very small particulate food in mucus secreted onto their oral surface. Benthic cnidarians are sessile animals, although some sea anemones are capable of detaching from a substrate and swimming temporarily to escape starfish predators. Asexual as well as sexual reproduction is common in this phylum.

Within the Phylum Cnidaria, the Class **Hydrozoa** includes the colonial hydroids (Figure 7.3), formed of unions of structurally and functionally different types of individuals. Although they are usually small and inconspicuous, a large part of the marine growth attached to rocks, shells, and wharf pilings and usually called 'seaweed' is actually composed of hydrozoan colonies. Some hydroids produce free-swimming medusae as part of their life cycle, but in the majority of species the medusa remains attached to the parent, where it functions as a sexually reproducing individual. The much larger Class **Anthozoa**, with over 6000 species, includes sea anemones and a variety of corals, as well as less familiar forms such as sea whips and sea fans. Sea anemones (Figure 7.5c) are common residents of intertidal and subtidal communities, but are also found at over 10 000 m depth; they are solitary animals, ranging in diameter from about 1 cm to more than 1 m. Included among the Anthozoa are a variety of taxonomically different forms called 'corals'; the important subgroup of stony corals that form massive reefs in tropical regions is considered in more detail in Section 8.6.

Benthic worms belong to a number of different phyla. The threadlike **nematodes** (Phylum Nematoda) constitute one of the most numerous and widespread groups of marine (and terrestrial) animals, although most of the species are inconspicuous inhabitants of soft sediments. A single square metre of bottom mud off the Dutch coast was reported to contain about 4 500 000 individual meiobenthic nematodes. Taxonomic problems have hampered ecological research on this abundant group, but it appears that there is a wide diversity in feeding types with some species being carnivorous, others feeding on plants, or on decaying material and associated microfauna. The Phylum **Nemertea** (Figure 7.5d) encompasses about 600 species of elongated worms, all characterized by having a long eversible proboscis that is used to capture food. Nemerteans are more abundant in temperate seas than in tropical areas, and they are more common in shallow zones. Free-living **flatworms** (Phylum Platyhelminthes) (Figure 7.5f) reside in sand or mud, under stones and shells, or on seaweeds, but they are seldom present in large numbers. **Sipunculids** (Phylum Sipuncula) (Figure 7.5i), also called peanut worms, are unsegmented worms ranging in length from about 2 mm to more than 0.5 m. Many of the 250 or so species burrow into sand or mud, using movements of their large proboscis to force their way through the sediments; others inhabit rock or coral crevices, or even empty snail shells. They are mostly deposit feeders. **Echiurids** (Phylum Echiura) (Figure 7.5h) are somewhat similar to sipunculids in size and general habit. Most species use their large nonretractible proboscis to forage for food contained in sediments. Although some species occur intertidally, most are found only in very deep water habitats. The majority of deep-sea echiurid species have dwarf parasitic males attached to the female, a mode of reproduction that is reminiscent of that of the deep-sea angler-fish (see Section 6.6).

More than 10 000 species belonging to the Phylum Annelida, Class Polychaeta, make up the largest and most diverse group of marine worms. **Polychaetes** (Figure 7.5e) are the segmented worms with multiple appendages called parapodia. Size ranges from a few millimetres to 3 m in length. Ecologically, polychaetes can be separated into those that move actively over the seafloor or burrow into sand and mud, and those that inhabit permanent tubes or burrows. Most crawling species and some of the active burrowers are carnivorous and feed on various small invertebrates that are captured with the predator's jaws. Some polychaetes also use their jaws to tear off pieces of algae. Many burrowers and some tube dwellers are deposit feeders that consume sand or mud directly by mouth. Other deposit-feeding species have developed special tentacle-like structures that extend onto or into the substrate; sediment particles adhere to mucous secretions on the surface of these structures, and this material is then conveyed to the mouth by cilia. As well, many of the sedentary species are filter feeders, using special head appendages to collect plankton and suspended detritus. This group, with both epifaunal and infaunal species, frequently forms a large fraction of the benthic biomass in many habitats.

The **pogonophora** (Figure 7.5j) are regarded as specialized annelids by some workers, or as a separate phylum by others. These sessile worms are most abundant in deeper areas, occurring down to 10 000 m. They secrete long leathery tubes that are attached to hard substrates. A cluster of tentacles projects from the tube, and the common name of 'beard worms' is derived from this feature. The pogonophora are highly unusual in lacking a mouth or gut, but they share these features with the **vestimentiferan worms** that are described in Section 8.9 and shown in Colour Plate 39. Both groups depend on symbiotic chemosynthetic bacteria for their nutritional requirements, although pogonophorans may also utilize dissolved organic substances absorbed through their tentacles.

The Phylum Hemichordata includes the **enteropneusts** (Figure 7.5g), or acorn worms, which occur intertidally as well as at deep-sea hydrothermal vents (Section 8.9) and in trenches (Section 8.8). The largest species reportedly attains lengths of over 1.5 m, but most are much smaller. Many live in burrows in mud and sand, others move sluggishly over the sediment surface, or form entanglements on firm substrates. Burrowing forms use their proboscis to plough through the sediments, and most ingest sand or mud from which the organic matter is digested. The amount of substrate consumed is indicated by the large accumulations of faecal castings that accumulate at the posterior opening of the burrow. Nonburrowing species and some burrowers are suspension feeders; plankton and detritus stick to the mucus-covered proboscis and then are transported in ciliated grooves to the mouth. Enteropneusts are difficult to collect because of their fragility, but deep-sea cameras have recorded their trails and tangled masses on the seafloor.

QUESTION 7.2 Table 7.1 lists seven phyla of different types of marine worms. What features do these animals share to classify as a 'worm'. and why do you think they are such a successful marine benthic group?

Members of the Phylum **Mollusca** include over 50 000 marine species, among them the familiar snails and related nudibranchs or sea slugs (Class Gastropoda), and the bivalved clams and mussels (Class Bivalvia). This phylum also includes the flattened chitons (Class Polyplacophora) (see Figure 8.2), with a shell divided into eight plates. Less well-known members are the burrowing scaphopods (Class Scaphopoda) (Figure 7.7a) with tusk-shaped shells, and the wormlike, shell-less aplacophorans (Class Aplacophora) found within sediments. Most of the octopus species (Class Cephalopoda) are also essentially benthic species, although they are capable of swimming. The great diversity in this phylum is expressed in the facts that molluscs inhabit all depths of the sea, are found both on and within sediments, have representative species in all trophic levels, and are present in all benthic communities.

Echinoderms (Phylum Echinodermata) (Figure 7.6) are exclusively marine animals. Although differing in external appearance, all echinoderms are characterized by having radial symmetry, with the body divided into five parts around a central axis; a skeleton composed of calcareous plates; and tube feet. The approximately 5600 species are divided into five classes. The Class **Asteroidea** includes roughly 2000 species of starfish (or seastars) (see Figure 8.2) whose habitats range from intertidal zones to about 7000 m. Many starfish are carnivorous, and they may have considerable ecological impact on cultivated shellfish beds as well as in natural habitats. Other starfish species are deposit feeders or, more rarely, suspension feeders. The Class **Ophiuroidea** comprises almost 2000 species of long-armed brittle stars (Figure 7.6c) and basket stars. Deep-sea photographs often show ophiuroids carpeting the seafloor, where they feed on deposited sediments, on small dead or living animals, or on suspended organic material. Some 800 species of spiny sea urchins (Figure 7.6d) and flattened sand dollars are placed in the Class **Echinoidea**. Urchins are conspicuous members of the macrobenthos of rocky shores, kelp beds, and coral reefs; they use a special chewing apparatus to feed on all types of organic material, but most shallow-water species are regarded as basically herbivorous and deep-sea species (to about 7000 m) are considered to be deposit feeders. Sand dollars are inconspicuous, but often numerous, infaunal species; some are capable of suspension feeding as well as deposit feeding. The Class **Holothuroidea** (with 500 species) includes the elongated sea cucumbers (Figure 7.6b), named because of their resemblance to the vegetable. The epibenthic species of holothurians can be either deposit or suspension feeders; infaunal species swallow the sand or mud in which they live. Although also found in shallow waters, the greatest number of abyssal echinoderms are sea cucumbers, and they feature prominently in deep-sea photographs. The Class **Crinoidea** is the most ancient echinoderm group, and presently includes about 650 species of animals known commonly as feather stars and sea lilies. Feather stars

long e thin e wiggle.

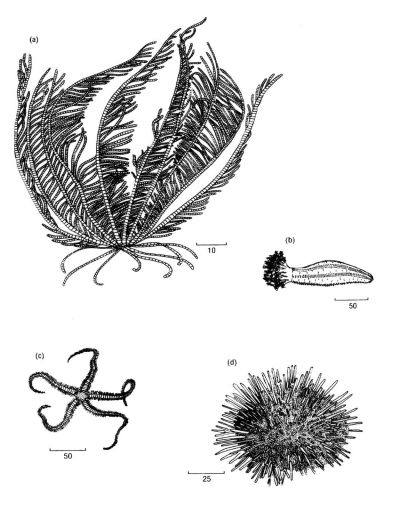

Figure 7.6 Representative benthic echinoderms: (a) feather star (crinoid); (b) sea cucumber (holothurian); (c) brittle star (ophiuroid); and (d) sea urchin (echinoid). (All scales in mm.)

(Figure 7.6a) live mostly in depths above 1500 m, and although they often cling to the seafloor, they are mobile animals that are capable of crawling as well as of swimming temporarily. Sea lilies are attached by stalks to the seabed, and they are typically deep-sea inhabitants that are most abundant between 3000 m and 6000 m. All crinoids are considered to be suspension feeders.

Bryozoa (Figure 7.7b), or moss animals, belong to the Phylum Ectoprocta. Like the hydroids, bryozoans are colonial and sessile animals that form inconspicuous encrustations or seaweed-like growths on intertidal rocks, shells, or artificial surfaces. A few species have also been recorded from depths of more than 8000 m. Each individual in a colony is small (usually <0.5 mm long) and, in the majority of species, is largely encased in an external calcium carbonate skeleton. A special food-trapping device, called a lophophore, consists of numerous ciliated tentacles, and it can be projected into the overlying water to capture small planktonic food or suspended detritus. Despite including almost 4000 marine species, the group has received little attention from ecologists.

Brachiopods (Figure 7.7c) constitute a phylum (Brachiopoda) of somewhat less than 300 marine species that superficially resemble molluscs in having a

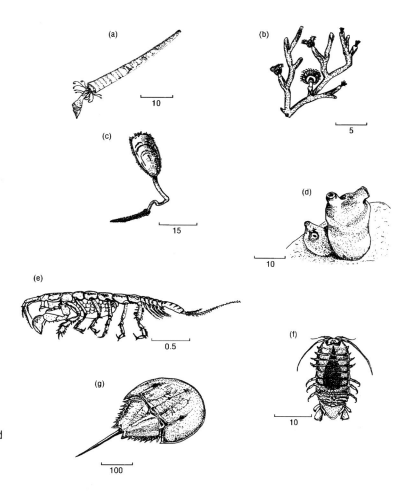

Figure 7.7   Representative benthic animals:
(a) scaphopod; (b) bryozoan colony; (c) stalked
brachiopod; (d) two tunicates; (e) tanaid;
(f) isopod; and (g) a horseshoe crab (not
related to true crabs). (All scales in mm.)

bivalved calcareous shell (5–80 mm in diameter), though their fundamental
body-plan is quite different. Most live in the upper 200 m, and most are
cemented to a hard substrate. However, some of the more common
brachiopods (e.g. *Lingula*) live in vertical burrows in sand or mud, and some
have been collected from 5500 m. Like the bryozoans, brachiopods possess a
lophophore that is employed in suspension feeding.

Tunicates (or **ascidians**) (Figure 7.7d) are benthic relatives of the pelagic
larvaceans and salps, described in Section 4.2. These sessile barrel-shaped
animals belong to the Phylum Chordata, Class Ascidiacea. Most of the
common tunicates are solitary organisms, but there are also many colonial
species that develop by asexual budding. Ascidians are commonly found in
intertidal waters, attached to rocks, shells, wharves, or other firm substrates,
but they also inhabit depths to at least 8000 m. The free end of a tunicate
has two siphons that provide passage for a current of water drawn through
the animal by cilia. Suspended particles are removed from the water by a
sheet of secreted mucus, and entangled food is conveyed to the gut by cilia.
Filtered water is forcibly expelled from the animal's excurrent siphon, giving
rise to the common name of 'sea squirt'. Deep-sea forms exhibit
modifications in the feeding apparatus, and they are thought to feed on
suspended sediments, or even directly on meiobenthos. In general, however,
the suspension-feeding mechanism of these benthic tunicates is similar to
that employed by their pelagic relatives, the salps. Although ascidians are
generally not major elements of benthic communities, they are capable of

removing significant amounts of plankton or suspended material from water in proximity to the seafloor; a single tunicate only a few centimetres long can filter about 170 litres of water daily.

The segmented **Crustacea** are well represented on the seafloor. Meiobenthic species include **ostracods** and the **cyclopoid** and **harpacticoid copepods**, all of which were described briefly in Section 4.2. Harpacticoids are an especially abundant group whose members crawl over or burrow through soft sediments. Also included in this size category are the **tanaids** (Figure 7.7e). These small (usually <2 mm long) crustaceans have slender, more or less cylindrical bodies, and they are burrowers or tube dwellers present at all depth levels down to at least 8000 m. Little is known about the biology of these 350 species.

Common macrobenthic crustaceans include the **isopods** (Figure 7.7f) and **amphipods**. Isopods usually have a flattened body of 5–15 mm in length; deep-sea species are generally larger, however, with one genus reaching 40 cm. Isopods are often observed running rapidly over rocks in the intertidal zone, but some species are burrowers, including one group that tunnels into wood. The majority of the 4000 marine species are omnivorous scavengers. Amphipods are closely related to isopods, but differ in that most have a laterally compressed body (see Figure 4.10 k for planktonic species). They range from a few mm to about 30 cm in size, with the largest species occurring in the deep sea. Depending on the species, amphipods are capable of crawling or burrowing, but many of the bottom-dwellers are also capable of swimming, even if infrequently. Many of the species construct temporary or permanent burrows or tubes. The depth distribution in this group ranges from the semi-terrestrial beach fleas living near the high tide level to species living in hadal trenches. Most amphipods are detritivores or scavengers, but a few are specialized filter feeders.

Barnacles (Cirripedia) (see Colour Plate 31) are familiar marine animals and the only sessile crustaceans. There are about 800 species, including a large number that are parasitic in other marine invertebrates. These shrimplike animals live within an external covering of calcareous plates. Some attach directly to substrates, others are stalked. The more familiar barnacles form crowded aggregations in rocky intertidal regions, but some species have become specially adapted to attach to mobile surfaces and live on the bodies of whales, sharks, sea snakes, manatees, fish, or crabs. Although most common in shallower depths, some species are present to at least 7000 m. Free-living (i.e. nonparasitic) barnacles feed by rhythmically sweeping their feathery appendages through the surrounding water. These animals often foul the bottoms of ships and the surfaces of buoys and wharf pilings. They have been the subject of extensive ecological studies since Charles Darwin's classic taxonomic monograph on this group.

Benthic **decapod** crustaceans include the familiar crabs, lobsters, and shrimp, and the group has both epifaunal and infaunal representatives. Decapods show their greatest diversity in shallower waters, but a few species live at depths of 5000–6000 m. The group includes predators, omnivores, and scavengers. Some are filter feeders (e.g. burrowing mud shrimp and mole crabs), but detritus rather than plankton is often the dominant food. Many species in this group are economically important as human food, and those species along with molluscs make up the shellfish industry.

There are a few other phyla of marine animals that have not been considered here, either because they consist of only a few species or because they are not usually abundant in benthic communities. In some cases, we still lack an understanding of the ecological role of certain obscure animal groups simply because present sampling gear does not adequately collect them. Many new species (as well as new families) have been discovered within the last 25 years, primarily in the deep sea (see Sections 8.8 and 8.9), and this list can be expected to continue to grow.

QUESTION 7.3 From information in the previous section, what are some of the mechanisms used by benthic animals for defence against predation?

7.2.2 SAMPLING AND PRODUCTION MEASUREMENTS

Benthic animals can be collected by a variety of sampling gear. In shallow waters less than about 30 m, direct observation, counting, and collection by scuba diving are effective techniques that permit a diver to sample any type of substrate and provide a greater appreciation of the natural conditions, including patchy distribution of the macrobenthos. In depths beyond safe diving limits, **grabs** or **box corers** are commonly used to sample soft sediment communities. Ideally, these devices remove a quantitative bottom sample from which the resident animals can be screened out on a series of sieves. **Hydraulic cores** of undisturbed samples can be obtained but because these are small, they are used only for quantitative sampling of meio- or microbenthos. **Dredges** are containers equipped with sturdy bags of fairly wide mesh that are pulled over the seafloor; they are designed to scoop up benthic animals, but they do not generally provide quantitative samples. Other types of sampling devices are mentioned in Chapter 8 in connection with special environments.

Standing stock of benthos is given in terms of numbers of animals per unit area (usually m^{-2}). Benthic biomass is expressed in terms of g m^{-2}; ideally, these data are presented in grammes of carbon or grammes of dry weight, but they may be given in grammes of wet weight (for conversions, see Appendix 2). Biomass changes constantly, and measures of secondary productivity take this into account. Biomass increases through growth and reproduction of the individuals in a population or community; existing biomass is removed by predation or other sources of mortality. Under certain circumstances, these changes can be monitored over successive intervals, and production can be determined from the methods given in Section 5.3.1 (see especially equations 5.3, 5.4, and 5.5).

Age classes may be distinguishable in populations that breed during a well-defined time period, or from age-size relationships, or in species that have age-related features such as growth rings on shells. If the biomass of each age class is determined and changes in biomass are followed by successive sampling, it is possible to calculate production during the interval between samples. Table 7.2 presents production data collected over a period of 616 days from a population of infaunal clams in the North Sea. In the first interval of 50 days, mortality was very high but the mean weight of the surviving individuals increased by about 4 mg; this resulted in a decrease in biomass, but a net production of 317 mg m^{-2} day^{-1}. Completing the calculations in the table will show that proportionally more biomass was produced, but then lost from the population due to predation or other sources, during the first two intervals than later in the study.

Table 7.2 Production data from a population of the clam *Mactra* in the North Sea.

t (days)	X (no. m^{-2})	Mean weight (mg)	Biomass (mg m^{-2})	Biomass loss (mg m^{-2} day^{-1})	Net production (mg m^{-2} day^{-1})
0	7045	1.416	9976		
				411	317
50	990	5.364	5310		
				27	18
225	378	9.910	3746		
				14	66
398	289	44.286	12799		
616	246	73.542	18091	12	36

QUESTION 7.4 In Table 7.2, biomass has been calculated only for the first two age classes, and loss of biomass and net production only for the first interval of 50 days. Complete the rest of the table using equation 5.3 to calculate biomass, and equation 5.4 to calculate biomass losses and net production. (Refer to Section 5.3.1 for help if necessary.)

If it is not possible to follow a single cohort of animals and there are numerous overlapping age classes, benthic secondary productivity becomes more difficult to assess directly. In such situations, laboratory studies can be used, if feasible, to determine typical growth rates of each arbitrary size class. The population is then divided into component size classes regardless of age, and production is calculated by multiplying the numbers present in each size class by the growth rate of that class.

7.3 DETERMINANTS OF BENTHIC COMMUNITY STRUCTURE

The numbers and types of organisms making up any particular benthic community are determined by a variety of physical and biological factors. Among physical factors, sediment type is of paramount importance in dictating the type of community that can become established in a particular locality. The relative proportions of epifauna and infauna will vary according to whether the seafloor surface is composed of rock, mud, sand, or some combination of these types. At the same time, however, the activities of the organisms themselves will alter the substrate. Burrowing and feeding activities of the infauna continually disturb and rework soft sediments (a process called **bioturbation**), and sessile organisms that attach to hard surfaces alter the topography of their environment. In shallow coastal waters, tidal levels, degree of wave action, and salinity and temperature variations will influence the composition and relative abundance of benthic species.

Biological determinants of benthic community structure include competition, predation, and the type of development leading to recruitment into the adult population. As in pelagic communities (Section 5.2), competition among individuals or among species for limited resources (e.g. food) may affect population densities and relative abundance of species. Competition for limited space in hard-substrate communities may influence the distribution patterns of sessile zoobenthos. Selective feeding by predators can alter the outcome of competition between prey species, and predation may affect total species diversity in benthic communities. The effects of competition and predation in specific types of communities are discussed in more detail in Chapter 8.

Community structure also can be influenced by the type of development that is exhibited by dominant species. Benthic animals may have direct development, in which there is no free-swimming larval stage and the young emerge as miniature adults, or more commonly, they may have indirect development with the production of meroplanktonic larvae (see Section 4.3). Two types of meroplanktonic larvae are distinguished. **Lecithotrophic larvae** hatch from relatively large eggs that are produced in small numbers by the parent. They spend a very brief time in the water column (hours to 2 weeks) and usually do not feed on planktonic food, depending instead on yolk in the egg for their growth and development. **Planktotrophic larvae** hatch from small eggs produced in large numbers; they rely on phytoplankton or bacteria for nourishment and remain in the pelagic environment for several weeks or months.

Lecithotrophic larvae, which spend a very short duration in the plankton, do not usually drift far from the adult population which produced them, but planktotrophic larvae remain in the water column for longer times and can be transported by currents over long distances. Consequently, planktotrophic larvae in particular are a means of dispersal for species that are immobile or slow-moving as adults. Both larval types are vulnerable to pelagic predators while they are planktonic, and planktotrophic larvae especially may be carried away by currents from suitable benthic substrates upon which they could successfully settle. Lack of abundant planktonic food may also contribute to the higher mortality rates of planktotrophic larvae. As meroplanktonic larvae develop and grow, they become photonegative and begin to move deeper in the water column, toward the seafloor. Many, if not all, pelagic larvae actively search for a suitable substrate before selecting one on which to settle and metamorphose into the adult form. Larvae seeking settlement sites rely on a variety of cues such as substrate type, suitable food, light intensity and, in many cases, on the presence of adults of their own species.

The relative success of larval recruitment into adult populations will influence community structure in terms of relative numbers and biomass of the resident species. Species that produce planktotrophic larvae often have variable recruitment success; in successful years, many larvae may survive to settle in the adult population, but in other years, larval mortality may be high and recruitment low. In comparison, populations of species with lecithotrophic larvae (or those with direct development) tend to have low but constant biomass over long periods.

QUESTION 7.5 Benthic invertebrates with direct development (no pelagic larval stage) may nevertheless be widely distributed. Can you think of any way(s) in which such species could disperse to new geographic localities?

on drift wood ships ete

7.4 SUMMARY OF CHAPTER 7

1   The benthic environment is divided into a number of distinctive ecological zones based on depth, seafloor topography, and vertical gradients of physical parameters. These are the supralittoral, littoral, sublittoral, bathyal, abyssal, and hadal zones.

2   Benthic plants include macrophytic angiosperms like mangrove trees, marshgrasses, and seagrasses. Macrophytic algae include green, red, and brown seaweeds, and the long-stemmed kelps, a type of brown algae. Microphytic algae include benthic species of diatoms, Cyanobacteria, and dinoflagellates.

3   The number of phyla and the number of species of benthic animals exceeds those of pelagic species, at least partly because of the greater physical variety of benthic habitats.

4   Benthic animals are separated into infaunal and epifaunal species, depending upon whether they live within sediments or on the surface of the seafloor, respectively. Size categories of the zoobenthos consist of the larger macrofauna (>1.0 mm), the small meiofauna which is characteristically found in sand and mud, and the microfauna which is made up mostly of protozoans.

5   Benthic primary productivity is measured by a variety of methods including the carbon-14 method for microphytic species, and harvesting and weight measurements of macrophytic plants. Methods of estimating benthic secondary production are similar to those employed for pelagic animals and described in Chapter 5.

6   The numbers and types of species making up any particular benthic community are determined by a variety of physical and biological factors. In shallow coastal communities, the types of species present and their relative abundance will be partly determined by tidal levels and degree of exposure to air, wave action, and range of salinity and temperature. At all depths, the type of sediment (e.g. sand, rock, mud) will dictate the relative proportions of epifauna and infauna. Biological factors that influence benthic community structure include competition for limited resources (e.g. food, space), predation, and type of development.

7   Benthic animals may have direct development, in which there is no free-swimming larval stage, or they may produce pelagic planktotrophic or lecithotrophic larvae. Planktotrophic larvae are relatively small and are produced in large numbers; they must feed on planktonic food, and they remain in the water column for several weeks or months. Although they are a means of dispersal for the species, planktotrophic larvae have high rates of mortality, and the adult populations have variable rates of recruitment from year to year. Lecithotrophic larvae hatch from relatively large eggs that contain large amounts of nutritive material and that are produced in small numbers; these larvae do not remain planktonic for long, and they do not feed while in the water column. Compared with planktotrophic larvae, mortality rates are lower for lecithotrophic larvae, and species with this type of development tend to have low but constant biomass because recruitment is less variable.

Now try the following questions to consolidate your understanding of this Chapter.

QUESTION 7.6   Compare Table 7.1 with Table 4.1 (a) How many phyla are represented in the plankton and in the benthos? (b) Why is there a difference in the numbers of phyla in these major marine environments?

QUESTION 7.7   Would you consider benthic species that produce lecithotrophic larvae to have predominantly *r*- or *K* -selection characteristics? Explain the bases for your answer. (Refer to Section 1.3.1 and Table 1.1.)   *K* .

QUESTION 7.8   Stromatolites (see Section 7.1) in Hamelin Pool, Western Australia, reach a maximum height of 1.5 m. Assuming no erosion or change in sea level, how old are these structures?

$$1.5 m = \frac{1500 mm}{0.5 mm \text{ per year}}$$

$$= 3000 \text{ years.}$$

This chapter considers different types of benthic communities, ranging from the highest intertidal levels to the deepest trenches. Although these benthic habitats are treated separately here, they are all coupled in a dynamic fashion with the overlying pelagic environment.

In shallow water, both phytoplankton and benthic plants contribute to the primary production of benthic communities. Phytoplankton and zooplankton are consumed by shallow-living filter feeders. Conversely, exudates and detritus from attached benthic plants supply necessary nutrients for the phytoplankton and planktonic bacteria, and bottom currents may cause resuspension of sediments, making benthic microphytes and bacteria-covered sediment particles available as food sources for zooplankton. Some fish and some marine mammals rely on shallow-water benthos for food.

The great majority of benthic communities, however, are located in the aphotic zone, and most are entirely dependent on organic matter that is photosynthetically produced in the euphotic pelagic zone. The only exceptions are certain communities in which the food chain begins with chemosynthetic production by bacteria (see Section 8.9). Part of the organic matter that sinks or is transported from the surface waters is the food source that supplies deep-water benthic communities. Sinking organic and inorganic particles also form the sediments in which the benthos live. Decomposition processes tend to take place in deep water or on the seafloor, and the nutrients that are released are eventually returned to the surface where they are used by the phytoplankton.

Marine organisms may exploit both the benthic and pelagic environments during different stages of their lives. Many invertebrates are benthic as adults, but disperse by producing planktonic larvae. Conversely, some planktonic organisms produce resting stages (spores, cysts, or eggs) that sink into sediments where they remain dormant until favourable water conditions cause them to 'hatch' into swimming or drifting stages.

The concept of **benthic–pelagic coupling** recognizes the many interactions between these two vast environments, and attempts to integrate the ecologies of the seafloor and the water column.

8.1 INTERTIDAL ENVIRONMENTS

The terms littoral and intertidal are synonyms for that coastal area which is periodically exposed to air by falling tides and submerged by rising water levels. Included in this general area are a variety of distinctive ecosystems such as rocky shores, sandy beaches, and mudflats, each supporting specially adapted assemblages of species. Rocky intertidal areas support a preponderance of epifauna, whereas the soft-substrate sand and mud communities have higher proportions of infaunal species. The intertidal regions mark the transition from land to sea, and although they make up only a very small part of the total world ocean, they support rich and diverse communities of marine plants and invertebrates as well as birds and inshore species of fish. Even some land mammals (e.g. mink, skunks, and raccoons) visit this area to feed on easily accessible shellfish, and many shorebirds depend on the rich food supply to be found in these habitats.

8.1.1 TIDES

Tides are the periodic rise and fall of sea level over a given time interval, and they are caused by the interaction between the gravitational attraction of the Moon and Sun on the Earth, and the centrifugal force resulting from the rotation of the Earth and Moon. On most coasts, **semidiurnal tides** result in the intertidal zone being exposed and covered by water twice each day. However, because of certain physical conditions, there is only one tide per day (a **diurnal tide**) in some regions such as the Gulf of Mexico.

Tidal range is greatest during **spring tides** which occur twice each month when the Earth, Sun, and Moon are aligned. At the other extreme, tidal range is minimal during the **neap tides** which occur at the first and the third quarters of the Moon, when these planetary bodies are not in alignment. The high water mark is the greatest height to which the tide rises on any day, and the low water mark refers to the lowest point to which the tide drops.

The extent of the littoral zone in any particular locality is governed by the slope of the shoreline and by local tidal ranges, which are partly determined by the configuration of coastlines. Tidal range varies from barely perceptible in places such as Tahiti and the Baltic Sea, to as much as 15 m in the Bay of Fundy in eastern Canada.

8.1.2 ENVIRONMENTAL CONDITIONS AND ADAPTATIONS OF INTERTIDAL ORGANISMS

The littoral regions experience the greatest variations in environmental conditions of any marine areas. Here, organisms are periodically exposed to air, and they encounter wide fluctuations in temperature and salinity. Rainfall and land runoff both contribute to lowering salinity. In cold climates, intertidal organisms are subjected to ice formation and ice scouring. In addition, many intertidal regions are exposed to heavy wave action and current motion.

Intertidal plants and animals show a variety of special adaptations to the changing conditions of their environment. Whereas inhabitants of sand and mud tend to burrow into the soft substrates to escape desiccation, temperature and salinity extremes, and wave action, organisms living on rocky shores exhibit more diverse adaptations to these environmental features. Rocky-shore species of bivalved molluscs (e.g. clams, mussels) and barnacles close their shells tightly during emersion, enabling them to retain moisture around the gills and thus preventing desiccation as well as exposure to freshwater. Many snails also retreat into their shells, sealing the shell aperture with a horny or calcareous operculum on the foot. On the other hand, many benthic plants and some of the intertidal animals have no particular mechanisms to avoid water loss. Algae like *Fucus* and *Enteromorpha*, for example, tolerate as much as 60–90% loss of water from their tissues.

Shells, or other types of rigid exoskeletons like those of sea urchins, also protect animals from mechanical injury in areas where wave action can be severe. In some sea urchins and molluscs, the shell is much thicker in populations exposed to heavy wave action than in populations which are sheltered. Strong attachment to rock surfaces or other firm substrates prevents plants and animals from being washed away by waves and currents. Benthic algae attach to rocks by special holdfasts. Barnacles, oysters, some tubeworms, and tunicates secrete cementing substances for firm attachment. Mussels secrete tough elastic byssal threads from a special gland in the foot, and these secure their positions. The broad, flattened foot of limpets, abalones and chitons provides a suction-like attachment, and their low, streamlined profiles also help to resist wave impact. Certain animals (e.g. some sea urchins and rock-boring clams) are equipped to bore into hard surfaces by mechanical abrasion, chemical secretions, or both. Many of the more mobile intertidal animals, like crabs and isopodes, seek out rock crevices where the wave action is reduced; this sheltering behaviour also permits them to remain in moist refuges at low tide. Rock pools form similar refuges for animals such as starfish, crabs, and small intertidal fish, all of which avoid desiccation by remaining in these pools at low tide.

8.2 ROCKY INTERTIDAL SHORES

Much is known about the inhabitants and ecology of rocky shores compared with other marine habitats. The accessibility of these densely-populated marine communities has permitted researchers to make long-term direct observations, and to conduct *in situ* experiments on factors determining community structure.

8.2.1 ZONATION

A striking characteristic of all rocky shores is that the resident plants and animals are grouped in distinctive bands, with some species living high in the intertidal zone and others being grouped at lower tidal levels. This vertical **zonation** of species applies to all rocky intertidal communities, although the specific pattern of zonation and the species composition of the zones varies according to geographic location, tidal range, and whether sites are exposed to severe wave action or are protected. Zonation is largely based on sessile species, like algae, barnacles and mussels, although some mobile animals also tend to be zoned but with less sharp demarcation. In general, many of the larger motile animals move with the tides and often remain in a relatively constant water depth, or retreat to rock pools at low tide.

On rocky shores, the supralittoral zone (see Chapter 7) is inhabited by encrusting black lichens (which are combinations of algae and fungi) and blue-green algae, certain species of *Littorina* (periwinkles) that graze on the vegetation, and relatively large (3–4 cm long) isopods (*Ligia*). Primitive insects (e.g. *Machilis*) may also be present.

Just below the supralittoral zone, periwinkles (*Littorina* species) are usually found in extraordinarily dense populations, with numbers ranging from several hundred to 10 000 snails per square metre. Lower in the intertidal, barnacles form a sharply demarcated belt, and these crustaceans may also have densities of thousands per m². In many localities, mussels crowd together in dense aggregations below the barnacle zone. There is intense competition for the limited space among the attached algae and sessile animals as shown in Colour Plate 31.

Gregariousness is an adaptive feature in many of these intertidal species. By crowding together, periwinkles create microhabitats in which more moisture is retained during exposure at low tides. They reproduce by internal fertilization, and crowding also increases their chances of finding mates. Mussels freely release gametes into the sea, where fertilization occurs; in this case, gregariousness increases synchrony of spawning among many individuals. Barnacles are hermaphrodites that cross-fertilize, and high population densities are necessary for reproduction. The penis of a barnacle can only reach about twice as far as the diameter of its exoskeleton, so it is essential that the animals be in close proximity.

Intertidal zonation of organisms is not determined simply by tidal levels, but results from a variety of physical and biological causes. The upper limit of a particular zone is often determined by physical factors and the ability of particular plants and animals to deal with exposure to air and with temperature and salinity variability. The upper limit of any one species may also be set by biological factors such as the absence of suitable food, or grazing or predation pressure. The lower limit of a particular zone is usually determined only by biological factors.

How can the causes of vertical zonation be determined?

Sessile animals, like barnacles and mussels, make ideal subjects for studies of vertical zonation because the same individuals can be monitored over long periods. Photographs taken at successive intervals record the size and position of individuals and enable researchers to follow growth, interactions with neighbouring individuals, and death. The accessibility of the intertidal zone also makes experimental manipulation possible, and individual animals may be removed from sites, or whole rocks with attached organisms may be relocated to different tidal levels. It is also possible to exclude predators by enclosing study populations within wire cages. The results of such techniques are described below.

Barnacle populations on rocky shores in Scotland are composed of two species (Figure 8.1). Adults of the small *Chthamalus montagui* form a distinct band high on the shore, above the mean high water mark of neap tides, with only a few adults being found down to the mean tide level. The larger *Balanus balanoides* has a much wider distribution for both adults and larvae, extending above and below the mean neap tide marks. Long-term observations indicated that the larvae of *Chthamalus* actually settle throughout most of the *Balanus* zone, but only survive to adulthood in the upper areas. This is because young *Chthamalus* are eliminated from areas below the mean high water neap mark by competition for restricted attachment space with the faster growing *Balanus*. *Balanus* either overgrows, undercuts, or crushes *Chthamalus*. The observation that the lower limits of *Chthamalus* zonation are dictated by biological competition was confirmed by experimental work in which all newly settled *Balanus* larvae were removed from rocks containing young *Chthamalus*; in such studies, *Chthamalus* survived well at all tidal levels. On the other hand, the upper zonation levels for both barnacle species are determined by physical factors, with *Chthamalus* being more tolerant to heat and/or desiccation than *Balanus*. Further, intraspecific competition may be an important mortality factor for *Balanus*; if larval recruitment is high, the growing barnacles begin to compete for space, and the younger or slower-growing individuals of the species are eliminated.

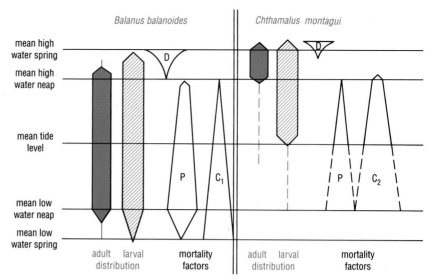

Figure 8.1   The effects of competition and predation on barnacle distribution in Scotland. C_1, intraspecific competition; C_2, interspecific competition between *Chthamalus* and *Balanus*; D, desiccation; P, predation by *Nucella*, a predatory snail. The widths of the distribution bars indicate relative abundance; the widths of the mortality bars indicate relative importance of the factors concerned. Note that the upper limits of distribution for both species are determined by physical factors (i.e. tolerance to dessication). Snail predation and intraspecific competition for space are the major causes of mortality for *Balanus*, and both factors become increasingly important at lower tidal levels. For *Chthamalus*, the major cause of mortality of newly settled larvae is intraspecific competition for space with the faster-growing *Balanus*. Few *Chthamalus* larvae settle below mean tide level, but those that do are eliminated by predation and interspecific competition.

Predation too may be an important determinant of zonation patterns. The whelk *Nucella lapillus* is a major predator of barnacles in Scotland. Like many predators, this snail prefers to eat larger prey, and thus prefers *Balanus balanoides* over *Chthamalus*. When cages were used to exclude *Nucella* from natural populations of barnacles, it became evident that snail predation was a major cause of mortality for larger (and older) *Balanus*, especially those living in the lower intertidal zone where *Nucella* was most abundant. Thus the lower limit of the *Balanus* zone is determined largely by predation.

Similar predator-prey relationships can be found on the west coast of North America between three barnacle species (*Chthamalus dalli, Balanus glandula* and *B. cariosus*) that are preyed upon by three different species of *Nucella*. In this area, most of the mortality of young *Balanus glandula* is the result of predation rather than crowding and competition for space, and *B. cariosus* attains an adult size that is too great to be eaten by *Nucella*. However, here too predation and competition for space act to set the lower limits on the barnacle distributions.

Physical and biological factors also act in concert or independently to set the zonation patterns of benthic algae. Algae compete for sunlight (see Section 7.1) and for restricted space with other plants and with animals, and these biological factors partly establish the position of plants on shores. Upper zonation limits of algal species are often set by tolerances for exposure and desiccation, and may also be determined by the grazing pressure of herbivores. As an example, the *Torrey Canyon* oil spill in 1967 killed the dominant grazing molluscs in the intertidal areas of parts of south-western England, and subsequently there was a rise in the upper zonation limits of several intertidal algae. With eventual recovery of the grazers, the higher reaches of the algae were once again grazed down and the original zonation pattern was re-established.

8.2.2 TROPHIC RELATIONS AND THE ROLE OF GRAZING AND PREDATION IN DETERMINING COMMUNITY STRUCTURE

Both benthic algae and phytoplankton are important primary producers supporting rocky intertidal communities, but production figures are relatively low. The intertidal zone is a difficult habitat for benthic algae in several respects. In tropical regions, heavy rainfall, high light intensities, and exposure to high air temperatures with resulting desiccation are major problems. Freezing and scouring by ice limit algal production in arctic and subarctic intertidal areas. In temperate climates, where benthic algae reach their full potential, there is competition between algal species for access to sunlight, and competition for attachment sites with other algae and with sessile animals. The average annual productivity of the rocky intertidal areas of the world is of the order of 100 g C m^{-2}. However, production rates of around 1000 g C m^{-2} yr^{-1} may occur in particularly favourable areas.

Attached algae are grazed by a variety of molluscs and sea urchins. Mussels, barnacles, clams, tunicates, tubeworms (polychaetes), and sponges are among the many filter feeders that are dependent on plankton. Intertidal carnivores include starfish, which eat limpets, snails, barnacles, mussels, and oysters; predatory snails, which consume a variety of prey including clams, mussels, and barnacles; and sea anemones, whose prey includes shrimp, small fish, and worms. Important scavengers include isopods and crabs. Shorebirds may also have considerable predation impacts on intertidal life (see Sections 6.5 and 8.5).

Experimental work has demonstrated that herbivores such as sea urchins, limpets, chitons, and littorine snails may control both the level of primary production and the species composition of benthic plants. For example, removal of limpets from experimental areas results in the appearance of different species of algae and in heavier algal growth compared with undisturbed sites. Removal of sea urchins from intertidal and subtidal regions also tends to create a greater initial diversity of algal species, although this may eventually change to lower diversity as other determinants of community structure come into play. The species composition of the algae in a community may also result from competition for space and light between different algal species, with the dominant species being those that are fastest growing in the particular locality.

Competition and predation are also important determinants of species composition and diversity among intertidal animals. Along the north-west Pacific coast of North America, the rocky intertidal community is dominated by mussels, barnacles, and the carnivorous starfish *Pisaster ochraceus*. *Pisaster* feeds on a variety of molluscs and barnacles, as illustrated in Figure 8.2. Experimental removal of the starfish from the community resulted in lowering species diversity from about 30 species to one dominant species, the mussel *Mytilus californianus*. When *Pisaster* is present, its feeding activities control the numbers of dominant sessile prey (barnacles and mussels) so that space is kept open and none of these species becomes dominant; at the same time, primary production is also enhanced by the provision of space for benthic algae. When the top predator is removed, competition for space is intensified and, in this region, *Mytilus californianus* overgrows and outcompetes all other macrobenthos to take over the available space throughout most of the mid-intertidal zone. *Pisaster* is referred to as a **keystone species** because its activities disproportionately affect the patterns of species occurrence, distribution and density in the communities in which

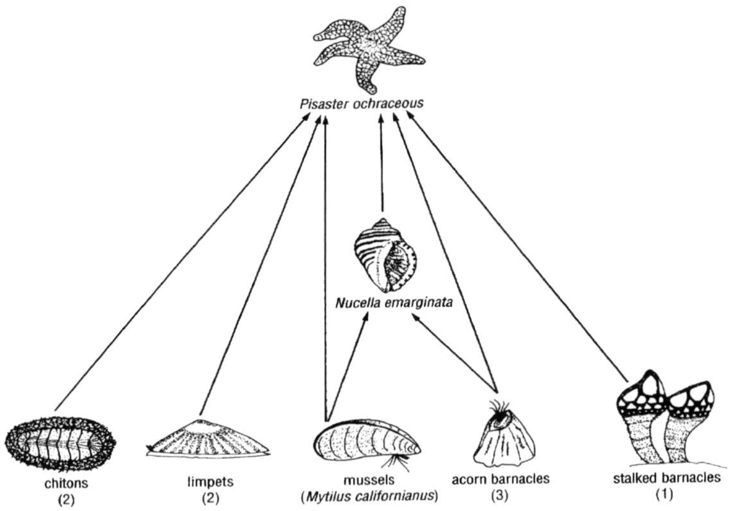

Figure 8.2 The food of the starfish *Pisaster ochraceus*, a keystone species of rocky intertidal communities along the Pacific coast of North America. Numbers in parentheses indicate the number of species in a particular group. (Animals are not to scale.)

High tidal areas as there is more nutrient & more space.

it lives. Similar types of interactions control rocky intertidal community structure in other areas of the world. In New England, for example, the mussel *Mytilus edulis* is the competitively dominant sessile species, whose numbers are usually kept in check by two species of starfish (*Asterias forbesi* and *A. vulgaris*) and by the snail *Nucella lapillus*.

QUESTION 8.1 Would you expect to find a greater biomass per unit area of benthic organisms in intertidal areas with a high tidal range (e.g. >2 m), or in those with a small tidal range (e.g. <0.5 m), and why?

8.3 KELP FORESTS

In cold temperate regions, intertidal rocky-shore communities merge subtidally into kelp forests. The term **kelp** refers to a variety of very large brown algae that are usually found only outside the limits of the 20°C summer isotherms. These algae form distinctive subtidal communities in areas of upwelling, fast currents, or heavy surf. Kelp require a hard substrate for attachment, and they grow off rocky shores to depths of 20–40 m, depending on water clarity. Kelp beds occur along the western coasts of North and South America, extending into subtropical latitudes in upwelling areas. In the western Pacific, extensive kelp beds are found off Japan, northern China, and Korea. In the Atlantic, large kelp beds occur off the Canadian east coast, and off the coasts of southern Greenland, Iceland, and northern Europe including the United Kingdom. The highest biomass levels of kelp are around subantarctic islands like the Falklands. New Zealand and South Africa are also localities that support kelp in quantities sufficient for exploitation.

Each kelp plant typically has a holdfast for attachment to the substrate and a flexible stipe (or stalk) (Figure 8.3). Large, thin blades (equivalent to leaves) are attached to the stipe. Kelp may also have gas-filled floats (or pneumatocysts) which keep the blades near the water surface where solar radiation is highest. Because of their large photosynthetic surface and the constant supply of nutrients in the surrounding turbulent water, kelp are highly productive plants. Common Pacific genera include *Nereocystis*, *Postelsia*, and *Macrocystis*. *Macrocystis pyrifera* is commonly known as the

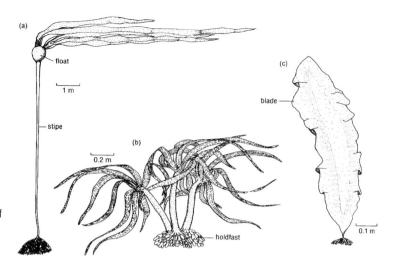

Figure 8.3   A number of kelp species illustrating diversity of structure in this group of brown algae. (a) *Nereocystis luetkeana*; (b) *Postelsia palmaeformis*; (c) *Laminaria saccharina*

giant kelp as it may exceed 50 m in length, and it forms aquatic forests off California. Various species of *Laminaria* (usually 3–5 m in length) are the dominant kelp along coastal areas of the North Atlantic, and they also occur as an understory species in the temperate Pacific.

Kelp not only are the largest of all algae, but some are considered to be among the world's fastest-growing plants. Growth rates of 6–25 cm per day are common, and *Macrocystis pyrifera* grows as much as 50–60 cm per day off California. Kelp can be either annual or perennial species, some regrowing new stipes and blades from the original holdfast either yearly, or every few years. All kelp reproduce via production of spores.

The high growth rates of kelp translate into productivities of from about 600 to more than 3000 g C m^{-2} yr^{-1} (compare with production values for rocky intertidal areas in Section 8.2.2). Off Amchitka Island in the Aleutians, annual kelp production is from 1300 to 2800 g C m^{-2}. This high production once supported populations of the giant Steller's sea cow (see Section 6.4). Off Nova Scotia, in Atlantic Canada, *Laminaria* forests produce about 1750 g C m^{-2} yr^{-1}. Kelp beds off South Africa have a production of about 600 g C m^{-2} annually. In some locations, kelp are harvested for fertilizer, iodine salts, industrial chemicals, and alginates used as food additives. The commercial harvest of the California giant kelp amounts to 10 000–20 000 tonnes dry weight per year.

Kelp communities provide spatial heterogeneity and diverse habitats and thus support a highly diverse association of animals. The large surface area of the kelp blades provides space for numerous epiflora and epifauna, including diatoms and other microflora, and colonial bryozoans and hydroids. A variety of molluscs, crustaceans, worms, and other animals live on the plants, or on the substrate between plants. In some areas, much of the primary production may be consumed by herbivores such as sea urchins. Some snails and sea slugs (e.g. *Aplysia*) also feed on kelp directly, but usually consume only very small amounts of the total production. This habitat also supports a variety of fish that feed on kelp-associated animals and that find protection from predators such as seals, sea lions, and sharks.

In many areas, however, as much as 90% of the kelp production is not consumed but enters the detritus food chain. The edges of the kelp are continually abraded by wave action, with small fragments being torn off, and there may be self-thinning with some plants losing the competition for available light, space, and nutrients. Annual kelp species (e.g. *Nereocystis* spp.) may attain a summertime biomass of more than 100 tonnes per hectare, all of which may be destroyed in the first winter storm. This material enters the detrital pool of the kelp bed or is exported to other areas. Kelp that is uprooted in storms may wash up in large quantities on beaches, where it is eaten by amphipods or isopods. Kelp also release considerable quantities of organic matter into solution, and this exudate is utilized by bacteria and thereby converted into particulate biomass (see Section 5.2.1).

The sea otter (*Enhydra lutris*) (Colour Plate 32) is considered to be a keystone species in North Pacific kelp forests. Otters eat a wide variety of prey including crabs, sea urchins, abalone and other molluscs, and slow-moving fish, and a single otter may eat 9 kg of food per day. Off Amchitka Island, otters at a density of 20–30 km^{-2} annually consume about 35 000 kg km^{-2} of prey. Otter predation on sea urchins regulates the ecological balance between kelp production and the destruction of the kelp by the herbivorous urchins.

Sea urchins (*Stronglyocentrotus* spp.) graze directly on living kelps, and they are capable of eating through the holdfasts that anchor the algae to the seafloor. The detached kelp is then swept away from the area in ocean currents. Otters, by direct predation, maintain relatively low densities of sea urchins and thus protect the kelp from overgrazing. The importance of otters in maintaining healthy kelp forests was revealed from ecological comparisons of different islands in the Aleutian chain. Certain islands off the Alaskan coast were found to have lush kelp communities with thriving populations of otters, seals, and bald eagles; other nearby islands had no kelp or otters, and few seals or bald eagles. The depauperate islands were the focus of extensive hunting expeditions during the eighteenth and nineteenth centuries, and historical records document the elimination of otters from many kelp beds during this time. Only those islands which had been repopulated with the few surviving otters had flourishing kelp forests. Since 1911, sea otters have been protected by law, and populations have recovered in some areas of Alaska and British Columbia. They have also been deliberately reintroduced into other depopulated regions, including along the California coast. Where otter populations have recovered, sea urchin numbers have decreased, and kelp production has increased. Despite protection, otters remain a vulnerable species. They are disliked by fishermen, who often perceive them as competitors for fish or shellfish (especially the valuable abalone). Otters are also particularly vulnerable to oil spills, as the *Exxon Valdez* experience of 1989 demonstrated when at least 5000 otters were killed by exposure to crude oil spilled off the Alaskan coast.

Wherever sea urchins occur in very high densities in kelp beds, they are capable of eliminating the kelp as illustrated in Figure 8.4. Prior to 1968 there were luxurious beds of *Laminaria* along rocky shores of the Nova Scotian coast (eastern Canada). The kelp beds extended to a depth of about 20 m, and they supported sea urchin populations of about 37 individuals m^{-2}. From 1968 onward, the urchins (*Stronglyocentrotus droebachiensis*) became more and more abundant, and barren areas developed where the kelps were eliminated. By 1980, urchin-dominated barren grounds extended along more than 400 km of coastline. The rocky substrate became encrusted with coralline red algae, which are not controlled by urchin grazing. In the early 1980s, the urchin population was decimated by epidemic disease and

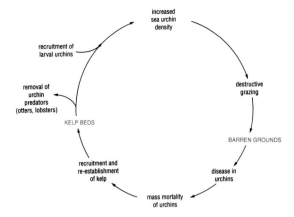

Figure 8.4 The biological events that alternate between productive kelp beds and barren grounds that are dominated by sea urchins.

kelp began to reappear along most of the coast. Within three years of the mass mortalities of urchins, luxuriant kelp beds had been re-established in some areas.

Anecdotal evidence from fishermen suggests that mass mortalities of sea urchins, and reciprocal fluctuations in kelp and urchin abundance, have occurred off the Nova Scotian coast since at least the turn of the century. The population explosions of sea urchins may be related to very high recruitment success of larvae in certain years, and this may be related to changes in local seawater temperature. As well, higher than average water temperatures have been linked with outbreaks of disease in urchins. These facts suggest that fluctuations in kelp and urchins may be natural events triggered by environmental change, and that they may have been occurring over a very long time. In any case, it is clear that urchin grazing and disease regulate the ecological dynamics of these subtidal *Laminaria* communities.

8.4 SAND BEACHES

The intertidal zones of sandy beaches appear barren when compared with rocky shores. In particular, beaches exposed to severe wave action often seem entirely devoid of life. This is because the nature of the substrate sets living conditions which are best met by infaunal organisms that usually remain hidden from direct observation. This makes sand beaches more difficult to study than areas where the activities of the resident organisms can be directly observed. Compounding the problem is the fact that many of the organisms in this environment are of very small size, making their separation from the sediment tedious, and their taxonomic identification difficult.

8.4.1 ENVIRONMENTAL CHARACTERISTICS

Beach sand grains are usually formed of irregularly-shaped quartz particles mixed with a high proportion of shell fragments, and with detritus derived from both marine and terrestrial sources. The particle size of sand varies from <0.1 mm to >2 mm and is determined largely by the degree of wave action; protected beaches have finer sand particles than exposed areas, where waves resuspend and transport small-sized grains. There is a gradient in particle size between sand and mud, with mud being composed of finer particles and mudflats being formed in areas of little water movement (see Section 8.5). Some substrates are difficult to characterize, giving rise to terms like muddy sand, or sandy mud. As particle size increases, sand grades into gravel or shingle. These large-particle substrates do not retain water because of their high porosity, and the shifting and abrasion of the large particles also contribute to an absence or low diversity of life on gravel and shingle shores.

Sandy beaches typically have a gradual slope, and this means that the sediment drains and dries out relatively slowly. Although oxygen is plentiful in the overlying water, oxygen content in the substratum diminishes with depth because of the respiration of micro-organisms and the oxidation of chemicals within the sand. Anaerobic conditions are marked by a black sulphide layer beginning at a depth of from a few millimetres to nearly a metre, depending on the organic content of the sand. Chemosynthetic bacteria are present in the sulphide layer (refer to Section 5.5).

Animals require special adaptations to live in an environment where the substrate is physically unstable in the sense that the sand grains are continually moved by turbulent water. The continual shifting of the surface layers of exposed beaches excludes large sessile species and most large epifauna in general. As well, sand beaches contain relatively low concentrations of organic matter. On the other hand, sand buffers against large temperature and salinity fluctuations, and organisms burrowing into sand are kept moist at low tide. Sand also acts as a protective cover from intense solar radiation. Although there are differences in the physical environment and in the distribution of species from high to low tidal marks, the zonation patterns are not as clearly obvious as on rocky shores. Zonation on sandy beaches is also dynamic and variable; as the tide rises, many populations change their positions on the shore, or enter the water column.

8.4.2 SPECIES COMPOSITION

Primary producers
There are no large attached plants below the high tide mark on sand beaches. The dominant benthic primary producers are diatoms, dinoflagellates, and blue-green algae. These are restricted to the near-surface layers of the sediment because light does not penetrate very deep in sand. The primary productivity of these benthic plants is very low (<15 g C m^{-2} yr^{-1}), and the system depends primarily on energy derived from primary productivity in the surrounding water and on organic detritus.

Macrofauna
There is a low diversity of macrofauna compared with rocky shores or mud communities. Burrowing polychaetes and bivalves plus crustaceans are usually the dominant members in terms of biomass. In temperate latitudes, the supralittoral zone is occupied by air-breathing amphipods (beach hoppers or beach fleas) and, sometimes, also by isopods. Amphipods and isopods burrow into the sand during the day, and feed at night on detritus like decaying seaweed that has washed ashore. On tropical sand beaches, the highest reaches are occupied by ghost crabs (*Ocypode*) which are also scavengers.

The mid- and lower tidal zones support a higher diversity of macrofauna. Small, fast-burrowing wedge-shaped clams (*Donax, Tellina*) are often present in vast numbers, some migrating up and down the beach with tidal changes.

Larger razor clams (e.g. *Ensis, Siliqua*) are also confined to sandy shores, and they too are rapid diggers. These mobile bivalves tend to have smooth, thin shells with slender profiles that ease passage through sand. Stouter thicker-shelled bivalves like cockles (e.g. *Cardium*) or *Macoma* also inhabit sand, but they tend to anchor themselves more firmly in the sediment and move less. The bivalves may be either suspension feeders or deposit feeders, with some species capable of taking advantage of both food sources. In general, deposit feeders tend to dominate in fine-particle sands, presumably because the concentration of organic material is higher than in coarser sands. The sand environment is also home to certain snails that plough through the sand; these include the olive shells (e.g. *Olivella*), and the larger moon snails (*Natica, Polinices*), all with smooth, undecorated shells. The majority of olive shells prey on small molluscs. Moon snails are also predators, especially of bivalves, and they gain access to their prey by drilling a hole through the shell. Where they are abundant, moon snails can have important effects on community structure; when *Polinices* is experimentally removed, the numbers of clams and other infaunal prey increase.

Few animals form permanent burrows because wave action and the relatively large particle size of sand cause their collapse. However, exceptions can be found in those polychaete species that line their burrows with mucus or membranous materials. Many of the sand-dwelling polychaetes are deposit feeders, a few are suspension feeders on plankton or resuspended organic material, and some (e.g. *Nephthys, Glycera*) are predators or scavengers that move through the sand actively seeking food.

Characteristic crustacean inhabitants of mid-tidal levels include mole crabs of the genus *Emerita*. They typically lie with their entire body buried and only their antennae projecting above the surface of the sand to capture small suspended food particles from receding waves. Despite being swift burrowers, they are commonly preyed upon by shore birds. Prawns and mysids are other sandy-shore crustaceans; they burrow temporarily but emerge to feed. Predators of this community can include larger epifaunal crabs.

Various types of echinoderms may be present at lower tidal levels, including burrowing sea cucumbers, heart urchins, and sand dollars, most of which are deposit feeders. The shortened spines of the sand dollars facilitate burrowing, and the young of some species ingest and selectively accumulate the heaviest sand particles containing iron oxide in their digestive tract. These increase the density of the small sand dollars and act as a weight belt, keeping them in the sediments when wave action is intense. Starfish are not common inhabitants in temperate communities, but some tropical species (e.g. *Oreaster*) are present in low intertidal or subtidal depths where they feed on organic matter contained in the sand.

Vertebrate members of sand beach communities include fish that are permanent residents, like sand eels that burrow into the sand at low tide, and temporary inhabitants (e.g. flatfish) that move into the area only at high tide to feed on smaller animals. Shorebirds are also important predators of the community, and mammals (e.g. rats, otters) may visit this area to obtain food.

Meiofauna

The meiofauna of sand beaches includes some of the most diverse and highly adapted species in this environment. The term **interstitial fauna** is also applied to those animals that live in the interstices, or spaces, between

208

the sand grains. They either attach to sediment particles, or move through the interstitial spaces without dislodging the grains. Many animal phyla are represented in this category, and some groups (e.g. gastrotrichs, see Figure 8.5c,d) are wholly or largely restricted to this particular environment. Biomass of the meiofauna usually ranges between 1 and 2 g m^{-2}, with the average number of individuals being 10^6 m^{-2} (from the sand surface down to the anoxic layer).

Figure 8.5 illustrates some of the characteristic meiofauna of sand. Many of their adaptations are morphological and can be seen in the figure; these include small size (only a few mm in largest dimension even among groups that are usually large, such as echinoderms and molluscs), elongate or wormlike shapes, and flattened bodies. As well, many have a strengthened body wall as protection against crushing in a physically unstable substrate. This may involve the development of spines or scales (e.g. in gastrotrichs), a well-developed cuticle or exoskeleton (as in nematodes or crustaceans), or an internal skeleton of calcareous spicules (some ciliates and sea slugs). Alternatively, soft-bodied animals like ciliates, flatworms, and hydroids have developed the ability to contract strongly to protect against mechanical damage. Many of the interstitial species have special adhesive organs for maintaining a hold on the sediment particles; these may be epidermal glands, hooks, or claws (note Figure 8.5e in particular).

The majority of the sand meiofauna are mobile, but some foraminiferans and hydroids (Figure 8.5h) remain firmly attached to sand particles. All feeding types are present, from animals like ostracods and harpacticoid copepods that graze on benthic diatoms and dinoflagellates, to detritus feeders (gastrotrichs, nematodes), to predators such as hydroids and flatworms. Suspension feeders are the rarest type, and these are sedentary animals like bryozoans and

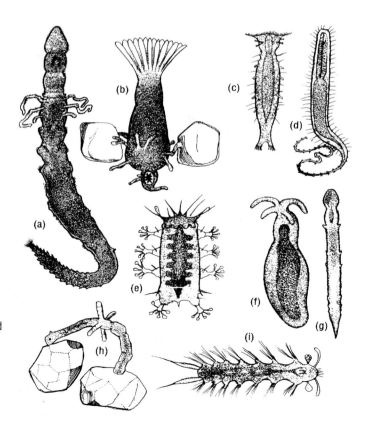

Figure 8.5 Representative meiofauna from sand, all between 0.1 and 1.5 mm in length. (a) *Psammodrilus* (a polychaete); (b) *Monobryozoon* (a bryozoan attached to sand grains); (c) *Dactylopedalia* (a gastrotrich); (d) *Urodasys* (a gastrotrich); (e) *Batillipes* (a tardigrade); (f) *Unela* (a gastropod mollusc); (g) *Pseudovermis* (a gastropod mollusc); (h) *Psammohydra* (a hydroid attached to sand particles); and (i) *Nerillidium* (a polychaete)

tunicates. The meiofauna fall prey to macrofaunal deposit feeders, shrimp, and young fish.

Fecundity of the meiofauna is low owing to their small sizes and the consequent physical constraints on producing large numbers of gametes. Many of the species produce only one to ten eggs at a time, and about 98% of the species lack pelagic larvae. The young are often brooded by the parent until they are able to live freely or, alternatively, eggs are attached to the sand and the young hatch as benthic juveniles. Dispersal is by passive transport of those eggs or adults that are caught in water currents when the sand is washed away, or by organisms attached to sand particles that adhere to the feet of wading birds.

QUESTION 8.2   About 98% of the meiofauna in sand do not produce planktonic larvae. What factors favour direct development and suppression of a pelagic phase in these species and in this environment?

8.5   ESTUARIES

Estuaries are partially enclosed regions where large rivers enter the sea. They rank among the most productive of marine ecosystems as they typically contain a high biomass of benthic algae, seagrasses, and phytoplankton, and support large numbers of fish and birds. Estuaries are enriched by nutrients from land drainage, but their high productivity is also the result of nutrient retention within the estuary. This is due to the water circulation pattern that is set up when less dense freshwater overlies heavier salt water. Figure 3.15 illustrates how estuaries tend to entrain nutrients from deep, saline water into the freshwater flowing seaward from the river, with the nutrient enrichment usually leading to a phytoplankton bloom seaward of the river mouth. Some of the bloom will sink out into the lower, more saline layers, and the decomposing phytodetritus will then be carried back toward the land. Thus the special circulation pattern of estuaries, combined with tidal flow, results in the sinking of particles and nutrients from seaward-flowing river water, and in these nutrients being carried back at depth in the saline water that flows inward and upwells to replace that carried away by the surface flow.

Each estuary has unique physical features that influence its ecology. These include the amount of river discharge, depth and general topography, specific circulation patterns, climatic regime, and vertical tidal range. Nevertheless, certain generalities emerge from the many comparative studies of life in estuaries. In several respects, the estuarine ecosystem is much more complex than open ocean ecosystems, and the plankton community at the seaward edge of the estuary is only one of several communities governed by different groups of primary producers. The major components that typically make up estuaries are illustrated in Figures 8.6 and Colour Plate 33; the relative area occupied by each of these communities depends on local tidal action and the topography of the estuary.

Starting from the upper reaches of temperate-latitude estuaries, there is firstly a sheltered, upper intertidal **saltmarsh community** dominated by a variety of marshgrasses (e.g. *Spartina, Salicornia*); this community is largely replaced by mangroves in tropical and subtropical latitudes (see Section 8.7). The marshgrasses, which are rooted flowering plants, may be as much as 2 m high, and they function as a trap for nutrient-rich sediment. Above-ground

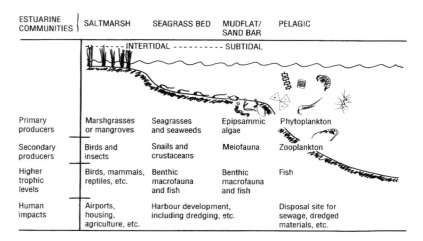

Figure 8.6   A schematic depiction of the communities composing the estuarine ecosystem, showing their dominant flora and fauna and potential human impacts.

primary production of marshgrasses ranges from 200 to 3000 g C m^{-2} yr^{-1}, and production by benthic mud algae contributes another 100 to 600 g C m^{-2} yr^{-1}. Thus saltmarshes rank among the most productive ecosystems on Earth. Most of the living plant material is not grazed directly, but enters detritus food webs either on the marsh or in adjacent waters. This plant debris decays slowly and, over long periods, the accumulation of debris and trapped sediment may create peat deposits that are several metres deep. This upward growth of saltmarshes results in changes in relative tidal level and drainage, and thus in changes in the species composition of plants; this process of marsh evolution eventually contributes to the infilling of estuaries.

The upper reaches of a saltmarsh mark the transition between the sea and land. This habitat has great variations in salinity and temperature, and relatively few species of plants and animals live here permanently. Terrestrial animals such as raccoons, rats, and snakes invade this area, and there are large insect and bird populations. Faunal diversity is greater in the lower intertidal areas of the marsh, and the saltmarsh macrobenthos may include deposit-feeding fiddler crabs (*Uca*) that build burrows in the mud, snails (e.g. *Nassarius, Hydrobia, Littorina*) that feed on the rich deposits of benthic diatoms, and mussels of the genus *Modiolus* that are specially adapted to live in or on mud and that can respire in both air and water. The leaves and stems of the marshgrasses serve as attachment sites for many small organisms, and significant numbers of micro- and meiobenthos live on or in the bottom sediments. Bacteria attain densities as high as 10^9 cm^{-3} in the sediments, and they are an important food for protozoans and meiofauna. Saltmarshes fulfil the important function of providing shelter and food for shrimp, juvenile lobsters, and the young stages of many species of marine and estuarine fish.

A seagrass community is located seaward of the saltmarsh, in the intertidal and subtidal zones. It may contain significant stands of seaweeds in addition to the seagrasses, but in general, seaweeds do not grow as well in muddy estuarine waters as they do in clear waters. The dominant plant of this estuarine community in temperate latitudes is *Zostera*, commonly called eelgrass; in tropical climates, it is replaced by *Thalassia*, or turtlegrass. The brown seaweed *Fucus* and green seaweeds, *Enteromorpha* and *Ulva*, may grow on patches of rock among the seagrass beds. Measurements of the productivity of seagrasses are complicated by the fact that many epiphytic diatoms may grow on the blades of the seagrass, and these may add to the total primary productivity. For example, on the eastern coast of the United States, *Zostera* may produce about 350 g C m^{-2} yr^{-1}, and associated plants contribute a further 300 g C m^{-2} yr^{-1}. Generally, the annual production of temperate seagrasses is about 120–600 g C m^{-2}, while tropical seagrass communities have higher net primary productivities of up to about 1000 g C m^{-2} yr^{-1}.

Numerous meiofauna, including protozoans and nematodes, are associated with the seagrass epiphytes which are grazed by snails, isopods, amphipods, and harpacticoid copepods. Sessile filter-feeding invertebrates (e.g. hydroids, bryozoans, and tunicates) attach to the seagrass leaves. Snails, bivalves, polychaetes, and various types of crustaceans dominate the mobile invertebrate fauna of seagrass communities. This estuarine zone, like the saltmarsh, serves as a nursery area for the young of many species of fish, including commercial species such as menhaden and salmon.

In both the saltmarsh and seagrass communities, little of the primary production is consumed by herbivores. Both communities are dominated by detritus-based food chains because marshgrasses and seagrasses contain large amounts of refractory material, such as cellulose, that is difficult for herbivores to digest. Less than 10% of the marshgrass is grazed by terrestrial herbivores, and usually only a small fraction of the seagrass production is eaten by such animals as sea urchins and migrant birds (e.g. geese). However, in some tropical regions, turtlegrass may be consumed in large quantities by dugongs or manatees, and by sea turtles. In general, though, by far the largest fraction of the net primary production in both communities dies and is colonized by fungi and bacteria, to be converted eventually into microbial biomass. The numbers of bacteria in estuarine water are much higher than in seawater, and bacterial densities in sediments may reach 200–500 × 10^6 per gramme of estuarine mud. Thus a large amount of plant detritus is produced, some of which is exported out of the estuary, and much organic carbon is recycled to re-enter the food chain through the microbial loop (see Section 5.2.1). Within the sediments, much of the organic matter is decomposed under anoxic conditions, with anaerobic bacteria using primarily inorganic sulphate as a source of oxygen.

QUESTION 8.3   Is the occurrence of hydrogen sulphide in sediments an indication of pollution?   *No.*

On the seaward side of the seagrasses, either a subtidal **mudflat** or **sand-bar community** will be present depending on the current and tidal regime. In fact, this community is continuous underneath both the intertidal seagrass and saltmarsh communities. The dominant primary producers of mudflats or sand bars are the **epipsammic** algae, which are generally species of benthic diatoms or dinoflagellates that are specially adapted to grow on sediment particles. The surface of mud is sometimes colonized by thick mats of filamentous blue-green algae of several types. The productivity of this region tends to be inversely correlated with the grain size of the sediment particles, so that mudflats are generally more productive than sand bars in the same location. The primary productivity of these communities (in the absence of a cover of marshgrasses or seagrasses) tends to be low. Sand bars have a primary productivity of about 10 g C m^{-2} yr^{-1} but mudflat production by benthic microphytes may be as high as 230 g C m^{-2} yr^{-1}.

212

The smaller the grain
size the greater the
surface area for algae
to attach to.

QUESTION 8.4   Why should the particle size of sand vs. mud affect the productivity of the epipsammic algae?

The mudflat community supports a wide range of animals, with crabs and flatfish being common epifauna, and bivalves, polychaetes, and mud shrimp dominating the infauna. There is a rich meiofauna of small copepods, nematodes, and polychaetes, and an equally rich microfauna of protozoans, especially ciliates. Detritivores are usually predominant in the community. In the shallower reaches of this community, large numbers of birds eat the detritivorous invertebrates. Food consumption by birds may represent a significant impact on the mudflat species. For example, each of several thousands of knots (*Calidris canutus*) on a large mudflat may eat as many as 730 small clams (*Macoma*) per day; a single redshank (*Tringa totanus*) may consume up to 40 000 burrowing amphipods (*Corophium*); and one oyster-catcher (*Haematopus ostralegus*) can eat 315 cockles (*Cardium*) daily. Overall, birds may take between 4% and 10% of the accessible invertebrate fauna.

The **pelagic community** located on the seaward edge of the estuary is controlled largely by the primary productivity of the phytoplankton, and this ranges from about 100 to 500 g C m^{-2} yr^{-1}, depending on water clarity. Although nutrients may be plentiful, turbidity of the water often restricts light penetration and limits phytoplankton production. In shallow estuaries, as much as half of the phytoplankton may be consumed by filter-feeding benthos, with the rest being eaten by zooplankton. Zooplankton also may feed on benthic diatoms and bacteria-covered sediment particles that are resuspended by intense mixing in shallow estuaries. In deep fjordlike estuaries, benthic plants are light-limited and most of the estuarine primary production is carried out by phytoplankton.

Although estuaries are highly productive and host many juvenile fish, as well as large numbers of crustaceans, molluscs, shorebirds and waterbirds, the number of species found in these areas is relatively small compared with other marine habitats. Few species are adapted to cope with the salinity, temperature, and turbidity variations present in this habitat. Salinity tolerance plays a major role in the distribution of any particular species in an estuary, although distribution is also determined by such factors as substrate type and degree of tidal exposure.

Figure 8.7 illustrates the typical distribution and relative diversity of freshwater, brackish, and marine animals in relation to estuaries. Estuaries support an essentially marine fauna, but the number of marine species declines as the water becomes less saline, and the species change from those that are stenohaline to those that are euryhaline (see Section 2.3.2). The majority of animals living in rivers do not tolerate salinities greater than about 0.5, and they do not penetrate further seaward than the uppermost reaches of the estuary. Only a few freshwater organisms (**oligohaline** species) can survive in water having a salinity of 0.5 to about 5; these include principally various insect larvae, oligochaete worms, snails, and some fish such as sticklebacks. There are relatively few brackish-water species that are restricted to estuarine conditions with salinities of about 5–20, and most are animals with marine affinities. Euryhaline marine organisms constitute the majority of species living in estuaries, and their distributions extend from the sea into the central regions of estuaries.

Stenohaline = Animals that can only tolerate a narrow range of salinity.

euryhaline = can tolerate a wide range of salinity.

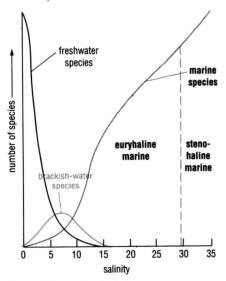

Figure 8.7   An idealized diagram of the distribution of freshwater, brackish-water, and marine animals relative to salinity. (Numbers of species given in relative units.)

Stenohaline marine species are unable to tolerate salinities lower than about 25–30, and they are largely excluded from estuaries. Some fish (e.g. salmon, eels) are transient residents of estuaries, and move freely from the sea to rivers and lakes, or vice versa (see Section 6.6.1). Overall, estuaries have fewer species than adjacent aquatic environments, but abundance within individual species as well as biomass are often markedly increased.

In general, the extent of penetration into estuaries by marine and, conversely, freshwater species is determined by the rate and magnitude of tidal change, rather than by the salinity gradient. That is, marine species penetrate farther upstream, and freshwater organisms reach much closer to the sea, in estuaries where tides are small and the salinity gradient is relatively stable. The minimum number of species occurs in that part of the estuary where the salinity variation is greatest. Finally, the distributions of benthic species within estuaries are also controlled by sediment type.

QUESTION 8.5 Can you offer any explanation(s) as to why species diversity declines in estuaries relative to adjacent environments, but numbers of individuals and biomass increase?

8.6 CORAL REEFS

Coral reefs are well known for their spectacular beauty (Colour Plates 34 and 35), and they are perhaps the most diverse and ecologically complex of marine benthic communities. They are unique in being formed entirely by the biological activity of certain corals belonging to the Phylum Cnidaria (see Table 7.1). These tropical reefs result from massive deposits of calcium carbonate laid down by the corals over aeons of geologic time. These are among the oldest of marine communities, with a geological history stretching back for more than 500 million years.

8.6.1 DISTRIBUTION AND LIMITING FACTORS

Living coral reefs cover about 600 thousand km^2, or somewhat less than 0.2% of the global ocean area and about 15% of the shallow sea areas within 0–30 m depth. The largest reef is the Great Barrier Reef that extends along the east coast of Australia for a distance of more than 2000 km and is as much as 145 km wide. Reefs are located exclusively within water bounded by the 20°C isotherms and so are virtually confined to the tropics (Figure 3.10). Reef-building corals cannot tolerate water temperatures of less than 18°C, and optimal growth usually occurs between 23° and 29°C, although some corals tolerate temperatures of up to 40°C. A number of other physiological demands further limit the distribution of reef-building corals. They require high salinity water ranging from 32 up to 42. High light levels are also necessary for reef-building (for reasons that will be explained below), and this restricts corals to the euphotic zone. Even in the clear oligotrophic water of the tropics, most reef-building species live in water that is shallower than 25 m. The upward growth of a reef is restricted to the level of lowest tides, as exposure to air for more than several hours kills corals. Corals are also absent in turbid waters, as they are very sensitive to high levels of suspended and settling sediment which can smother them and clog their feeding mechanisms. High turbidity also affects reef-building by decreasing the depth of light penetration. New reefs are initially formed by

[handwritten margin note: This is a niche habitat only specially adapted animals can live here. But there will be fewer predators.]

[handwritten margin note: oligotrophic = low in nutrient content]

a. upwelling cold water
b. too muddy.

the attachment of meroplanktonic coral larvae to a hard substrate, and for this reason reefs always develop in association with the edges of continents or islands.

QUESTION 8.6   Refer to Figure 3.10. (a) Can you explain why coral reefs are generally absent on the west coasts of the Americas and Africa between 30° S and 30° N? (b) What might prevent reef formation off north-eastern South America, northward from the mouths of the Amazon and Orinoco rivers?

8.6.2   CORAL STRUCTURE

Corals are closely related to benthic sea anemones (both are in the Class Anthozoa) and are more distantly related to planktonic jellyfish, benthic marine hydroids, and the freshwater *Hydra*. Not all corals are reef-builders; some are solitary or colonial animals that are capable of living in deeper and/or colder water and are found throughout the world's oceans. Reef-building stony corals are colonial animals, and each reef is formed of billions of tiny individuals called polyps (Figure 8.8; Colour Plate 36). Each polyp secretes a calcium carbonate exoskeleton around itself that generally measures about 1–3 mm in diameter. Each polyp is equipped with tentacles containing batteries of nematocysts (see Section 4.2), and these stinging cells can be used to capture prey and for defence. The polyps can produce a large colony by asexual division, or budding, and all the polyps in a colony remain connected to each other by extensions of their tissues. Corals also reproduce sexually, producing planktonic larvae that disperse, settle, and establish new colonies.

Individual coral colonies vary in size, but some are very large, weighing up to several hundred tonnes. The form of a colony, whether it is branching, massive, lobed, or folded, depends on the species and also on the physical

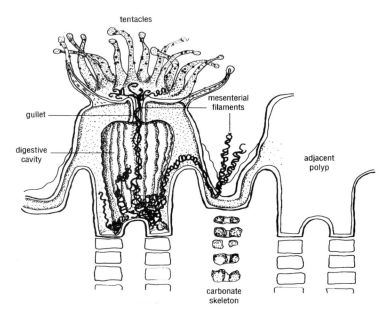

Figure 8.8   Anatomy of a coral polyp. The animal is basically a contractile sac housed in a carbonate skeleton. The central mouth is surrounded by six, or a multiple of six, tentacles equipped with batteries of nematocysts. The tiny zooxanthellae live in cells in the lining of the central digestive cavity. Each polyp secretes a protective carbonate exoskeleton consisting of a radial arrangement of vertical plates; as it grows upward, the polyp deposits new layers under itself.

environment in which the coral is located. The same species may have a very different form when it grows in areas exposed to wave action as opposed to calm conditions, or when it grows in shallow versus deeper waters.

8.6.3 DIVERSITY

The diversity of life on a coral reef is extraordinarily rich. Figure 8.9 illustrates only a very few dominant types of the coral-reef fauna. The Great Barrier Reef is composed of about 350 species of hard corals, and is home to more than 4000 species of molluscs, 1500 species of fish, and 240 species of seabirds. In addition, there are many more species of macrobenthos, and the numbers of micro- and meiofauna remain unknown. Representative species of almost all phyla and classes can be found in the reef ecosystem.

Figure 8.9   A coral reef habitat illustrating some of the many inhabitants of this diverse ecosystem.

1	petrel	16	snail
2	jellyfish	17	nudibranch (sea slug)
3	angelfish	18	sponges
4	lobed corals	19	colonial tunicate
5	sea whips (gorgonian corals)	20	giant clam (*Tridacna*)
6	triggerfish	21	pseudochromid fish
7	sea fans (gorgonian corals)	22	starfish
8	tube anemone	23	soft corals
9	stone coral	24	cleaner shrimp
10	bryozoans	25	sea anemones
11	brain coral	26	clownfish
12	butterfly fish	27	worm tubes
13	moray eel	28	snail (cowry)
14	cleaner fish	29	sea fan (gorgonian)
15	tube corals		

Reefs in the Indo-Pacific have a high diversity of coral species, with at least 500 reef-building species throughout the entire region. Atlantic reefs are impoverished in comparison, with only about 75 species of reef-building corals. The number of species in other animal groups associated with reefs is also generally lower in the Atlantic sector than in the Indo-Pacific. The number of mollusc species is estimated at about 5000 in the Pacific versus 1200 in the Atlantic, and there are about 2000 versus 600 fish species in these respective reef areas. The differences in species diversity may result from differences in the age of the oceans, and the respective geologic times over which reefs have evolved. Geologically, the Atlantic is a more recent ocean, and its reefs were also more severely influenced by decreased temperatures and lowering sea levels during ice ages. Most Atlantic reefs are only 10 000–15 000 years old, these dates corresponding to the last glacial age. In contrast, the Great Barrier Reef is about 2 million years old, and some Pacific atolls date back about 60 million years.

The reef itself provides food and shelter for many plants, invertebrates, and fish. For sessile species, the reef offers a site of attachment. Surface irregularities in the reef limestone create a variety of microhabitats like crevices and tunnels, and these contribute to the faunal diversity of the system. Areas of rubble and sand also accumulate between coral heads, and these sediment types require different sets of adaptations and develop different communities from that associated with the hard-substrate reef. A reef is also differentiated into regions distinguished by physical differences in wave action, depth, and degree of tidal exposure. This wealth of different habitats is a major factor in supporting the many species of a reef.

Coral polyps usually dominate the living biomass of a reef, but other reef organisms also contribute to the carbonate reef structure. These include the hard, coralline red algae that grow in thin layers over the surface of the reef. These encrusting algae precipitate $CaCO_3$ and play a role in cementing the reef fragments together. Some green algae also secrete calcium carbonate, other green algae do not. In addition to encrusting algae, there are benthic algae that are erect species, and some that live within the spaces of the coral framework. Seagrasses often grow in the sandy areas within or surrounding the reef. All of these plants provide food for herbivorous species of invertebrates and fish. However, the algae are generally inconspicuous inhabitants of the reef, and animal life is visually dominant.

In addition to the reef-building stony corals, other types of cnidarians are prominent reef members (see Figure 8.9 and Colour Plate 37). These include several types of non-reef-building corals, including fire corals, pipe corals, and soft corals. Sea whips and sea fans are also common reef inhabitants; they are close relatives of stony corals and have internal skeletons composed of calcareous spicules. Other major invertebrate groups in a reef community include echinoderms (starfish, sea urchins, and sea cucumbers), molluscs (limpets, snails, and clams), polychaete worms, sponges, and crustaceans (including spiny lobsters and small shrimp). Some of the invertebrates are encrusting species, like bryozoans; some build calcareous tubes, like certain polychaete worms; and some snails attach tube-like shells to the reef. All of these activities serve to cement the limestone reef framework together. In the Pacific, giant clams belonging to the genus *Tridacna* are also important structural components of reefs (Figure 8.9). These molluscs contribute an astonishing biomass to the reefs, as they grow to over 1 m in length and may exceed 300 kg in weight.

Fish comprise the dominant vertebrates on a reef. Many of the reef fish are brightly coloured and visually conspicuous. About 25% of the world's species of marine fish are found only in reef areas. These diverse species of fish show a high degree of feeding specialization and food selection. Some are herbivores, feeding on algae or seagrasses; some specialize in being plankton-feeders; and some are piscivorous, or are predators of benthic reef invertebrates. Fish not only play important ecological roles in grazing or predation, but the faeces of these abundant animals contribute an important source of nutrients to the reef ecosystem.

The very large number of reef species, and the abundance of life on the reef, lead to intense competition between species and between individuals for limited resources. The high degree of food specialization observed in many reef species is a reflection of the high species diversity of the reef, and every available food resource is efficiently utilized. There is also intense competition for space on the reef, and every microhabitat is occupied by organisms adapted to their particular site. Experimental work has revealed that the mesenterial filaments (Figure 8.8) of some corals contain substances that kill polyps of adjacent colonies. Aggressive, slow-growing corals can thus avoid being overgrown by less aggressive, but faster-growing species.

8.6.4 NUTRITION AND PRODUCTION IN REEFS

Reef-building corals are also called **hermatypic** corals. They are distinguished from non-reef-building (**ahermatypic**) species by having a special symbiotic association with certain algae. Each hermatypic coral polyp contains masses of photosynthetic dinoflagellates, called **zooxanthellae**. These are a vegetative form of free-living dinoflagellates; when cultured under laboratory conditions, they develop into motile flagellate forms identical with planktonic dinoflagellates (see Section 3.1.2). The zooxanthellae in all corals belong to a single genus, *Symbiodinium*, with different species or strains being specific to particular coral species. The zooxanthellae live within cells in the lining of the gut of corals, reaching concentrations of up to 30 000 cells per mm^3 of coral tissue. Under stressful environmental conditions, the symbiotic algae can be expelled from the coral. Because much of the colour of corals is due to the pigmentation of the zooxanthellae, this expulsion is referred to as 'bleaching'.

The algal-coral relationship is beneficial to both species. The coral provides the algae with a protected environment, but it also provides certain chemical compounds that are necessary for photosynthesis. Carbon dioxide is produced by coral respiration, and inorganic nutrients (ammonia, nitrates, and phosphates) are present in waste products of the coral. In return, the algae produce oxygen and remove wastes; but most importantly, they supply the coral with organic products of photosynthesis that are transferred from the algae to the host. These chemical products include glucose, glycerol, and amino acids, all compounds that are utilized by the coral polyps for metabolism or as building blocks in the manufacture of proteins, fats, and carbohydrates. The symbiotic algae also enhance the ability of the coral to synthesize $CaCO_3$. Rates of calcification are significantly slowed when zooxanthellae are experimentally removed from corals, or when corals are kept in shade or darkness. The relationship between the two independent processes of CO_2 fixation by photosynthesis and CO_2 fixation as $CaCO_3$ is complex and not fully understood. However, the symbiotic association with photosynthetic dinoflagellates explains why hermatypic corals require clear, lighted water. This association also leads to intense competition for space within areas of sufficient light to support the zooxanthellae.

QUESTION 8.7 Can any corals grow below the euphotic zone?

Yes ahermatypic types can.

The coral–zooxanthellae symbiosis is maintained over time and distance because the algae are already contained in coral larvae before they are released from the parent polyp. This relationship is not unique on the reef, however. Zooxanthellae are also present in other reef inhabitants, including the majority of other cnidarians, some tunicates, some shell-less snails, and in the giant clam *Tridacna*.

The symbiotic arrangement between algae and corals or other invertebrates results in nutrients being tightly recycled within coral reefs. This internal nutrient cycling is of primary importance in maintaining the productivity of the reef in oligotrophic tropical water.

Symbiotic algae do not supply all the nutritional requirements of their hosts. All the animals harbouring zooxanthellae are mixotrophic and capable of meeting their additional nutritional needs in other ways. Corals are true carnivores that capture zooplankton, employing their nematocysts to paralyse the prey. Many coral species can also feed on suspended particles by producing mucous nets or mucous filaments to entangle food that is then drawn to the mouth by rows of cilia. Ciliary-mucus feeding extends the size range of potential food items to include even bacteria. Corals may also directly absorb dissolved organic matter.

The relative importance of zooxanthellae versus captured particulate food to the nutrition of any particular coral probably depends on the particular species, and it will be influenced by the specific chemical that is produced and translocated from the symbiotic algae to the host. It should also be influenced by various environmental parameters such as depth, light intensity, abundance of zooplankton, etc.

8.6.5 PRODUCTION ESTIMATES

Primary production in the coral reef system is carried out by the benthic algae attached to or associated with the reef, by the suspended phytoplankton, and by the zooxanthellae living within the animals of the reef. This ecological fractionation of primary producers makes accurate measurements of primary productivity extremely difficult because different techniques must be employed for each. With the exception of the phytoplankton, it is also difficult to assess the standing stock of primary producers. To do so requires determining the plant/animal proportions of coral polyps and the relative contributions of various types of benthic algae to total reef biomass. Until these have been determined, the size of the primary producer trophic level remains uncertain.

Production studies of coral reefs suggest that gross primary productivity ranges from about 1500 to 5000 g C m^{-2} yr^{-1}, values that are much higher than those of open tropical oceans (see Section 3.5 and 3.6). In fact, they represent some of the highest rates of primary production of any natural ecosystem. However, many of the nutrients contributing to this production are recycled (i.e. the f-ratio <0.1, see Section 5.5.1). Symbiosis between primary producers and dominant animal species of the community, with nutrients prevented from being washed away, is a dominant controlling feature of the biological production, just as it is in the deep-water, sulphide-communities which will be described in Section 8.9.

Net primary production on reefs is lower than might be expected because respiration of the primary producers is high, with gross production to respiration ratios (P/R) usually ranging from 1.0 to 2.5. In comparison, healthy phytoplankton have a P/R ratio of about 10. In addition, the coral-reef food chain is much longer than in upwelling zones (see equation 5.2), so that respiration losses throughout the entire ecosystem are high. This results in lowering the production of top-level predators relative to the high gross primary productivity.

8.6.6 FORMATION AND GROWTH OF REEFS

During the voyage of the *Beagle* in the 1830s, Charles Darwin observed that there were three basic types of coral reefs, and he formulated an hypothesis of reef formation that linked these types. His ideas are summarized below and illustrated in Figure 8.10.

Reef formation is initiated with the attachment of free-swimming coral larvae to the submerged edges of islands or continents. As the coral grows and expands, a **fringing reef** is formed as a band along the coast or around an island. This type of reef is predominant in the West Indies (Caribbean Sea). It is also the first stage in the process of forming atolls.

If the fringing reef is attached to the edges of a volcanic island or other land mass that is slowly sinking, while the coral continues to grow upward, a **barrier reef** will eventually form. Barrier reefs are separated from the land mass by a **lagoon** of open deep water. The Great Barrier Reef of Australia is the best known of this type, but it is in fact an aggregation of many reefs.

Atolls mark the last stage in this geological process. When a volcanic island subsides below sea level, the coral reef is left as a ring around a central lagoon. Continued coral growth maintains the circular reef, but calm conditions and hence increased sedimentation in the central lagoon prevent development of a reef in this area. Hundreds of coral atolls are found throughout the South Pacific Ocean, all of them located far from land but attached to underwater **seamounts** (volcanic elevations rising from the seafloor) which have subsided with age.

Darwin's ideas on atoll formation were not substantiated until the 1950s, when drilling programmes on coral atolls encountered volcanic rock hundreds of metres below the surface. His hypothesis has been further supported by the discovery of seamounts, submerged far below the sea surface, that still have attached remnants of shallow-water corals.

QUESTION 8.8 Excluding pollution influences, would you expect to find a difference in total biological production between a barrier reef located offshore of a continent and a mid-oceanic atoll? Explain your answer.

The rate at which a reef develops depends on a balance between the growth rates (budding) and calcification of the coral polyps and the rates of destruction of the limestone framework. Corals always grow upward, toward light, as each polyp deposits new carbonate layers under itself (Figure 8.8). Growth of the coral skeleton is much faster in sunlight than in darkness (and

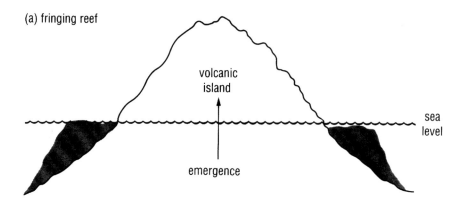

(a) fringing reef

volcanic
island

emergence

sea
level

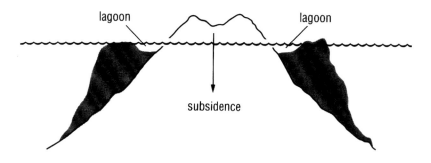

(b) barrier reef

lagoon

lagoon

subsidence

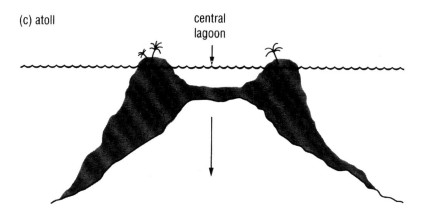

(c) atoll

central
lagoon

Figure 8.10 The formation of coral atolls
according to Darwin's theroy of subsidence.

therefore also faster in shallower water) and, not surprisingly, the rate of
growth can be decreased if photosynthesis of the zooxanthellae is reduced by
sediment-laden water or chemicals (see Section 8.6.4). Growth rates may
also decline with age and increasing size of a colony. In general, corals are
regarded as slow-growing, with measured rates of growth usually varying
from <1 to 10 cm yr^{-1}.

However, growth rates of individual coral species do not necessarily describe
the rate of growth of an entire reef system. This is partly because different

species of corals have different growth rates, but also because growth and expansion of the reef is regulated by many other factors such as predation, competition for space with other organisms, and light intensity, to name only a few. Further, the limestone framework is continually being destroyed by biological activities and physical events (see below). Estimates of total reef growth can be made from measured changes in reef topography over several years, or from geological information on the thickness of reef limestone deposits. These estimates of net vertical upward growth of reefs vary greatly, from only a few millimetres per year, to 30 cm per 11 years under favourable conditions.

In order to obtain better estimates of the rate at which entire reef systems grow, it is also necessary to know something about the factors that destroy the reef and the rate at which the limestone is broken down. Reefs are subject to physical erosion by wave action and currents, and tropical storms can cause extensive damage. Reefs are also subject to continual **bioerosion**, or breakdown of the calcium carbonate skeleton by reef inhabitants. Some organisms associated with the reef remove part of the coral skeleton by boring into the reef, using chemical dissolution or mechanical abrasion; these include certain species of algae, clams, sponges, sea urchins, and polychaete worms. Some animals (e.g. herbivorous limpets and snails, parrotfish) remove pieces of the reef skeleton inadvertently during grazing. Small coral fragments are consumed by deposit-feeders such as sea cucumbers, and thus become further reduced in size. These destructive activities eventually break down reef material to fine-grained carbonate sand. Much of the fine-grained detritus is flushed away from the reef by waves and currents, but some accumulates in pockets between coral heads.

8.6.7 ZONATION PATTERNS ON REEFS

All reefs exhibit zonation patterns resulting from a combination of bottom topography and depth, and different degrees of wave action and exposure. The patterns differ according to locality and type of reef, with atolls having the most complex zonation. The major divisions are illustrated in Figure 8.11 and discussed below, but depending on locality, the zones may be subdivided into as many as a dozen.

The **reef flat** (or back-reef) is located on the sheltered side of the reef, extending outward from the shore or coastline to the reef crest. This area is only a few centimetres to a few metres deep, and large parts may be exposed at low tide. The width of the reef flat varies from a few tens to a few thousands of metres. The substrate is formed of coral rock and loose sand. Beds of seagrasses often develop in the sandy regions, and both encrusting and filamentous benthic algae are common. Because it is so shallow, this area experiences the widest variations in temperature and salinity, but it is protected from the full force of breaking waves. The reduced water circulation, accumulation of sediments, and periods of tidal emersion combine to limit coral growth. Although living corals may be scarce except near the seaward section, this area of many microhabitats supports a great number of species in the reef ecosystem, with molluscs, worms, and decapod crustaceans often dominating the visible macrofauna.

The **reef crest** (or algal ridge) lies on the outer side of the reef, with its exposed seaward margin marked by the line of breaking waves. As the name implies, the reef crest is the highest point of the reef, and it is exposed at

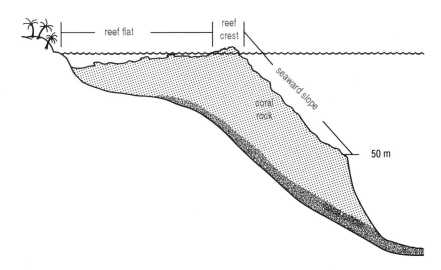

Figure 8.11   A generalized cross section of a typical Caribbean fringing reef, illustrating the major ecological zones.

low tide. The width of this zone varies from a few to a few tens of metres. In some localities, encrusting red coralline algae are dominant; in other reefs, brown algae predominate in this zone. Living corals are very scarce where wave action is severe; usually only one or two robust coral species dominate in this region.

The outermost **seaward slope** (also called fore-reef) extends from the low tide mark into deep water. The upper part of this zone is broken by deep channels in the reef face, through which water surges and debris from the coral reef leaves. Large corals dominate here, and there are many large fish. The maximum number of coral species tends to occur at 15–25 m, then declines fairly rapidly with increasing depth. At 20–30 m depth, there is little wave action and the light intensity is reduced to about 25% of that at the surface; here, corals tend to be smaller branched forms. At 30–40 m, sediments accumulate on the gentle slope and coral becomes patchy in distribution. Sponges, sea whips, sea fans, and ahermatypic corals become increasingly abundant and gradually replace hermatypic corals in deeper and darker water. At 50 m, the slope steepens into deep water. The depth limit for reef-building corals is about 50–60 m in the Pacific, and about 100 m in the Caribbean; the difference is probably related to differences in light penetration.

8.7   MANGROVE SWAMPS

Mangrove swamps, also called **mangals**, are a common feature covering 60–75% of tropical and subtropical coastlines. These forests of trees and shrubs that are rooted in soft sediments occur in the upper intertidal zone. They produce a marine system that is similar to a saltmarsh in having aerial storage of plant biomass and in harbouring both terrestrial and marine species. The euryhaline plants making up this specialized community are tolerant of a wide range of salinities and are found both in fully saline waters and well up into estuaries, but they are restricted to protected shores with little wave action. The distribution of mangroves overlaps with that of coral reefs, but extends farther into subtropical regions. In many areas, mangrove swamps border coastlines protected by barrier reefs.

8.7.1 WHAT ARE MANGROVES?

The term 'mangrove' refers to a variety of trees and shrubs belonging to some 12 genera and up to 60 species of flowering terrestrial plants (angiosperms). Dominant genera include *Rhizophora, Avicennia*, and *Bruguiera*. They have in common the following features:

(a) They are salt-tolerant and ecologically restricted to tidal swamps.

(b) They have both aerial and shallow roots that intertwine and spread widely over the muddy substrate in an impenetrable tangle (Colour Plate 38). The substrate is oxygen-poor, and the aerial roots allow the plants to obtain oxygen directly from the atmosphere. Many of the mangrove species also have special prop roots extending down from the trunk or from branches to serve as extra support.

(c) Mangroves have special physiological adaptations that prevent salt from entering their tissues, or that allow them to excrete excess salt.

(d) Many mangrove plants are **viviparous**, producing seeds that germinate on the tree. Young plants drop from the tree into the water, and the floating plants are dispersed by water. The life cycle of these long-lived plants is illustrated in Figure 8.12.

Some Indo-Pacific mangrove forests may contain 30 or more species of mangroves. There are fewer in Atlantic areas; a total of 10 species is distributed throughout the New World, and mangrove swamps in Florida, for example, support only three species.

Figure 8.12   The life cycle of viviparous mangrove trees.

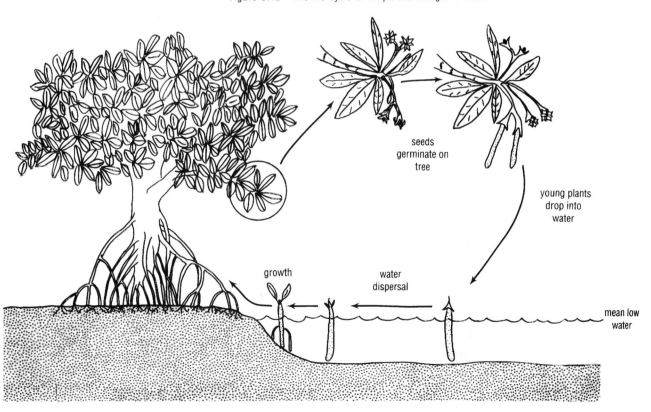

8.7.2 ECOLOGICAL FEATURES OF MANGROVE SWAMPS

The physical environment of mangals is characterized by considerable fluctuation in salinity and temperature. This is also a region that is strongly influenced by tidal action. Water exchange transports nutrients into mangrove areas, and exports material out. Tidal flow also results in an inflow and outflow of animals, such as fish and shrimp, into the tidal area. Animals living high in the intertidal zone are subjected to the greatest environmental variation and to potential desiccation. Nevertheless, the plants and animals are adapted to tidally-induced fluctuation, and the largest mangrove swamps are in areas with a large vertical tidal range.

Mangroves are found in regions of little wave action, and the intertwining roots of the plants further reduce water velocities. This results in trapping of suspended sediments and organic material (particularly leaves) which settle on the bottom to form black mud. The sediments tend to be anoxic because of high bacterial activity and because of poor circulation within the fine-grained substrate.

There is a progressive change from marine to terrestrial conditions from the seaward side of a mangrove area to the landward edge. There is a corresponding zonation of different mangrove species, based at least partly on their respective salt tolerances.

Ecologically, a mangrove community can be divided into (a) the above-water forest, (b) the intertidal swamp, and (c) the submerged subtidal habitat. These distinctive zones support unique combinations of species which are described below.

The **above-tide forest** formed by the trunks and leaf canopy of the mangroves is an arboreal environment inhabited by terrestrial species. Birds, bats, lizards, tree snakes, snails, land crabs and mangrove crabs, spiders, and insects are all common residents, with insects being the most diverse and most abundant. The birds and bats are mostly insectivores or are piscivorous, feeding on small fish. The crabs are detritivores or omnivores and may feed on marine prey during low tide. In some areas, domestic animals (cattle, goats, or camels) may graze on the mangrove leaves. A study of Florida mangroves showed that about 5% of the total leaf production was consumed by non-mammalian terrestrial grazers, the rest entering the aquatic system as debris and becoming available for marine detritivores, either fish or invertebrates.

The **intertidal swamp** offers a variety of different substrates and different microhabitats to support a more diverse community of marine species. Some organisms attach to the mangrove roots, others reside in or on the mudflat or mudbanks. Barnacles and oysters are conspicuous epifauna on the roots, with the latter often the dominant contributor to community biomass. Certain species of isopods bore into the woody prop roots, and their activities may sever the roots, although the total impact to the mangrove swamp is usually limited. Periwinkles (snails) are found in abundance crawling over the roots in the upper intertidal zone. Some polychaete worms are also associated with the root system, with some tube-building species attaching to this hard surface. In this area, combined densities of snails, nematodes, and polychaetes commonly exceed 5000 m^{-2}.

The intertidal mudflat is the home of numerous burrowing fiddler crabs (*Uca*), and sea cucumbers commonly are present on the surface of the mud. Both of these groups feed on detritus. Red and green benthic algae are grazed by amphipods and some species of crabs. Pacific mangroves are frequented by large-eyed mud-skippers (genus *Periophthalmus*), fish that burrow into the mud but spend much time out of water, using modified fins to crawl on the mud flat or, in the case of one species, to climb the mangrove roots. Various species of shrimp and fish move in and out of this region with the tides.

Leaf fall is a major source of nutrients and energy in the intertidal swamp, and many of the residents are detritivores. Some remove suspended detritus by filter feeding (e.g. oysters), others feed on organic material in the sediments by deposit feeding (e.g. burrowing polychaetes), and others like crabs, shrimp, and amphipods capture larger particles of debris with their claw-like appendages. Most animals in the community probably consume detritus in addition to living plant and animal tissue.

The **subtidal** zone also has sediments of fine-grained mud with a high organic content, and sand patches may be present as well. The subtidal mangrove roots support a rich epiflora and epifauna of algae, sponges, tunicates, anemones, hydroids, and bryozoans, and their crowded numbers indicate that competition for space is intense on this substrate. In some areas, turtle grass (*Thalassia*) may be the dominant benthic plant, and it serves to stabilize the mud bottom. Burrowing animals (e.g. crabs, shrimp, worms) are common, and their burrows facilitate oxygen penetration into the mud and thus ameliorate anoxic conditions. Fish are most common in this zone, and many of them are plankton-feeders. The fish, as well as crabs, lobsters and shrimp, form the basis for local fisheries.

The primary producers in this system include not only the mangroves themselves, but benthic algae, seagrasses, and phytoplankton. Few production studies have been conducted in mangroves because they are a particularly difficult environment in which to work. However, it is clear that mangrove swamps are rich in recycled nutrients. Although large quantities of detritus may be exported from a mangrove community, the roots also trap organic-rich detritus which is broken down and decomposed in the sediments; the recycled nutrients then become available to be taken up by the roots of the mangroves. Thus the mangrove system is not solely dependent on nutrients dissolved in the surrounding oligotrophic seawater. Mangroves are also located in regions of intense solar radiation, and the combination of high nutrients plus high light should lead to high gross primary production rates. Plant respiration is variable and is possibly related to the degree of salinity stress in particular localities. However, it is estimated that mangrove swamps contribute between 350 and $500 \text{ g C m}^{-2} \text{ yr}^{-1}$ net production to coastal waters.

QUESTION 8.9 How do values of net primary productivity in mangrove swamps compare with those for phytoplankton production in tropical nutrient-deficient oceanic waters? Refer to Table 5.1.

8.7.3 IMPORTANCE AND USES OF MANGROVES

Mangroves figure importantly in the livelihoods of peoples living within or adjacent to these habitats. The trees themselves have traditionally been used for firewood and charcoal. The timber is water-resistant, so it is also used to construct boats and houses. The leaves are used for roof thatching and as fodder for cattle and goats. Even cigarette wrappers are manufactured from the young, unfolded, leaf sheaths of a certain mangrove species.

Most of these tropical coastal communities have long-standing fisheries based not only on fish like mullet, but also on the abundant populations of shrimp, crabs, bivalves, and snails. Fish nets and traps are often constructed, at least in part, from parts of the mangrove trees, and tannin extracted from the mangroves is used to increase the durability of fishing nets and sails.

Mangroves also have great importance in non-commercial aspects. They form protective barriers against wind damage and erosion in regions that are subject to severe tropical storms. In some areas, mangroves may facilitate the conversion of intertidal regions into semi-terrestrial habitats by trapping and accumulating sediment. For example, mangroves have spread seawards at rates of between 100 m and 200 m per year in Indonesia. The root system also serves as a protective nursery ground for many species of fish, shrimps, juvenile spiny lobsters, and crabs. The forest canopy not only supplies food for many of the arboreal and marine inhabitants, either directly or as detritus, but it is utilized for nesting sites for a variety of tropical birds.

8.8 DEEP-SEA ECOLOGY

The vast majority of the seafloor lies permanently submerged below tidal levels yet, relative to the intertidal regions, comparatively little is known about life in the bathyal, abyssal, and hadal zones (see Figure 1.1). This, of course, is due to their relative inaccessibility. Although it is possible to dive to several thousand metres in submersibles or to employ remote-controlled cameras, the number of hours of direct observations in the deep sea has so far been extremely low. Most information on deep-sea ecology comes from indirect inferences based on animals contained in benthic samples obtained from ships. Whatever the method, expense is a limiting factor in deep-sea research. Few countries or institutions have submersibles to use for basic research, and few have large research ships equipped to obtain deep-sea samples. Collecting a sample from 8000 m depth with towed gear, for example, requires a very large winch with at least 11 km of cable in order to allow for the towing angle. It may take up to 24 hours to let out that much wire, obtain a sample, and then retrieve it. With large ship costs easily exceeding tens of thousands of dollars per day, a single sample containing a few benthic animals can be beyond the budget of most oceanographic research facilities. Compounding this problem is the fact that animal life is just not very abundant in many deep-sea areas, so that it is desirable to have large numbers of samples. Nonetheless, new techniques for collection and observation, combined with accumulating numbers of analysed deep-sea samples, permit a general assessment of benthic life in deeper water.

The deep-sea environment has been generally regarded as stable and homogeneous in terms of many physical and chemical parameters. Water temperatures are generally low (from $-1°$ to $4°C$) and salinity remains at slightly less than 35. Oxygen content is also constant and is rarely limiting, with the exception of areas beneath upwelling zones or in some basins where water exchange is slight (e.g. the Cariaco Trench in the southern Caribbean Sea). Soft bottom sediments, originating from land and/or from the sinking of dead planktonic organisms, cover most of the deep seafloor. Hard substrates are largely limited to mid-ocean ridges and seamounts that jut up from the sea bottom. Relative to surface currents, bottom currents in the deep ocean basins are slow (generally <5 cm per second) but more variable than once believed. Some areas experience abyssal (or benthic) storms lasting up to a few weeks, during which bottom currents increase in speed and may reverse direction. Deep boundary currents that move along continental margins may have velocities of up to 25 cm s^{-1}, and these may cause sediment resuspension and thus influence sediment redistribution. Deep-sea environments may experience seasonal variability in the amount of organic material that sinks from the euphotic zone to the seafloor.

8.8.1 FAUNAL COMPOSITION

Most animal phyla are represented in this dark environment of low temperatures, high pressures, and predominantly soft substrates. It has been known since the time of the *Challenger* expedition that there are, however, changes in the relative abundance of different types of zoobenthos with increasing depth. Figure 8.13 is based on benthic samples taken from the Kurile-Kamchatka Trench in the North Pacific in the 1950s. It shows, for example, that sponges form a dominant component of the macrobenthic biomass between 1000 m and 2000 m, but they are small and scarce below 2500 m. Starfish are important members of the trench community down to 7000 m, at which depth they disappear. Holothurians (sea cucumbers), however, increase in relative abundance in deeper areas; one species (*Elpidia longicirrata*) makes up about 80% of the total biomass in the trench.

Globally, holothurians frequently dominate the biomass in depths over 4000 m where sediments are relatively rich in organic matter. Numerically,

Figure 8.13　The percentages of different animal groups in the biomass of macrobenthos at different depths in the Kurile-Kamchatka trench.

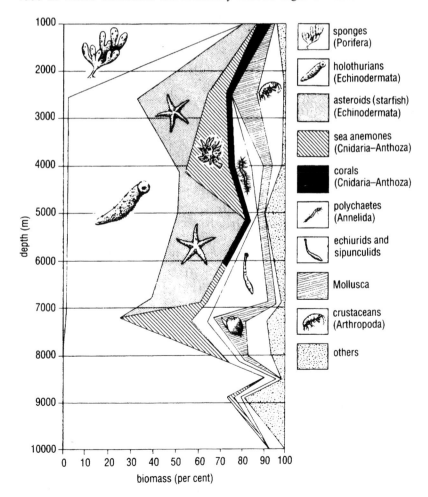

small burrowing polychaetes commonly make up 50–75% of the macrofauna in many, widely scattered, soft-bottom deep-sea sites. Small crustaceans (amphipods, isopods, tanaids) are also common deep-sea macrobenthic species, followed by molluscs (especially clams), and a variety of worms (sipunculids, pogonophora, echiurids, and enteropneusts). Brittle stars (ophiuroids) can be abundant in some areas; for example, they make up over 60% by numbers of the macrobenthos in the Rockall Trough west of Ireland.

Certain groups of animals attain their greatest abundance and diversity in the deep sea. The soft-bodied or calcareous sponges that are common in shallow water are largely replaced by glass sponges with siliceous spicules in deep water. Cnidaria are principally represented in the deep sea by sea anemones that live in burrows, and by sea pens and gorgonian corals that may form densely populated beds under eutrophic waters where there is sufficient suspended material for feeding. Slender, branching colonies of black corals have been found in the greatest depths. The Pogonophora are mostly a deep-water group found down to 10 000 m, and echiurid worms become more common in depths exceeding 5000 m. Some echiurids reach body lengths of 1 m, and they can occur in dense aggregations in organically-rich sediments where they form a large proportion of the biomass. The more primitive crinoids, the stalked sea-lilies, are mostly restricted to deep-sea habitats.

Benthic foraminiferans and related protozoans, the giant xenophyophores (see Section 7.2.1), increase in importance in deep water, both in terms of abundance and biomass. Unlike the shallow-living species with calcareous tests, deep-water foraminiferans have proteinaceous tests or exoskeletons made of agglutinated sediment particles. In certain areas, 30–50% of the seafloor may be covered by foram pseudopodia and, in the Aleutian Trench, forams comprise 41% of the meiofauna. Xenophyophores are known to occur in nearly all areas of the deep-sea basins at depths below 1 km. They may occur in densities of up to 20 m^{-2}, and they constitute up to 97% of the total benthic biomass in some areas of the South Pacific.

Some animal groups show a tendency toward gigantism in the deep sea (see also Section 4.4); these include the benthic foraminifera and the xenophyophores, as well as certain amphipod species that attain lengths of about 28 cm. However, there is a reverse tendency in some groups toward miniaturization, and the deep-sea meiofauna is numerically dominant over the macrobenthos. Nematodes are ubiquitous in marine soft substrates and make up 85–96% of the deep-sea meiofauna. Harpacticoid copepods and ostracods are also common deep-sea members of this size category, the former group constituting 2–3% of the meiofauna in abyssal zones. Tanaids are extremely diverse in the deep sea, and many of the species are meiofaunal; in the north-west Atlantic, they occur in densities of about 500 m^{-2}.

Certain animal groups are poorly represented in deep water. Decapod crustaceans (e.g. crabs, shrimp, lobsters), sea anemones, and echinoid echinoderms are absent or uncommon below about 6000 m. Fish are also rare in very deep waters; one of the deepest captured fish came from 7230 m in the Kurile-Kamchatka Trench. These generalizations are largely based on collections made with dredges or trawls, both of which are difficult to use over rocky substrates or in relatively steep-walled trenches, and both of which can be avoided by swimming animals. It is well to keep in mind that Jacques Piccard and Lieutenant Don Walsh, who together made the deepest dive in a bathyscaphe, observed flatfish and shrimp at over 10 000 m; neither of these groups of animals have been collected by conventional gear from such great depths.

QUESTION 8.10 What explanations can you give for why large sponges are so successful in shallower areas, and why sea cucumbers often dominate the macrobenthos of deep water?

[handwritten margin notes: sponges are epifaunal e suspension feeders; sea cucumbers are... infaunal or epifaunal c. deposit or detritevores]

Some deep-sea residents have a cosmopolitan distribution and are found in all the major oceans; other species are restricted to relatively small areas. In general, species become more limited in geographic range as water depth increases. Only about 20% of the species present below 2000 m in the Atlantic Ocean are also found in the Pacific or Indian oceans.

Table 8.1 Percentages of species living below 6000 m depth that are endemic to the hadal region.

Taxonomic group	Number of hadal species	% endemic
Foraminiferans	128	43
Sponges	26	88
Cnidaria	17	76
Polychaetes	42	52
Echiurid worms	8	62
Sipunculid worms	4	0
Crustaceans		
barnacles	3	33
cumaceans	9	100
tanaids	19	79
isopods	68	74
amphipods	18	83
Molluscs		
aplacophorans	3	0
snails	16	87
bivalves	39	85
Echinoderms		
crinoids	11	91
holothurians	28	68
starfish	14	57
brittlestars	6	67
Pogonophorans	26	85
Fish	4	75

Many species found in areas deeper than 6000 m are endemic to the hadal region, and many are restricted to a particular trench. Table 8.1 lists the number of hadal species known in particular invertebrate groups, and the percentages of these that are endemic to this deep-sea region. Endemic species constitute up to 75% of the benthos in certain Pacific trenches. The high degree of endemicity suggests that trenches are fundamentally isolated habitats which are centres for the generation of new species.

Trenches often have relatively high abundances of aplacophorans (wormlike, shell-less molluscs), enteropneust worms, and echiurid worms, all of which are poorly represented elsewhere. In the Aleutian Trench at depths of 7000–7500 m, the macrofauna is dominated by polychaetes (49%), bivalves (12%), aplacophorans (11%), enteropneusts (8%), and echiurid worms (3%).

The meiofauna is dominated by benthic foraminifera (41%), followed by nematodes (36%) and harpacticoid copepods (15%). Some of the zoobenthos of trenches exhibit pale coloration and are blind, characteristics that are shared with cave fauna. Very large size is also a characteristic of some hadal isopods, tanaids, and mysids.

Deposit-feeding infaunal animals are dominant in the soft, organically-rich sediments of the deep sea, usually comprising 80% or more of the fauna by numbers. At one site in the Atlantic, at a depth of 2900 m, 60% of the polychaetes, >90% of the tanaids, 90% of the isopods, >50% of the amphipods, and 45% of the bivalves are deposit feeders. Other groups, like sea cucumbers and sipunculids, also ingest the detritus and small organisms contained in surface or subsurface sediments. Because bottom currents are usually slow and do not disturb compacted sediments, the topographic features produced by these animals persist for long periods. Faecal mounds, burrows, trails, and tubes are some of the biological features that are commonly recorded in the deep sea by remote cameras.

Animals that feed on suspended particles are also found in the deep sea, but they are much less abundant and are usually restricted to particular localities. This is because most of these epifaunal animals require relatively firm substrates for attachment as well as high concentrations of suspended food particles. As a result, many types of epifaunal suspension feeders (e.g. common sponges, sea anemones, barnacles, mussels) show a marked decrease in abundance with increasing depth and distance from shore. They do flourish, however, on the rock substrates found on mid-ocean ridges and seamounts, and some may dominate in deep-sea sulphide communities (see Section 8.9). Relative proportions of deposit feeders and suspension feeders vary throughout the deep sea according to the degree of organic enrichment of the sediments and the supply of suspended food.

It is useful, in discussing deep-sea distribution of epifauna, to distinguish between two types of suspension feeding. **Active suspension feeders** (e.g. shallow species of sponges and tunicates) use their own energy to pump water through a filtering structure. These animals are successful in environments where suspended particle concentrations are high enough to repay the energetic costs of pumping. **Passive suspension feeders** (e.g. crinoids, some polychaetes, and most benthic Cnidaria like sea anemones and sea-fans) rely on external water currents to convey food to feeding appendages that are held into the flow. The passive feeders succeed only in environments where flow conditions are predictable and fast enough to supply them with sufficient particulate food. Suspended particle concentrations decrease with depth, but flow conditions become more predictable. Active suspension feeders disappear as suspended loads diminish, and the large filter feeders in the deep sea are passive feeders. Depending on the current speed, the friction of water moving over the seafloor may create physical mixing of the bottom water; this **benthic boundary layer** extends from 10 to several 100 metres above the bottom. Turbulence in this layer can result in resuspension of bottom sediments; heavy inorganic particles remain close to the seafloor, but suspended light organic particles will reach maximum concentrations some distance above the bottom. Typical deep-sea passive suspension feeders, such as sea-lilies (stalked crinoids) and bryozoan colonies, are found in highest densities where there is moderate current flow and resuspension of sediments. In contrast to their shallow-dwelling relatives, these animals are often supported by long stalks which hold them well above the seafloor where concentrations of suspended organic material may be optimal.

Some deep-sea groups have very different feeding mechanisms from those used by shallow-water related species. Glass sponges, for example, have an extremely porous body wall and water currents can enter passively as well as by active pumping. Members of one family (Cladorhizidae) of small-sized sponges occur in deep water to about 9000 m depth; at least some of them are highly modified carnivores that passively capture small swimming prey by means of filaments provided with hook-shaped spicules. Whereas shallow-water tunicates are active suspension feeders and many are colonial species, deep-sea representatives tend to be solitary forms that may supplement active pumping with mucous nets held into currents to capture food. Some deep-sea tunicates are even more highly modified and have become carnivorous.

In addition to deposit feeders and suspension feeders, the deep-sea food chain includes many scavengers. Cameras have recorded the speed with which a variety of swimming animals approach bait placed on the seafloor in deep water. These include giant amphipods, isopods, fish, and shrimp. Brittle stars and some polychaetes are among the slower-moving scavengers. Strict predators appear to be rare in very deep waters. However, diets of deep-sea benthic animals are not well known; feeding type is usually inferred from anatomical structure and gut contents.

8.8.2 SPECIES DIVERSITY

The number of species of many types of macrobenthos (e.g. snails, clams, polychaete worms) and fish tends to increase with depth from about 200 m to 2000 or 2500 m, then declines rapidly with further depth. Based on these observations, it was believed for many years that deep-sea species diversity was low compared with that of shallow-water communities. However, the development and use of a new collection device, called an **epibenthic sled**, changed this perception. The epibenthic sled (Figure 8.14) was designed to

Figure 8.14 An epibenthic sled designed to collect animals living on or just above the seafloor. The mesh-size is small enough to retain meiofauna, and the sampler can be closed during retrieval so that the entire sample is retained.

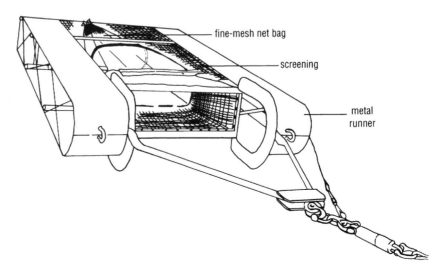

collect and retain smaller animals, in a size category that was previously not well sampled. When this gear was first employed in the 1960s, a single collection sometimes contained more animals than were collected by all the combined expeditions of the previous 100 years. To further demonstrate the effectiveness of this apparatus, one paper reported that over 120 new species of cumaceans (small crustaceans) had been collected. It soon became evident that the diversity of many smaller organisms increases with depth. For example, the number of species of meiobenthic copepods increases to at least 3000 m, and maximum diversity of benthic foraminiferans is found in depths exceeding 4000 m.

It is now established that there is high species diversity in the deep sea, especially among the small infaunal deposit feeders. As additional samples are obtained from the deep sea and more new species are described, the more diversity in this area seems to approach that of highly diverse terrestrial environments, such as the tropical rain forest. Some researchers estimate that there may be more than one million species of marine benthic animals, most of them living in deep-sea sediments. Species diversity does, however, vary in different oceanic areas. For example, in the North Atlantic, species diversity declines from the tropics toward north polar regions; but in the Southern Hemisphere, zoobenthos species diversity in the Weddell Sea (Atlantic sector of the Antarctic) is of the level normally associated with tropical regions. Deep-sea diversity also may vary according to different levels of surface primary production. In some areas, zoobenthos diversity is depressed under areas of upwelling and high surface productivity, probably as the result of reduced oxygen concentrations from decomposition of large amounts of organic material.

As more areas of the deep sea are surveyed with increasingly sophisticated gear, it is becoming apparent that the environment itself, in terms of substrate features and/or current regime, is more diverse than was once thought. Environmental diversity in the form of microhabitats (small areas having slightly different environmental characteristics) can itself lead to higher diversity in animals. Indeed, the deep-sea benthos is patchily distributed, with significant aggregations of animals having been detected in different taxonomic groups on scales ranging from centimetres to metres to kilometres. This patchy distribution underscores the importance of obtaining representative samples when assessing biomass and species diversity of deep-sea animals.

8.8.3 BIOMASS

Although the number of species is high in the deep sea, communities occupying the typical soft-sediment seafloor are characterized by low population densities and low biomass. Numbers of benthic individuals (both macrofauna and meiofauna) per unit area tend to decrease roughly exponentially with increasing depth and, to a lesser degree, with distance from shore. Under the central oceanic gyres, the total density of macrofauna ranges from 30 to 200 individuals per m^2. With a few exceptions, the dominant infaunal species tend to be small as well as sparse. In the central North Pacific, meiofauna and microfauna dominate the benthos in numbers (0.3% and 99.7%, respectively), and in biomass (63.8% and 34.9%, respectively). Deep-sea biomass values do not include the larger demersal species, which are more difficult to capture and to quantify on an areal basis; their inclusion would undoubtedly increase the biomass values given here.

Table 8.2 Average biomass values of benthic animals at different depths.

Depth range (m)	Mean biomass (g wet weight m^{-2})
Intertidal	3×10^3
to 200	200
500–1000	<40
1000–1500	<25
1500–2500	<20
2500–4000	<5
4000–5000	<2
5000–7000	<0.3
7000–9000	<0.03
>9000	<0.01

Average biomass values for different depth zones are given in Table 8.2. Benthic biomass is highest in shallow coastal areas within the euphotic zone, and it is lowest under oligotrophic, central, oceanic regions. Keeping in mind that the average depth of the world ocean is 3800 m, most of the seafloor supports less than 5.0 g wet weight m^{-2} of living organisms.

However, benthic biomass at any particular depth varies according to the amount of organic material delivered to the seabed. This is reflected, for example, in the different levels of biomass found in various trenches, all of which are deeper than 6000 m. Trenches are located in regions of frequent seismic activity, and thus are subject to brief episodes of high sedimentation caused by slumping of sediments down the trench walls. This results in the deposition of organically rich sediments from shallower depths, but also in the burial of benthic communities. The biomass of hadal fauna may be very high in trenches that lie near large land masses as they receive land-derived sediments and organic matter, as well as organic material sinking from the overlying nutrient-enriched and highly productive surface water. Under these conditions, benthic biomass may range from 2 to 9 g wet weight m^{-2} at depths of 6000–7000 m in the Kurile-Kamchatka Trench (North Pacific) and in the South Sandwich Trench (South Pacific). Trenches far from land masses and under oligotrophic water (e.g. Mariana Trench) have very low biomass values of about 0.008 g m^{-2}.

Benthic productivity cannot be assessed directly from biomass values, but many deep-sea species grow relatively slowly and their small biomass must indicate low productivity. Various estimates suggest that annual secondary production over most of the deep ocean floor is between 0.005 and 0.05 g C m^{-2}.

QUESTION 8.11 How does the decline in benthic biomass with depth compare with the vertical distribution of zooplankton biomass? Refer to Section 4.4 and Figure 4.14. *Both decline with Depth.*

8.8.4 FOOD SOURCES

Except for localized chemosynthetic production around deep-sea hot springs (see Section 8.9), there is no primary production in the dark, deep areas of the sea. Food availability, not low temperature nor high pressure, limits benthic biomass in the deep sea. The deep-sea food chain is dependent on surface production, and only a small percentage (1–5%) of the food produced in the euphotic zone is transferred to the abyssal seafloor. The

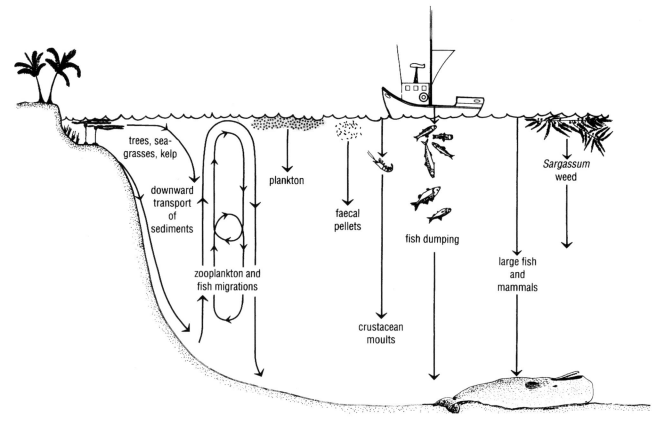

Figure 8.15   Potential food sources for deep-sea organisms.

percentage diminishes with increasing depth because of the increasing probability that organic particles sinking from the euphotic zone will be consumed or will decay before reaching the bottom.

A variety of potential food sources sink from the productive surface zone and thus may become available to deep-sea benthos (Figure 8.15). The relative contribution of each of these sources depends on their sinking rates and attrition in intermediate depths.

1. Dead phytoplankton, zooplankton, fish, and mammals. In many areas of the ocean, much of the phytoplankton is consumed by herbivorous zooplankton in the euphotic zone. That fraction which is not eaten sinks very slowly because of its small size and, in very deep areas, it is lost through predation, disintegration, or decomposition at intermediate water depths. In some regions like the North Atlantic, however, the phasing between seasonal phytoplankton blooms and zooplankton growth is such that much of the phytoplankton dies without being consumed (see Section 3.6), and the sinking phytodetritus may reach the seabed at depths down to 4000 m. The fate of most uneaten epipelagic zooplankton is similar, although sinking rates are somewhat faster because of their larger size. The corpses of large fish, squid, or marine mammals sink rapidly enough so that they may reach the seafloor in deep waters, where they become available to benthic scavengers; however, some are no doubt consumed by large animals at intermediate depths. In any event, the arrival of large animal carcasses on the sea bottom is generally an unpredictable and rare event, except perhaps under the

seasonal migration routes of some fish and mammals. This is not the case, however, in heavily fished regions where tonnes of unwanted, incidentally captured fish ('trash' fish) are dumped back into the sea. The dumping of unmarketable by-catch may amount to a very significant fraction of the total reported fish catch, but it will only locally increase benthic food supplies.

2. Faecal pellets and crustacean moults. Compact faecal pellets of some zooplankton (about 100–300 μm in size) are collected in mid-water traps designed to capture sedimenting particles, and their settling rates are such that they may reach the seafloor relatively intact. Faecal material of fish may also reach the seabed almost undegraded. Although some animals ingest faecal pellets, these wastes generally contain large fractions of indigestible materials. Moulted exoskeletons of planktonic crustaceans occur in the benthic boundary layer, but they may result from deep-water species. Moults are also low in nutritional value as they are composed primarily of chitin which cannot be digested directly by most animals, but it is broken down by **chitinoclastic** bacteria living in the guts of many species. Faecal pellets and moults are colonized by bacteria during their descent in the water column, and they are eventually converted to bacterial biomass. Bacteria are important nutritional intermediates in the food cycle of the deep sea, and they comprise a major food source for benthic deposit feeders. There is, in fact, an increase in bacteria in the bottom sediments of the deep sea, with numbers exceeding one million per gramme of sediment between depths of 4000 and 10 000 m.

3. Macrophyte detritus. A certain amount of organic material enters the sea near coastal zones in the form of wood from terrestrial plants, or from dislodged seagrasses and kelp. Some of this is carried well offshore in currents before becoming waterlogged and sinking. *Sargassum* is also a potential source of organic material in areas where it occurs (see Section 4.4). Larger plant particles sink rapidly enough to reach the seafloor more or less intact. Panels of wood placed at a depth of 1830 m and recovered 104 days later were riddled with burrows made by wood-boring clams, some of the very few animals that can utilize wood for food. These bivalves convert woody plant material to foods that are available to other animals. They produce faecal pellets that can be consumed by detritivores; those larval or adult clams that become exposed by the disintegration of the wood can be eaten by predators; and their dead remains become available for scavengers. Bacteria decompose and convert other types of macrophyte detritus into biomass available to benthic animals.

4. Animal migrations. The vertical migrations of zooplankton and fish result in a downward transfer of organic materials. Food that is captured in shallower depths is converted to animal biomass that may be consumed by predators at deeper levels, and faecal pellets may be released by migrators when they return to deeper water (see Section 4.5). Some deep-sea fish (e.g. angler-fish) spend their larval stages near the surface and then migrate into the depths as juveniles or adults, where they become potential food for deep-sea predators. All of these events accelerate the pace at which food enters the deep sea.

There is seasonal variability in the amount of organic material reaching the seafloor in temperate and high latitudes due to seasonal differences in surface production. The sinking of large amounts of macrophyte detritus also may be linked to seasonal storms that dislodge seagrasses or trees. The dumping of trash fish also is largely restricted to seasons of relatively calm weather and availability of fish schools. Some fish die after spawning (e.g. mesopelagic blue whiting), and their carcasses deliver a seasonal signal to the underlying deep-sea fauna.

Although some deep-sea species have seasonal reproduction, and growth bands in mollusc shells and echinoderm skeletal plates reveal seasonal growth in certain species, many deep-sea animals have continuous reproduction and their production does not appear to be linked to seasonal surface events. Secondary production may be linked, however, to the sporadic nonseasonal occurrence of an adequate food supply.

Although life in most of the deep sea is dependent on surface production, the deep-sea environment is spatially and temporally separated from the euphotic zone. In general, it is estimated that 75–95% of the organic matter in particles sinking from the euphotic zone is decomposed and recycled in the upper 500–1000 m of the water column, above the permanent thermocline. In the Sargasso Sea, surface production is >100 mg C m^{-2} day^{-1} and the flux to the bottom at over 3000 m varies from 18 to 60 mg m^{-2} day^{-1} depending on season. However, organic matter constitutes only about 5% of the total sedimenting material, the remainder being mostly inorganic carbonate and silicate. In the north-east Atlantic, seasonally deposited phytodetritus has an even lower organic carbon content of less than 1.5%; however, deep-sea animals have been observed to feed on this material despite its low nutritive value. In general, only about 5–10% of organic matter produced in the euphotic zone will reach depths of 2000 to 3000 m, and progressively less in abyssal and hadal zones. Thus food is very scarce in the deep sea compared with other ocean regions. Food limitation is one of the most important factors governing biological processes and community structure of the deep-sea benthos.

$$\frac{5700 \text{ m}}{60 \text{ m d}^{-1}} = 95 \text{ days}$$

QUESTION 8.12 Using the maximum sinking rates for diatoms (from Section 3.1.1), how long would it take a single dead *Chaetoceros* cell to sink to 5700 m depth?

8.8.5 RATES OF BIOLOGICAL PROCESSES

Accumulating evidence suggests that various biological processes in deep-sea animals, such as metabolism, growth, maturation and population increase, are slow in comparison to such processes in shallow-water environments. One of the first pieces of evidence resulted from an accident at sea. In 1968, the research submersible *Alvin* slipped from its launching cradle after the pilot and scientists were on board, but before the ports were secured. All three men managed to exit safely, but the *Alvin* sank in 1540 m of water and carried their packed lunches with it. The submersible was not recovered until over 10 months later, at which time it was discovered that the scientists' lunches, although waterlogged, were still in good condition and edible. Placed for three weeks in a refrigerator at 3°C (the same temperature as at 1540 m), the food spoiled. This unexpected observation stimulated experiments in which organic substrates were exposed *in situ* at depths down to 5300 m; when these were recovered, they confirmed very low rates of bacterial metabolism in the abyss. The metabolic rate of abyssal bacteria living in sediments is from 10 to over 100 times slower than that of equivalent bacterial densities maintained in the dark, at the same low temperature, but at atmospheric pressure. Bacterial productivity is thought to range from about 0.2 g C m^{-3} day^{-1} (at 1000 m) to 0.002 g C m^{-3} day^{-1} (at 5500 m).

Low metabolic rates have also been reported for some benthopelagic animals, including teleost fishes. However, studies of large deep-sea epibenthic decapods and echinoderms indicate respiration rates comparable to those of related forms in shallow water when measured at the same temperature. When oxygen uptake by benthic communities is compared, that in the deep sea is two to three orders of magnitude lower than that of shallow-water communities. This reduction in oxygen uptake is partly due to a lower density of organisms per unit area, but it also reflects the lower metabolic activity of deep-sea organisms.

Another indication of slower biological processes came from studies of *Tindaria callistiformis*, a small (<9 mm long) clam inhabiting soft sediments of the North Atlantic at 3800 m depth. Radioactive dating of the shells of this little deposit feeder suggested that it grows extremely slowly, with the shell increasing in length at about 0.084 mm yr^{-1} (see Figure 8.16 for a comparison with other molluscs). This would mean that *Tindaria* requires 50 years to reach sexual maturity, and that its life span would be about 100 years. However, the technique used to obtain these results has since been criticized, and faster growth rates have been estimated for other deep-sea benthos.

The wood-boring clams (see Section 8.8.4) that rely on ephemeral falls of wood in the deep sea are notable exceptions to slow growth rates. In order to search out and colonize new sources of wood, they have evolved opportunistic life strategies that involve rapid growth, early maturity, and production of many young.

QUESTION 8.13 Would you consider *Tindaria callistiformis* and wood-boring clams to be primarily examples of *r*- or *K*-selected species? (Refer to Section 1.3.1 and Table 1.1.)

Many deep-sea species have low fecundity. The number of eggs produced per individual is generally much lower in many deep-sea residents when compared with their shallow-water relatives; this can be related to the miniaturization that occurs in many groups. One small (< 1 mm long) deep-sea clam, *Microgloma*, produces only two eggs at a time, and even larger deep-sea clams may produce only a few hundred eggs at any one time. In contrast, shallow-water clams typically spawn tens or hundreds of thousands of eggs. Although type of development is not known for many deep-water species, the dominant mode appears to be production of lecithotrophic larvae.

Low fecundity and therefore low dispersal suggest slow rates of recolonization in the deep sea, and this has been confirmed by experimental studies. Boxes of sterilized azoic sediment were placed at depths of 10 m and 1760 m, and examined after 2 months and after 26 months. The shallow boxes were colonized rapidly by invertebrates; after only 2 months, the boxes contained 47 species and 704 individuals (35 714 individuals m^{-2}). In comparison, the deep-sea boxes yielded only 14 species and 43 individuals (160 individuals m^{-2}) in the same period of time. Even after 26 months, recolonization of the deep-sea boxes was such that they contained 10 times fewer individuals and species than samples taken from the surrounding sediment at the same depth.

The low rates of metabolism and production shown by many deep-sea animals can be correlated with low food supply in their environment, but it is also likely that the food requirements of deep-sea species are much lower than those of surface-dwelling animals. Only a limited number of biochemical studies have been done on deep-sea animals, but these suggest that body protein concentrations and caloric content decrease with depth in fish and crustaceans. On a weight for weight basis, the food requirements of a slow-moving rat-tail fish living in the deep sea are likely to be about 20 times less than those of an active epipelagic salmon. Low rates of metabolism may also result from the physical–chemical constraints on enzyme kinetics that are known to occur at high pressures and low temperature.

8.8.6 FUTURE PROSPECTS

Much remains to be explored in the deep sea. Only a very small percentage of the seafloor has been examined using dredges, cores, submersibles, or remote cameras. By 1995, less than 100 m^2 of the deep seafloor had been *quantitatively* sampled. Certainly we can expect that many more new species will eventually be discovered. More detailed physical studies, particularly of water flow over the seafloor, may further alter our perception of homogeneity in this environment. For example, the effect of abyssal storms on benthos is not known.

We need to learn more about the biology of individual species in terms of their food and feeding patterns and energy requirements; their patterns of reproduction, development, and growth; and their interactions with other species. Only then can we approach an understanding of the factors that establish community structure, that determine production, and that maintain high species diversity in the deep sea.

8.9 HYDROTHERMAL VENTS AND COLD SEEPS

In 1977, scientists working off the Galapagos Islands discovered unusually high seawater temperatures at about 2500 m depth in an area where new seafloor is being formed. The hydrothermal activity accompanying this process manifests itself in the release of mineral-laden fluid either emitted as a warm (5–100°C) diffuse flow from cracks and crevices in the seafloor, or emerging as plumes of superheated (250–400°C) water from chimneylike vents. As the hydrothermal fluid mixes with the surrounding seawater, temperatures are moderated to between 8° and 23°C. Chemical analyses of the water in the vent areas revealed low concentrations of oxygen but very high concentrations of hydrogen sulphide (H_2S), a compound that is usually highly toxic to animals even in much lower concentrations.

QUESTION 8.14 What is the normal ambient seawater temperature at a depth of 2500 m? Refer to Section 2.2.2. 2 - 4°C

One of the most exciting events in benthic marine ecology occurred when scientists, diving in a submersible, discovered extremely dense concentrations of benthic animals living in this hydrothermal vent area. Since that time, similar deep-sea communities of animals have been found in other localities around the world, all of them in areas of tectonic activity. Sites at which biological investigations have been conducted include the Mid-Atlantic Ridge and spreading centres along the rim of the Pacific Ocean basin.

8.9.1 CHEMOSYNTHETIC PRODUCTION

Many of the animals in the densely populated vent communities are of extraordinarily large size. The occurrence of very high benthic biomass in deep waters, far removed from surface photosynthetic production, immediately raised the question of how these animals obtained sufficient food.

Further studies revealed the existence of a food chain driven entirely by geothermal (terrestrial) energy, and not dependent on solar energy. Vent communities are dependent on the presence of hydrogen sulphide, a reduced sulphur compound that is released in hydrothermal fluid. This compound is utilized by sulphur-oxidizing bacteria (e.g. *Thiomicrospira* and *Beggiatoa*), and the energy released by the oxidation is used to form organic matter from carbon dioxide by the same biochemical pathway that is employed by photosynthetic organisms. The reaction requires molecular oxygen which is provided by the surrounding seawater. The biochemistry of the chemosynthetic production can be generally summarized as:

$$CO_2 + H_2S + O_2 + H_2O \rightarrow \underset{\text{carbohydrate}}{CH_2O} + H_2SO_4$$

In vent communities, chemosynthetic bacteria are the primary producers of the food chain, and the bacterial biomass becomes available for consumption by higher animals. Mats of filamentous bacteria (up to 3 cm thick at some sites) can be grazed by animals like limpets, and bacteria suspended in water can be filtered by suspension feeders. In some cases, the bacterial production proceeds within tissues of host animals in special symbiotic relationships.

At present, the sulphur-oxidizing bacteria have received the most attention and are believed to comprise most of the bacterial biomass in vent areas. However, it is probable that other types of bacteria, utilizing different reduced materials (e.g. methane, ammonia) as sources of energy, also contribute to chemosynthetic production in these regions. In any event, bacterial production at hydrothermal vents is estimated to be two to three times that of photosynthetic production in the overlying water.

8.9.2 VENT FAUNA

Approximately 95% of the animals discovered at hydrothermal vents have been previously unknown species. To date, about 375 new species have been described, many requiring the establishment of new taxonomic families because they are so different from related species.

Spectacular giant, red, tube-dwelling worms found initially at the Galapagos vents proved to be a new genus and species, *Riftia pachyptila* (Colour Plate 39). These **vestimentiferans** (see Section 7.2.1 and Table 7.1) are encased in leathery tubes, with only a plume of many tentacle-like respiratory filaments protruding from the open end. They are highly unusual in lacking a mouth or digestive tract, but they are free-living and not parasitic. The largest vestimentiferans from the Galapagos site measure 1.5 m long and 37 mm in diameter, and have tubes of up to 3 m in length. These animals have extraordinarily high growth rates of up to 85 cm year^{-1}.

Densities of *Riftia* can be as high as 176 individuals m^{-2}, and biomass of *Riftia* alone ranges from 6800 to 9100 g wet weight m^{-2}. When combined with the wet weight of large bivalves living in the same site, biomass of this particular vent community can exceed 20–30 kg m^{-2}.

QUESTION 8.15 How does the biomass of the Galapagos vent community compare with typical biomass values at 2500 m depth? Refer to Table 8.2.

× 2000 – 3000 larger.

Riftia has a special internal organ known as a trophosome (meaning 'feeding body'), which contains masses of symbiotic bacteria. These bacteria make up to 60% of the dry weight of a *Riftia* individual, and it becomes a semantic question as to whether this organism is more a bacterial colony than a worm. The haemoglobin of *Riftia* is unique in being able to carry both oxygen and hydrogen sulphide simultaneously. The bacteria obtain energy from hydrogen sulphide brought to them in the blood system of the worm; the bacteria utilize CO_2 and the energy derived from the oxidation of the sulphide to form organic carbon. Some of this organic carbon is, in some way, passed into the tissues of the worm. Whether this is the sole source of nutrition for these large worms, or whether they are also able to absorb dissolved organic matter (e.g. amino acids) from the surrounding seawater, remains unanswered.

Another conspicuous and dominant animal at the Galapagos hydrothermal vents is a clam, *Calyptogena magnifica*, that reaches lengths of 30–40 cm (Colour Plate 40). The soft body parts of this bivalve are red, as they are in *Riftia*. In both animals, the colour is derived from haemoglobin in the blood. Most molluscs contain the blood pigment haemocyanin; its replacement in *Calyptogena* by haemoglobin, a more efficient oxygen-carrier, may be an adaptation to the low and variable oxygen concentrations in the surrounding water. The gills of *Calyptogena* contain masses of attached sulphur bacteria and the clams, like *Riftia*, benefit nutritionally from this symbiotic relationship.

Growth rates of *Calyptogena* have been calculated to be from 10 mm yr^{-1} to as high as 60 mm yr^{-1}; these rates are compared with growth rates of various other species of molluscs from different localities in Figure 8.16. Note that clams and mussels from vent areas have rates of growth that are comparable to those of shallow-water relatives, but they are approximately three orders of magnitude greater than those estimated for another deep-sea clam, *Tindaria* (see Section 8.8.5). Metabolic rates of *Calyptogena* and other

Figure 8.16   Growth curves for different species of shallow- and deep-water clams and mussels.

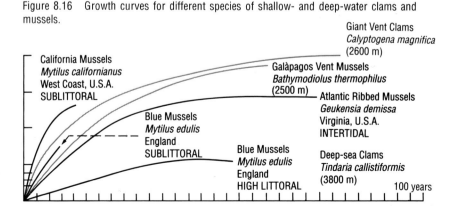

large vent animals are also similar to those of shallow-water relatives, and are orders of magnitude higher than those of related animals in other parts of the deep sea.

Related species of vestimentiferans and large clams similar to *Calyptogena magnifica* occur at other hydrothermal vent sites. Other vent molluscs include *Bathymodiolus thermophilus*, a giant mussel that has symbiotic bacteria on its gills but also is capable of suspension feeding, and limpets (>30 species) and snails which graze directly on mats of free-living bacteria covering hard surfaces. Several types of suspension-feeding polychaete worms are associated with vents, including tube worms (Family Alvinellidae) that attach in large numbers to vents emitting superheated water.

Deposit-feeding polychaetes are found in sediments around the vents. Long, thin enteropneusts, commonly known as spaghetti worms because of their appearance, can be abundant. Various types of crabs are found at most vents; some are scavengers, at least part of the time, and some prey on *Riftia*, small mussels, or polychaetes. Shrimp in densities of up to 1500 m^{-2} surround smoking vents on the Mid-Atlantic Ridge, and they apparently feed on bacterial mats. Sea anemones are abundant at certain sites, but other types of cnidarians are absent in these communities. A primitive type of barnacle dominates some Pacific hydrothermal localities. Fish are not usually important members of these communities; only five species have so far been recorded from vents.

Zooplankton are found in higher densities around vents than in surrounding waters. Copepods, amphipods and other planktonic crustaceans have been described, but few have been studied in detail. A new copepod genus (*Stygiopontius*), with seventeen species, occurs at every vent site. Vent meiofauna is dominated by nematodes and benthic copepods, as it is in other deep-sea regions. As more hydrothermal vents are discovered and sampled, the list of new species of animals grows rapidly.

8.9.3 SHALLOW VENTS AND COLD SEEPS

All of the explored hydrothermal vent sites are in waters deeper than 1500 m, but geothermally-driven chemosynthetic production is not restricted to the deep sea. There are, for example, hydrothermal vents that release high concentrations of sulphides in intertidal areas off southern California. There, benthic mats of sulphide-oxidizing bacteria contribute to the total primary production of the area along with photosynthetic production by benthic plants and phytoplankton. Limpets living near these vents are reported to graze on the bacterial mats, whereas limpets living in non-vent areas typically graze on photosynthetic algae encrusting rocks.

In 1984 a community of exotic organisms was discovered in the Gulf of Mexico, at the base of the Florida Escarpment. This massive limestone cliff rises some 2000 m above the sea bottom; at a depth of 3270 m, hypersaline waters containing high concentrations of sulphides and methane seep out onto the seafloor. Although the water temperature is low, the organisms in this cold sulphide seep area are remarkably like those found in hydrothermal vents. White bacterial mats cover exposed substrates. There are dense concentrations of 1-m-long tube worms (a new genus and species of vestimentiferan) as well as thick patches of large mussels and clams (the latter a new species of *Calyptogena*). Snails, limpets, and crabs are also conspicuous inhabitants of this particular seep.

Cold-water seeps result from a variety of causes, and they have been found along continental margins and in subduction zones where oceanic crust is carried back down into the Earth's mantle. Those that have been explored support similar assemblages of animals. Seep communities are also dependent upon chemosynthetic production by sulphide bacteria, and not on photosynthesis. The discovery of cold seeps demonstrates that the most important component necessary for high biological production in the deep sea is a source of reduced inorganic compounds, not heat. The existence of hydrothermal vents and cold seeps in deep waters indicates that low temperature and high pressure do not limit activities of deep-sea organisms. In deep areas where benthic ecology is dependent on photosynthetic production at the surface, biological processes and benthic production are limited by low food. In areas where sulphide-based food chains are possible, biological production in the deep sea may exceed that in the euphotic zone.

Although each hot vent or cold seep site has distinctive physical features and distinctive fauna, the communities of animals associated with high sulphide concentrations are similar in some respects. The dominant species are often ecologically similar, if not taxonomically related. Large vestimentiferans commonly occur at many sites, as well as similar species of limpets, clams, mussels, and crabs. All of the communities are characterized by having high population densities, high biomass, and rapid growth rates. These are unique concentrations of life in depths that usually are characterized by low density and low productivity; as such, these sulphide communities are appropriately referred to as 'oases'. Bacterial chemosynthesis is the major source of food in all of these communities. No sunlight is necessary and no photosynthetic production is needed from the surface; the populations of organisms are sustained entirely by inorganic materials that are converted into bacterial biomass, which then becomes available for consumption by higher animals.

8.9.4 UNIQUE ENVIRONMENTAL FEATURES OF SULPHIDE COMMUNITIES

Although hydrothermal vents and cold seeps support communities with high biomass, species diversity is low compared to that at other deep-sea localities. Endemic species predominate; over 90% of the animals found at vents and seeps do not occur outside their special habitats. The environments of vents and seeps possess certain attributes that place physiological constraints on animals and require special adaptations. Many animal groups are not represented at these sites, presumably because they have not evolved the ability to cope with the special conditions. With a few exceptions, cnidarians other than sea anemones, echinoderms of all types, sponges, xenophyophores, brachiopods, bryozoans, and fish are uncommon or absent. Molluscs, polychaetes, and crustaceans account for more than 90% of all vent species.

Hydrothermal vents, in particular, are transient environments that undergo large and rapid changes. The geological processes that create vents are dynamic events, and new vents are being formed as others close. Vent communities probably persist only for several years to several decades. Old, inactive vents are surrounded by the shell remains of clams and mussels that died when their energy source disappeared.

Animals living around vents are subjected to high temperature variance and to an oxygen concentration that can switch rapidly from anoxic to oxic conditions. There are also short-term fluctuations in H_2S concentrations. Salinity in vent plumes varies from about one-third to twice that of normal seawater. Hydrothermal fluid contains many inorganic substances which precipitate upon contact with seawater. The chimneys that emit superheated water are formed from the precipitation of sulphide deposits containing copper and zinc. This also means that animals are subjected to a rain of inorganic precipitates that coat their surfaces, and that they are exposed to potentially toxic concentrations of dissolved and precipitated heavy metals.

Both vents and seeps contain very high concentrations of H_2S (to 19.5 mM in hydrothermal fluid), and this raises the question of how the animals escape being poisoned by the high levels of H_2S. Hydrogen sulphide, even at concentrations less than one-thousandth of those found in some vent animals, poisons aerobic respiration. Only certain bacteria have the appropriate enzyme systems to oxidize this molecule in order to obtain chemical energy. Preliminary studies on *Riftia* indicate that it has special biochemical adaptations that protect the worm from H_2S toxicity. These include a special sulphide-binding protein in its blood, and enzyme systems in its body wall that oxidize any free sulphide entering the cells. Other vent species may be similarly equipped with sulphide-detoxifying systems.

Hydrothermal vent communities are quite small, usually only about 25–60 m in diameter, and they (as well as cold seeps) may be separated from other similar communities by as much as hundreds to thousands of kilometres. Vents are also ephemeral, lasting only on scales of years.

Given these features, how can animals maintain populations in widespread localities and how are new vents populated?

In order to succeed in short-lived, scattered habitats, animals of these communities could be expected to grow rapidly to sexual maturity, produce many young, and have efficient means of dispersal (see Section 1.3.1). Such *r*-selected traits would allow them to reproduce within the time span of their habitat and to continually colonize new vent areas. Although reproductive studies of vent fauna are few, zoogeographic studies suggest that vent species do rely on larval dispersal. On-going research into types of larval development and dispersal mechanisms is seeking information on these aspects.

8.10   SUMMARY OF CHAPTER 8

1   Relative to most other marine habitats, intertidal areas are characterized by great fluctuations in environmental conditions. Littoral plants and animals are specially adapted to cope with variable temperatures and salinity, and to withstand periodic exposure to air.

2   Rocky intertidal regions support dense communities with a high proportion of epiflora and epifauna that may compete for limited space. Many of the sessile species are arranged in distinct vertical zones. The upper boundary of any particular zone is often set by physiological limits of the species, such as tolerance to desiccation and temperature change. The lower limits of zones are generally established by biological factors such as predation and competition.

3   Intertidal areas of sand beaches support communities in which the primary producers are benthic species of diatoms, dinoflagellates, and blue-green bacteria, and the resident animals are predominantly infauna and

meiofauna. The meiofauna are specially adapted to live on sand grains, or in the interstitial spaces between sand particles, by their small sizes, elongate shapes, protected integuments, and adhesive organs.

4   Annual primary production averages about 100 g C m^{-2} in rocky intertidal areas, with a maximum of 1000 g C m^{-2} in particularly favourable areas. Benthic primary productivity in sand beaches is less than 15 g C m^{-2} yr^{-1}, and this system relies on energy derived from detritus and from primary production in the surrounding water.

5   Subtidal kelp forests occur on rocky substrates in cold temperate regions. Kelp are among the fastest growing of any plants, and the productivity of kelp beds ranges from about 600 to more than 3000 g C m^{-2} yr^{-1}. Much of this production is not consumed directly, but enters the detritus food chain. Sea urchins are dominant components of kelp communities, and their feeding activities greatly influence the community structure. In some North Pacific kelp beds, otters are the top predators that act as keystone species.

6   The circulation pattern of estuaries results in entrainment of nutrients and makes them some of the most productive of marine ecosystems. The upper reaches of estuaries are occupied by saltmarsh communities with total annual primary production generally ranging between 300 and >3000 g C m^{-2}. Seagrass beds typically form in intertidal areas of the middle reaches of estuaries, and total primary production by the seagrasses and associated epiphytes is about 600 to 1000 g C m^{-2} yr^{-1}. Both communities are dominated by detritus-based food chains. Estuaries also have subtidal mudflats or subtidal sand banks in which annual primary production (mostly by epipsammic algae) ranges from 10 to >200 g C m^{-2}.

7   The high productivity of estuaries supports dense populations of animals in some areas. However, many animals are excluded from living in estuaries because of the fluctuating salinity, and thus species diversity is low.

8   Coral reefs are formed by stony corals that contain symbiotic dinoflagellates called zooxanthellae. The algae utilize carbon dioxide and waste products of the coral in photosynthesis, and in return the coral is provided with organic compounds such as glucose and glycerol. Photosynthetic fixation by the zooxanthellae provides only part of the energy required by corals; the remainder is supplied by predation on zooplankton and bacteria, and by absorption of dissolved organic matter.

9   Primary producers of coral reefs include phytoplankton, benthic algae, and zooxanthellae. Gross primary productivity is very high, ranging from about 1500 to 5000 g C m^{-2} yr^{-1}, but production to respiration ratios usually are between 1.0 and 2.5, and very little new nutrient material enters the system (i.e. f-ratio is <0.1). The high production of this system supports a community with very high species diversity.

10   Corals grow relatively slowly, at rates of <1 to 10 cm yr^{-1}. Growth of a reef is also controlled by bioerosion and physical events (e.g. storms) that destroy the carbonate framework. Net vertical upward growth of reefs varies from a few to almost 30 mm yr^{-1} under favourable conditions.

11   Mangrove swamps occur along 60–75% of tropical and subtropical coasts. The major primary producers of these communities are salt-tolerant terrestrial plants that can live in oxygen-poor muddy substrates. The roots of the mangroves provide attachment sites for epifauna, and leaf fall is a major

source of nutrients and energy for the detritus-based food chain. Net primary production is estimated to be between 350 and 500 g C m^{-2} yr^{-1}.

12 The bathyal and abyssal zones together constitute over 90% of the benthic environment. Deposit-feeding infauna predominate in organically-rich sediments, and the meiobenthos are particularly diverse. Benthic biomass diminishes rapidly with increasing depth, and annual secondary production is between 0.005 and 0.05 g C m^{-2}. Most areas of the deep sea depend upon the fall-out from production in the euphotic zone, but only a small proportion of sinking organic matter reaches the seafloor in depths over 2000 m, and food limitation greatly influences biological processes and community structure of the deep-sea benthos. In general, typical deep-sea inhabitants exhibit slow metabolic rates, slow growth rates, and low fecundity.

13 Ocean trenches characteristically have a high proportion of endemic species. The biomass of hadal animals ranges from about 0.008 g m^{-2} in trenches far from land and underlying oligotrophic water, to as much as 9 g m^{-2} in trenches that lie near land under eutrophic water.

14 Hydrothermal vents and cold seeps support unique communities that are independent of solar energy and photosynthesis. Instead, the food chain in these environments is based on the presence of hydrogen sulphide that is utilized by chemosynthetic bacteria to form organic compounds from carbon dioxide. The bacteria are the primary producers in these communities, and they are either consumed directly by animals or they are found in symbiotic relationships with animals.

15 Deep-sea vents and seeps support extremely dense concentrations of large animals, and biomass may be as much as 30 kg m^{-2}. Although these environments have plentiful food and, in the case of vents, temperatures that are higher than usual in deep water, relatively few animals have developed the ability to live in high concentrations of H$_2$S and species diversity is low.

Now try the following questions to consolidate your understanding of this Chapter.

QUESTION 8.16   Apart from their high primary productivity, what important biological property do marshgrass and seagrass communities have in common that benefits animals?   *Refuge for small animals.*

QUESTION 8.17   If there is intense competition for space in the coral reef ecosystem, can you think of any reason(s) why the fastest growing species don't overgrow or crowd out other species and thus reduce the species diversity of the system?   *grazing by fish*

QUESTION 8.18   If sea level rose by 2 m in the next 100 years because of climate warming, would coral reef growth be able to keep up with the rise of coastal waters?   *Yes*

QUESTION 8.19   Although they are characteristic of different latitudes, tropical mangrove swamps and temperate saltmarshes share certain ecological features. What are these common characteristics?

QUESTION 8.20   In terms of quantity, how important might crustacean moults be in the sinking of organic materials into the deep sea? Refer back to Section 4.2 for help with your answer.

QUESTION 8.21   What are some of the biological or ecological advantages conferred on organisms that live at great depths?

QUESTION 8.22   Of the two types of marine food chains described in the deep sea, one based on bacterial chemosynthesis and one on algal photosynthesis, which would be evolutionarily older? (Refer to the Geologic Time Scale in Appendix 1.)

QUESTION 8.23   (a) Which benthic communities are characterized by having many endemic species? (b) What reason(s) explain the high degree of endemicity in these communities?

CHAPTER 9 | HUMAN IMPACTS ON MARINE BIOTA

Humans change marine environments and affect marine organisms in many ways. We harvest marine plants and animals, carry out mariculture, reclaim land, dam rivers that run to the sea, and dredge harbours. Marine organisms are transported around the world, and are deliberately or accidentally introduced into new areas. The sea has long been regarded as a convenient dumping site, and various pollutants are released into the marine environment from domestic or industrial outfalls or from accidental spills. Coastal ecosystems become enriched from nutrients contained in sewage, in discharged detergents, and in agricultural runoff. All of these activities, and others, may change the species composition of marine communities, result in loss of marine organisms or loss of marine habitats, or disrupt whole marine ecosystems. A summary of some examples of human impacts is given in Tables 9.1 and 9.2.

Table 9.1 Major impacts of industrial activities on marine environments.

Activity	Impact location	Effects
Harvesting fish	World-wide	Changes in the species composition of pelagic and benthic communities
		Changes in size structure of targeted fish populations
Fishing methods	World-wide (depending upon specific types of fisheries)	Benthic trawling destroys bottom habitat
		Dynamite fishing destroys corals
		Unselective fishing increases discarded by-catch
Discard of by-catch	World-wide	Increase in scavenging species
		Acceleration in delivery of nutrients to deeper water
		Possible increases in benthic biomass
Dam construction	Rivers running to sea	Loss of habitat for anadromous fish
Urban development	Estuaries; coral reefs; mangroves	Land reclamation leads to loss of habitat
		Sewage disposal and agricultural runoff may cause eutrophication
		Industrial runoff may pollute coastal waters
Commercial shipping	World-wide	Introduction of species into new environments
Coral mining	Tropical reefs	Destruction of corals

9.1 FISHERIES IMPACTS

What human activity has caused the greatest changes in the ocean?

The greatest and most serious human impact on marine ecosystems is caused by the annual removal of more than 100 million tonnes of fish and shellfish (reported catch plus by-catch, see Section 6.7.1). This harvest affects the species composition of pelagic communities as well as nutrient concentrations in surface waters (see discussion of f-ratio, Section 5.5.1). Mid-water and benthic communities may also be impacted by the dumping of dead by-catch which delivers a rich source of nutrients to deeper waters. There are also disruptive habitat changes caused by bottom trawling.

Advances in fishing technology have made it easier to locate fish schools and to catch more fish more effectively. At the same time, the world fishing fleet has increased rapidly, doubling to about 1.2 million vessels between 1970 and 1990. Long lines with thousands of baited hooks may extend more than 125 km from a ship, and some mid-water trawl nets with a mouth gape of 130 m and length of 1 km are large enough to encompass the Statue of Liberty or to extend around 12 jumbo jetliners. In some cases, more than 80% of a commercially lucrative stock is removed each year. These facts are reflected in Figure 9.1 which shows the increase in marine fish catch from less than 20×10^6 tonnes in the late 1940s to about 85×10^6 tonnes in 1993;

Figure 9.1   The marine fish catch (excluding by-catch) from 1947 to 1993, based on FAO statistics.

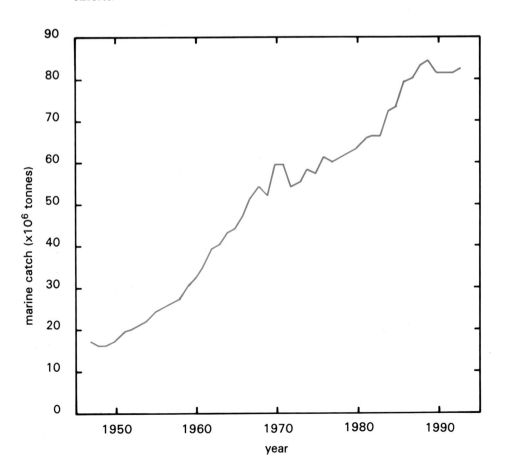

these figures do not include by-catch. However, it has become increasingly clear that many fish stocks throughout the world are now dwindling and that the **catch-per-unit-effort** (**CPUE**) of fishing has decreased. In the last two decades, fish catches have declined in all the major oceans except the Indian Ocean, where modern fishing fleets only began operating intensively in the late 1980s. A report issued by the United Nations Food and Agriculture Organization in 1995 concluded that 70% of the ocean's fish stocks are either fully exploited, overfished, or recovering from being overfished.

Although ocean climate change may be responsible for declines and changes in some fish stocks (see Section 6.7.2), overfishing is clearly responsible for the declines in many commercially favoured species. The 200-year-old fishery for cod and haddock stocks off eastern Canada and New England essentially ceased in the 1990s, although it may recover in time. Figure 9.2 shows how declines in these preferred species have resulted in changes in the abundance of other fish; skate and dogfish populations have increased by 35–40% since 1965, while cod, haddock, and hake have decreased by 45%. At the same time, less predation by cod and haddock on young lobsters has increased this lucrative shellfish harvest. Similar changes in community species composition have been documented in the North Sea, where sandlance greatly increased and became a target fish following declines in herring and mackerel stocks. In the Antarctic, commercial whaling and dramatic declines in the numbers of whales resulted in increased numbers of

Figure 9.2   Changes in the relative abundance of different fish species on Georges Bank, U.S.A., between 1963 and 1992, following overfishing of cod, haddock, and hake stocks. (Data from the National Marine Fisheries Service, U.S.A.)

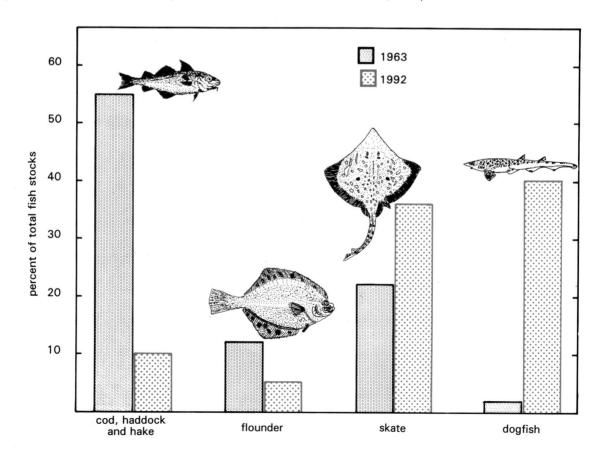

other animals that were dependent upon krill for food (see Table 5.2). In the 1960s, commercial fisheries in the Black Sea targeted 26 species of fish, many of them large predators with long life cycles. Overfishing, construction of dams, and pollution have reduced this to only five commercially viable species, all of small size. On the other hand, the total biomass of the Black Sea fishery harvest has actually increased due to the greater abundance of smaller fish species whose populations are no longer kept in check by predation, and to increased fishing effort. Overfishing may also cause changes in size structure of fish populations; for example, the total weight of spawning Atlantic swordfish fell by 40% between 1978 and 1989. There are also examples of declines in shellfish populations due to a combination of overexploitation and coastal pollution. Within 30 years, oyster catches in Chesapeake Bay, in the eastern U.S., fell from 20 000 to 3000 tonnes.

Anadromous fish, such as salmon, are adversely affected by the dams that block access to spawning areas. Despite expensive efforts to enhance stocks through hatchery rearing and construction of runways around dams, salmon stocks in the eastern North Pacific Ocean have declined in some rivers (e.g. the Columbia River). Loss of coastal spawning grounds and nursery areas through land development and/or pollution is another increasing problem for many fish species. Fishing activities may also destroy habitats. Heavy fish trawls can penetrate 6 cm or more into the seabed, thereby disrupting the natural substrate and releasing nutrients into the water column, and destroying zoobenthos that may be food for the demersal fish stocks. In particularly rich fishing grounds, more than 70% of the sediment can be ploughed by trawls.

Commercial fisheries discard about one of every four animals caught; the percentage of this unwanted by-catch may in fact be larger because much by-catch goes unreported. The discarded catch includes species with no economic value and young fish that are too small to market. In some cases, the by-catch may exceed the target catch. Shrimp fisheries have an exceptionally large by-catch. The shrimp fishery in the Gulf of Mexico catches and discards at least 5 million juvenile red snapper annually, or 4.2 kg of fish for every 1 kg of shrimp. World-wide, the total by-catch of shrimp fisheries may be as much as 17 million tonnes per year; much of this is made up of small fin-fishes, most of which do not survive capture and release. Discarded by-catch increases the number of scavengers in fishing grounds. In the North Sea, it is estimated that the annual discarded by-catch of about 90 000 tonnes of whitefish can potentially support about 600 000 gulls.

Unfortunately, as a valuable fishery species becomes more scarce, its economic value tends to increase. Thus it often remains profitable for fleets to continue to take an overfished species. Spawning populations of bluefin tuna (the world's most valuable fish) have declined by about 80% in the western Atlantic since 1970, and by 90% in the Gulf of Mexico since 1975, but a single large specimen may fetch more than US $80 000 (about $265 per kg in 1996) on the market.

Declining world fish catches have alerted nations to the fact that it is indeed possible for man to deplete fish populations and to alter oceanic ecosystems over vast regions. Some progress has been made to alleviate the problems of overfishing; regulations have been set concerning allowable sizes and total catch for some species, driftnets have been largely banned, tuna fishers have adopted new methods to avoid capturing and killing dolphins, and whales are no longer the targets of commercial fishing. But international regulations are difficult to establish and to enforce on the open seas, and economic issues, not scientific management, continue to drive the industry. Hopefully the next decade will see the resolution of global fisheries problems before entire fish stocks and pelagic ecosystems are irrevocably changed.

9.2 MARINE POLLUTANTS

Marine pollution has been defined by the Intergovernmental Oceanographic Commission as the introduction by humans, directly or indirectly, of substances or energy sources into the marine environment resulting in deleterious effects such as harm to living resources; hazards to human health; hindrance to marine activities, including fishing; impairment of the quality of seawater; and reduction of amenities. The number of different pollutants entering the sea is very large, and new substances are added every day. Some of the substances regarded as pollutants, like heavy metals and petroleum hydrocarbons, occur naturally in the sea and human introductions add to natural concentrations. Some introduced pollutants will decompose in time or will be attenuated by the very large volume of the oceans, so that their effect will not be noticeable. Other pollutants may have significant impacts. Some of the major anthropogenic pollutants and their effects are summarized in Table 9.2 and discussed below.

Table 9.2 Some major forms of marine pollution and their effects.

Pollutant	Location	Effect
Petroleum hydrocarbons	Local oil spills	Mass mortality of benthos and seabirds
	World-wide seas	Low-level concentration effects unknown
Plastics	Beaches	Aesthetically disturbing
	Floating debris	Entanglement of animals; ingestion by animals
Pesticides and related compounds	Local point-source inputs	Acute toxicity
	World-wide seas	Long-term sublethal effects largely unknown
Heavy metals	Industrial outfalls	Mostly sublethal effects causing growth abnormalities
Sewage	Local outfalls; agricultural runoff	Eutrophication and altered community structure; introduction of pathogens
Radioactive wastes	Local power plants; historical at-sea dumping sites	Generally considered to be below harmful levels
Thermal effluents	Local power plants	Warming leads to altered community structure

9.2.1 PETROLEUM HYDROCARBONS

Petroleum hydrocarbons have probably attracted the most attention as marine pollutants because the impact of an oil spill is visually very apparent. Table 9.3 lists some of the major spills, the largest having occurred during the Arabian Gulf war when approximately one million tonnes of oil were spilled into the Gulf of Arabia. The largest spill from an oil tanker occurred when the *Amoco Cadiz* went aground off Brittany in 1978, releasing 220 000 tonnes of crude oil (Colour Plate 41). More than 300 km of shoreline were affected, causing the elimination of at least 30% of the marine benthic fauna and the death of some 20 000 birds. The effects of the *Exxon Valdez* spill (ca. 30 000 tonnes of oil) on Alaskan populations of birds and otters were noted in Sections 6.5 and 8.3. Although such large spills are devastating within localized areas, the natural recovery time of shoreline communities, under moderate conditions of wave action, is usually within 5 to 10 years for most organisms, although bird and otter populations may take longer to recover because of their slower reproductive rates.

Table 9.3   Some major oil spills in the ocean. (Italicized names are of oil tankers.)

Location	Source	Amount of oil spilled (tonnes)	Date
Arabian Gulf	Gulf War	1 000 000	1990–91
Gulf of Mexico	Oil well	440 000	1981
Brittany, France	*Amoco Cadiz*	220 000	1978
Cornwall, U.K.	*Torrey Canyon*	117 000	1967
Wales, U.K.	*Sea Empress*	70 000	1996
Japan Inland Sea	Storage tank	8–40 000	1974
Alaska, U.S	*Exxon Valdez*	37 000	1989
Northwest Atlantic	*Argo Merchant*	30 000	1976
North Sea	Oil well	15 000	1979

Tanker accidents are responsible for only a small percentage of the oil entering the sea. The production and transportation of oil, conventional shipping, waste disposal, and runoff are all additional sources of oil in the marine environment. There are also natural seeps, where oil deposits close to the Earth's surface leak into the sea. It is estimated that between 2.5 and 5 million tonnes of petroleum hydrocarbons enter the ocean each year from all sources. Over time, very small amounts (ppb) of petroleum hydrocarbons have accumulated world-wide in the oceans. The experimental toxic effect of such hydrocarbons is generally at concentrations of parts per million (ppm), and therefore the present accumulated background is not considered harmful to marine organisms, although effects of long-term chronic exposure to such concentrations are not fully known.

QUESTION 9.1   In what type of marine environment would you expect to find the slowest rate of recovery from a large oil spill?   *Cold carm water*

9.2.2 PLASTICS

Discarded plastic materials in the oceans range in size from large nylon drift nets (see Section 6.2) to pellets of less than a millimetre in diameter which can be distributed by the wind over the whole ocean. These materials are not biodegradable; although plastics do break down as a result of physical and chemical weathering, this process is slow and therefore plastics accumulate over time in the sea. Loose driftnets (or 'ghost nets') or other discarded fishing gear, for example, can continue to entangle marine animals for years before washing ashore or sinking. Plastic bags or small plastic pellets are often mistaken for prey and ingested by marine turtles and seabirds, respectively. Pellets have been found in at least 50 species of marine birds. A survey of shearwaters (*Puffinus* sp.) in the North Pacific Ocean revealed that more than 80% of 450 birds had plastic particles in their stomachs. Ingested plastic bags are known to kill turtles; although one can speculate that ingested pellets may be harmful to birds or other marine life, there is presently no direct evidence to support this.

QUESTION 9.2   Can you think of any ways to reduce the amount of plastic material that enters the sea each year? *recycle it.*

9.2.3   PESTICIDES AND OTHER BIOLOGICALLY ACTIVE ORGANIC COMPOUNDS

The most common pesticides entering the oceans are various forms of chlorinated hydrocarbons. These man-made compounds do not occur naturally, they are not readily degraded by chemical oxidation or by bacterial action, and they accumulate in animal fat tissues because they are lipid-soluble.

The best-known insecticide, DDT, was first employed in 1940, and within 20 years it and its residues could be found throughout the biosphere. Because DDT was often sprayed from aircraft, it was easily carried by winds into the oceans. Eventually even Antarctic penguins, living several thousand kilometres from any place where DDT had been used, were found to contain ppb traces of DDT. During the 1960s, there was increasing evidence that marine organisms, particularly seabirds, were being adversely affected in marine areas where DDT concentrations were exceptionally high. One example occurred off southern California, where a pesticide company had released DDT for 20 years into the coastal environment. DDT entered the ocean food chain, and its effects on the marine biota could be detected for 100 km along the coast. Fish in this area contained >3 ppm DDT, and pelicans and sea lions that fed on the fish accumulated even higher tissue concentrations of DDT and were unable to breed successfully. Even following a dramatic reduction in DDT emissions after 1971, DDT levels in fish remained high for years. Harmful effects were found in the Los Angeles Zoo in 1976, with the death of all the cormorants and gulls that had been fed inadvertently for several years on locally caught fish contaminated with DDT. Autopsy results found DDT concentrations ranging between 750 and 3100 ppm in liver tissues of these birds.

DDT usage has now been banned in some countries, but continues to be used in tropical areas as an effective control against mosquito populations that carry malaria. At present, most of the chlorinated hydrocarbon in the sea, and 80% of that in marine organisms, is in the form of DDE, a chemical derived from the breakdown of DDT. In most surface waters, DDT/DDE concentrations are between 0.1 and 1 ng l^{-1} (or less than one part per trillion), and these levels are not considered to be harmful. It is now recognized that heavy use of synthetic pesticides in agriculture is associated with significant undesirable side effects, and efforts are being made to find alternatives through biological or genetic controls of pests.

In addition to pesticides, several toxic chlorinated hydrocarbons are used industrially and may be present in seawater. These include dioxins and PCBs (polychlorinated biphenyls), both of which may have deleterious effects on marine life. PCBs are stable compounds that tend to persist in the environment and to be concentrated through biological processes. These characteristics were underscored by the discovery of exceptionally high PCB concentrations in several beluga whales that died in the St. Lawrence River in 1985; they contained up to 575 ppm in lipid tissue and 1750 ppm in the milk. The recognition of environmental problems has resulted in a ban on production and usage of PCBs in the United States. In the mid-1960s, the organo-tin compound tributyl tin (TBT) was found to have exceptional antifouling properties, and was consequently applied to boat hulls and fishing nets to prevent the settlement and growth of marine organisms. Unfortunately it leaches into surrounding waters, where concentrations between 0.1 and 100 μg l^{-1} are toxic to the larvae of many benthic invertebrates, and levels as low as 0.001 μg l^{-1} may affect reproduction in some marine snails. Ship traffic has spread TBT globally and the compound has accumulated in sediments near harbours and ports. Several countries have now established regulations designed to curb TBT usage.

9.2.4 HEAVY METALS

Heavy metals such as mercury, copper, and cadmium occur naturally in seawater at low concentrations, and they enter the sea through natural erosion of ore-bearing rocks and subsequent transport in rivers or via dust particles in the atmosphere, and through volcanic activity. All of these metals can be poisonous to organisms in high concentrations, and thus potential health problems exist where heavy metals accumulate in the sea around industrial outfalls, or at marine sites used to dispose of some types of mine tailings. At such localities, the local benthos may accumulate metals in levels exceeding permitted concentrations for marketable marine products (the permitted level for mercury is 1 ppm). Generally, heavy metals have acute toxic effects, but accumulations of these substances in marine animals may also cause chronic effects such as growth abnormalities, including cancers.

A serious case of heavy metal poisoning in humans occurred in Minamata, Japan, where a plastics factory discharged an estimated 200–600 tonnes of mercury over a period of 36 years into the local bay. Illnesses began to appear in the early 1950s, and by 1956 these were diagnosed as mercury poisoning derived from eating contaminated shellfish and fish from the bay. Effects included severe neurological damage, paralysis, and birth deformities. By 1988, 2209 victims had been verified, of whom 730 died. Following this tragic discovery of the dangers to human health from eating mercury-containing seafood, regulations were adopted in many countries to limit mercury discharges and to limit the tissue concentrations allowed in seafood products.

What are normal concentrations of mercury in the muscle tissue of pelagic fish?

Most species of fish in uncontaminated oceanic waters contain about 150 μg kg^{-1} (0.15 ppm) of mercury in their muscles. However, some large pelagic species such as sharks, swordfish, black marlin, and tuna may have tissue concentrations as high as 1–5 ppm, but these levels are not indicative of anthropogenic pollution. These long-lived fish are large carnivores at the end of food chains, and their high mercury levels are acquired by bioaccumulation over their life spans.

Although other heavy metals, such as cadmium, copper, and lead, may accumulate in marine organisms exposed to high concentrations at waste discharge sites, they are not known to have caused serious human health issues. Problems arising from the use of another metal, tin, have been discussed in Section 9.2.3. International agreements now regulate marine discharge and usage of some of the more dangerous metals.

9.2.5 SEWAGE

Sewage disposal is a major form of coastal pollution throughout the world. Sewage outfalls near coastal communities release human waste as well as other organic matter, heavy metals, pesticides, detergents, and petroleum products. Nutrients from organic waste material may cause eutrophication; local waters may also be nutrient-enriched by detergents that contain phosphate and by agricultural and horticultural products entering from runoff. In addition, human sewage delivers pathogenic bacteria and viruses that are not necessarily killed by exposure to seawater; high concentrations of these microbes make local seafood unsafe to eat and contaminated waters unsafe for bathing. The chief health risk from sewage is through eating contaminated seafood, particularly filter-feeding clams or mussels which accumulate human pathogens on their gills. The cholera virus is a particular problem in some countries, and may be transmitted in just such a manner.

In urban areas of developed countries, sewage may receive special treatment to degrade organic matter or to remove nitrates and phosphates, but these processes are expensive. Usually no more sewage is treated than is deemed necessary, and in many places sewage is released into the sea without treatment. Generally the immediate area (within 100 m) around a large sewage outfall may be anoxic and dominated by anaerobic bacteria. At some greater distance from the outfall (within several km), nutrient enrichment typically leads to increased production of green macroalgae (*Enteromorpha* or *Ulva*) that form thick mats along the shoreline. A few opportunistic animals, such as the polychaete *Capitella*, are also indicative of sewage enrichment and may dominate affected benthic communities. At some ten kilometres from a major domestic outfall, there is usually sufficient attenuation of pollutants that community species diversity is not affected.

9.2.6 RADIOACTIVE WASTES

Radioactive wastes enter seawater from nuclear testing, from nuclear power plants or reprocessing reactors, or from deliberate dumping of waste materials. Heavy radionuclides have low solubility in water and tend to be adsorbed onto particulate matter; they therefore accumulate in sediments. Isotopes with long half-lives (e.g. caesium-137, strontium-90, and plutonium-239) are especially hazardous and are usually monitored in areas where they may escape from nuclear facilities. Barring major accidents, background levels in the marine environment around radioactive outfalls are generally regarded as safe. The potential for reaching high concentrations of radioactive materials exists in certain localized areas, notably around the several known sunken nuclear submarines, from nuclear dump sites at sea (which are now prohibited), and from nuclear testing that has been carried out within coral atolls (most recently in the South Pacific by France). However, it is predicted that leakage from such sources would occur at a slow rate and that there would be dilution of soluble radionuclides and adsorption of others on to bottom sediments.

Some marine organisms (e.g. seaweeds and bivalves) may accumulate radionuclides from surrounding water. For example, the alga *Porphyra umbilicalis*, growing in the vicinity of a reprocessing plant in England, accumulated 10 times the concentration of caesium-137 found in the ambient water and 1500 times the concentration of ruthenium-106. The experimental consequences of low-level doses of radiation on marine organisms are the same as those for terrestrial species and may include increased incidences of cancers, impaired immune systems, and genetic defects causing growth deformities. However, present levels of radiation in the sea have not produced any measurable environmental impact on marine biota.

9.2.7 THERMAL EFFLUENTS

Power plants may discharge several hundred thousand cubic metres of cooling water per hour into coastal waters, and this thermal effluent may raise the local seawater temperature by 1–5 C°. In some areas, this warmed water can be used beneficially to enhance growth rates of organisms grown in mariculture. However, in many cases elevated temperatures cause unwanted changes in the natural fauna and flora. For example, a persistent elevation of the ambient temperature by 5 C° along the subtropical Florida coast resulted in the replacement of natural algae and seagrasses by mats of cyanobacteria. In another example, the increase in water temperature of a temperate area of the U.S. allowed the entry of warm water wood-boring bivalves which caused damage to boats and wharves. In most cases, the area affected is limited to the plume of hot water and its immediate surroundings, an area that may range from less than one hectare to about 40 hectares.

There may be other effects from thermal effluents. They often contain chlorine, which is added to intake water to prevent fouling organisms from blocking pipes, and as little as 0.1 ppm of chlorine remaining in the effluent can be toxic to some organisms. Water flow of the plume also mechanically scours the seabed and so influences the fauna.

QUESTION 9.3   At what time of the year would you expect to have the greatest impact from thermal effluents released from power plants located on a subtropical coast?   ~~hottest~~ Summer.

9.3   INTRODUCTIONS AND TRANSFERS OF MARINE ORGANISMS

The movement of species from one region to another occurs naturally through larval drift, rafting, and other means, but humans have accelerated these movements and eliminated natural ocean barriers by accidental or deliberate introductions of species into new areas. In many cases, introduced species fail to develop reproducing populations, but in some instances an exotic species encounters favourable conditions and causes a significant impact on its new environment.

Marine organisms have often been introduced deliberately into new environments for mariculture purposes. For example, the Japanese oyster (*Crassostrea gigas*) and the east coast oyster (*C. virginica*) were brought to the north-west coast of North America because they were larger and grew more rapidly than the local oyster (*Ostrea lurida*); simultaneously, predatory snails that bore into oysters and other molluscs were accidentally introduced and both the oysters and this predator are now firmly established in this region.

Increased demand from the mariculture and aquaria industries and increased ship traffic have accelerated the rate at which species are transported to new environments. Each day, approximately 3000 species of marine animals and plants are being carried across oceans in the ballast tanks of ships that take on seawater in port and then release it at later ports of call. Examination of ballast water released in Oregon from a Japanese vessel contained 367 taxa of both holoplankton and meroplankton. Thus entire planktonic communities, including the larvae of benthic organisms, can be transported across natural oceanic barriers.

The release of the ctenophore *Mnemiopsis leidyi* (see Section 4.7) into the Black Sea is one example of an invasive species that encountered favourable conditions and no natural predators, and rapidly became so abundant in the Azov and Black Sea that it caused drastic declines in the resident zooplankton and subsequent declines in the anchovy fisheries of bordering countries. More recently, two species of jellyfish that are native to the Black Sea have appeared in San Francisco Bay (Figure 9.3); it is too early to assess their impacts on the ecosystem. Asian species of copepods are now present in Californian harbours, and the Chinese clam *Potamocorbula amurensis* has become one of the most abundant marine animals in the estuary of San Francisco Bay. The zebra mussel *Dreissena polymorpha*, originally from the Mediterranean, entered the North American Great Lakes from ballast water in 1986 and almost immediately began to grow in such

Figure 9.3 *Maeotias inexspectata*, a jellyfish native to the Black Sea that has recently been introduced into San Francisco Bay, California.

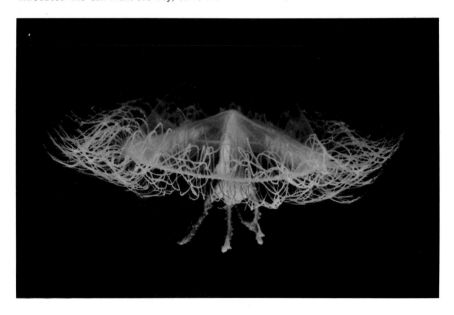

profusion that it blocked water intakes, resulting in an expense of ten billion dollars to deal with its destructive effects.

Australia has had over 35 successful invasions in recent years; the predatory starfish *Asterias amurensis* has invaded the coast of south-eastern Australia, and a common Japanese kelp, *Undaria pinnatifida*, is spreading along parts of Australia at a rate of 55 km a year. It is believed that the appearance of red tides (see Section 3.1.2) for the first time in Tasmanian waters resulted from toxic dinoflagellates being transported in ballast water of cargo ships; one such vessel contained more than 300 million toxic dinoflagellate cysts in the sediment accumulated in its tanks.

How could the spread of marine organisms by transport in ballast water be reduced?

Ships take on ballast water in coastal seaports. The problem of global transport of marine species could be ameliorated by requiring ships to exchange ballast water at sea. Coastal organisms are unlikely to survive in oceanic water, and oceanic organisms are much less likely to survive release into low salinity coastal water.

9.4 IMPACTS ON SPECIFIC MARINE ENVIRONMENTS

9.4.1 ESTUARIES

Many of the world's largest cities are located on estuaries; London, Shanghai, and New York are but a few examples. It is not surprising, therefore, that estuaries have suffered more from human impact than most other marine environments (see Figure 8.6). The most severely affected area is usually the saltmarsh community because this upper intertidal area is easily reclaimed for housing and industrial activity and, sometimes, for airport construction. In some localities, more than 90% of this community has disappeared. The deeper seagrass and mudflat communities are often disrupted by dredging operations to make large harbours and deepen shipping channels.

The same features that make estuaries productive also make them especially vulnerable to pollution. Just as nutrients are retained within the system (see Section 8.5), so are pollutants like petroleum byproducts, heavy metals, fertilizers, and pesticides. Both the plankton and benthic communities are affected by domestic and industrial outfalls releasing organic pollutants that may cause unwanted eutrophication. Pathogenic organisms, heavy metals, and pesticides that enter estuaries may all eventually work their way up the food chain into edible products for humans.

Because estuaries are naturally productive, they are often favoured fishing grounds and are frequently used for harvesting shellfish or for developing mariculture. Recently, however, many estuarine fisheries and shellfish beds have been closed due to high numbers of coliform bacteria from domestic sewage, or from the accumulation of pesticides or heavy metals in fish products. All of these forms of pollution may also alter the structure of the estuarine ecosystem so that traditional spawning grounds and nursery areas for fish may be lost. In extreme cases of pollution and eutrophication, anoxic zones may occur where only bacteria can survive.

Because of the dangers of human health risks from pollution of estuarine waters, monitoring programmes have been established in developed countries. Many pollutants occur in very low concentrations (ppb level) in seawater and are therefore extremely difficult to measure, even using sophisticated equipment and techniques. For this reason, a special programme called **Mussel Watch** was established to monitor concentrations of marine pollutants in mussels. This programme uses mussels (*Mytilus*) because they are abundant world-wide in coastal regions, because they are sessile animals that are exposed to any pollutants contained in the water flowing over them, and because these filter-feeding bivalves are known to accumulate a variety of pollutants. They have also been extensively studied, both experimentally and ecologically. Tissue levels of pollutants that have accumulated in mussels from contaminated areas can be more easily measured than pollutant concentrations in ambient waters, and these values can be compared with standards in mussels from uncontaminated regions. Since 1986 in the United States, mussels (or oysters) have been collected from over 200 localities once each year. Tissues are analysed for several heavy metals, chlorinated hydrocarbons including DDT and PCBs, and tributyl tin. Analysing these pollutants in mussel tissues is much easier and cheaper than analysing the same substances in trace concentrations in seawater.

QUESTION 9.4   Drawing on your general knowledge of geography, can you think of any large estuaries in the world that do not have adjacent cities and are not affected by human populations?

9.4.2   MANGROVE SWAMPS

Mangrove swamps (see Section 8.7) suffer from many of the same environmental disturbances that are experienced by estuaries. Dredging, land reclamation, garbage and sewage dumping are all disturbances that can have significant impacts on mangroves near populated areas. In these tropical and subtropical ecosystems, insect control (particularly of malaria-carrying mosquitoes) has resulted in accumulations of pesticides in estuarine sediments and in mangrove food chains. During the Vietnam war, spraying of herbicides on mangrove swamps defoliated and destroyed as much as 100 000 hectares. Oil spills smother both algae and invertebrates, and disrupt the oxygen supply to the root system. Where river water has been diverted into irrigation systems, the reduction in freshwater discharge and the resulting elevated salinities may be detrimental; for example, a considerable area of mangrove swamp has been destroyed by diversion of water flow from the Indus River in Pakistan.

Overcutting of mangroves is, and has been for centuries, a serious problem in many areas. Mangroves once existed along the shores of the Persian Gulf, where they were a much-needed source of firewood for humans and of green fodder for camels in a desert environment, but they were eventually eliminated by overcutting. Some efforts had been made to re-establish mangroves along north-eastern Saudi Arabia, but these were destroyed by the Gulf War. Other countries, recognizing the benefits of mangroves, have also developed afforestation programs, reintroducing mangroves with varying degrees of success. Globally, however, destruction of mangroves is progressing faster than reintroduction. Almost half of the world's mangroves have been eliminated in recent years in order to build shrimp farms or rice paddies. In countries like Bangladesh, removal of this buffering zone has led to intensified coastline inundation and erosion from tropical storms.

Typhoons and hurricanes remain perhaps the greatest destructive agents of mangrove swamps, as they affect very large areas and occur frequently. Not only do they uproot trees, but severe storms alter the salinity of both water and soil, and they cause massive sedimentation. It is estimated that recovery of mangrove forests from very violent storms takes at least 20 to 25 years. Whereas little can be done to reduce damage from natural events, it is possible to develop management policies for the exploitation of mangrove resources, including replanting. The rational utilization of mangrove areas depends ultimately on increasing public awareness of the importance of this unique marine community to local populations in developing tropical countries.

9.4.3 CORAL REEFS

Coral reefs not only have great beauty and support a very high natural diversity with many endemic species, but they also have wider biological and economic importance. Because corals remove large amounts of bound CO_2 from the oceans during calcification, the reefs play a role in the global CO_2 budget. They are of benefit in protecting coastlines and providing sheltered harbours. And as air travel has become cheaper and more available, reefs have brought in more and more 'tourist dollars' to boost the economy of human populations living in their vicinity.

However, coral reefs are extremely vulnerable to disturbance, and reefs are presently regarded as declining. The World Conservation Union and the United Nations Environment Program (UNEP) reports damage or destruction of significant amounts of reefs in 93 out of 109 countries. Much of this destruction results from human activities, some from changes in ocean climate.

Expanding human populations near reefs often result in the addition of various types of pollutants to near-shore waters. These can include agricultural runoff, pesticides, industrial pollutants, and sewage from beachfront hotels or coastal communities. Many of these sources increase the nutrient concentrations in the seawater, and this eutrophication triggers outbursts of benthic algae that can outcompete corals for available space. Often the algae overgrow the corals, smothering and killing the reefs by cutting off the sunlight required by the zooxanthellae. Just such an event followed a burst of land construction and development in an area of the Hawaiian Islands. After thick mats of algae overgrew and killed large areas of the reef, the decomposition activities of bacteria led to a lowering of oxygen concentrations in the seawater. The final result was a dramatic decrease in diversity, with a particular sea cucumber becoming the dominant animal.

Various types of coastal development also result in increased land erosion, which then increases the amount of sediment in water overlying the reef. The suspended silt decreases light penetration, thus reducing photosynthesis of the zooxanthellae and diminishing a nutritional source for the corals. Although coral polyps are capable of removing some settling sediment, using mucus trapping and cilia to cleanse their surfaces, excessive quantities of silt will clog this apparatus and smother the polyps. Deforestation, leading to increased runoff and excessive sedimentation, is a major cause of coral reef destruction. Logging in one area of the Philippines (Bacuit Bay) has increased erosion and killed 5% of the area's reefs. Dredging to deepen harbours or open ship channels through the reef has similar effects on adjacent reefs.

A coral reef offers a number of resources used directly by humans. In some localities, coral is cut and used as a favoured building material (Figure 9.4). In the Maldive Islands (Indian Ocean), about 200 000 m^3 of coral rock have been mined in the past 20 years; this represents about one-third of the available coral in shallow water. This practice has also destroyed large areas of reefs in French Polynesia and Thailand.

Local inhabitants have traditionally relied on fish from nearby reefs as an abundant protein source, but demands have increased for this resource as populations have grown in size and as expanding numbers of tourists enter the communities. At the same time, private and public aquaria have increased in number throughout the world, and the capture of exotic fish for sale to the aquarium trade has become an extremely lucrative business, worth about $40 million per year (in 1996). To meet these increased demands, traditional fishing methods often have been replaced by much more detrimental techniques. It has become common to blast with dynamite to stun fish, which are then easily collected when they rise to the surface. At the same time, portions of the reef are destroyed, and it is estimated that it may take some 40 years for areas destroyed by blast fishing to recover to 50% live coral cover. Cyanide is also used to stun fish for live collection, although it may result in the death of the fish and certainly kills other reef species. In the Philippines alone, an estimated 150 tonnes of sodium cyanide is used annually for fish capture. Increased fishing, whether by traditional or non-traditional methods, has resulted in overexploitation of many species in many regions. Where destructive fishing techniques have been banned, regulations often are not enforced.

Fish are not the only reef inhabitants that are removed for consumption or trade. Corals are removed for ornamental purposes; about 1500 tonnes of such coral was imported into the United States alone in 1988. Spiny lobsters,

Figure 9.4 Mined coral that has been cut for building purposes in the Maldive Islands, Indian Ocean.

sea cucumbers, and sea urchins are among some of the reef animals that are considered delicacies in many parts of the world. Snails and clams are collected for food, or to sell to tourists and shell collectors. Removal of large numbers of animals from reefs may alter the ecology. For example, sea urchins are responsible for removing part of the reef framework during grazing; this bioerosion may be intensified when their natural predators (some fish and molluscs) are overfished.

Tourism may strengthen the economy of reef areas, but it often has the effect of damaging or destroying that very resource which people pay to see. Tourism leads to increased development, which may increase erosion and siltation over the reef, and it almost certainly increases the amount of sewage entering the water. As more tourists arrive, more fish is sold to restaurants and hotels, and more shells and coral are removed as mementos. Even seemingly innocent activities can have repercussions on the reef ecology. Reefs off Florida in the United States have been seriously damaged by amateur boaters colliding with submerged coral heads or anchoring to corals which are easily broken in the process. Even walking on reefs at low tide is destructive to coral. Tropical countries that wish to have long-term economic benefits from their local reefs would do well to avoid overexploitation, and to educate both local populations and visiting tourists.

In the 1970s, attention was drawn to Pacific reefs that were being destroyed by *Acanthaster planci*, the crown-of-thorns starfish. This large (30–40 cm diameter) echinoderm feeds on coral polyps, but is normally present in low enough numbers that reef damage is slight. However, *Acanthaster* began to undergo population explosions on many reefs in the western Pacific, including the Great Barrier Reef. Tens of thousands of starfish were found on some reefs, and the impact of so many large predators was devastating, with entire reefs being destroyed in some areas. For example, in less than three years, *Acanthaster* destroyed approximately 90% of the coral along 38 km of reef off Guam. Parts of the Great Barrier Reef have experienced major damage from predation during two separate starfish outbreaks, one in the 1960s and again in the 1980s.

Many causes have been advanced to explain the outbreaks of *Acanthaster*. These include increases in waterborne pollutants or increases in sedimentation due to dredging or other activities. It also was suggested that shell collecting was to blame. Many tritons had been removed from reefs, as the shell of this large snail is a prized ornament. Tritons are one of the few natural predators of the crown-of-thorns starfish, and the decline of these snails could result in lessened mortality of *Acanthaster*. It seems, however, that no one explanation applies to all the reefs damaged by starfish predation. There have been suggestions that *Acanthaster* density fluctuations may be natural cycles that are linked to ocean climate change. As with the similar kelp–urchin interactions discussed in Section 8.3, the debate continues about whether starfish population explosions are a contemporary phenomenon linked with human activities, or whether they are an ecological pattern that has persisted for thousands of years. In any case, restoration of coral cover takes place within 10 to 20 years, but it may take much longer to re-establish the original species diversity.

Storms, exposure during exceptionally low tides, and other natural events may cause widespread coral damage. In 1982–83, a rapid 2–4 C° rise in seawater temperature to nearly 30°C was caused by a particularly strong El Niño event, and this damaged or killed 95% of the coral in the Galapagos

Islands, and 70% to 90% of the corals in the Gulf of Panama and in Indonesia. In 1987, reefs throughout the Caribbean were affected by a similar event. Corals are particularly sensitive to elevated temperatures as they live so close to their upper temperature tolerance limits. Any sustained temperature increase usually results in bleaching due to loss of zooxanthellae, and eventually to death if the thermal stress continues. In Hawaii, bleaching of coral has been correlated with discharge of heated effluent from a power plant, and bleaching can be experimentally induced by elevated temperature. On a world-wide basis, global warming poses a serious threat to coral reefs.

Reef-building corals similar to modern species have a geological record dating back about 250 million years, but other types of coral reefs occurred as long ago as 500 million years. More than 5000 species of extinct corals are known, compared with the present number of less than 600 reef-building species. The pace of species extinction may be accelerating in many areas. Some countries, recognizing the benefit of adjacent reefs, have developed governmental policies to protect them. About 65 countries now have almost 300 protected areas that include coral reefs; these include marine reserves and underwater parks. The economic benefits of controlled tourism may push other countries to develop conservation policies.

9.5 SUMMARY OF CHAPTER 9

1 The annual fish harvest of more than 100×10^6 tonnes has had a greater impact on the ocean than any other human activity. By 1995, fish catches were declining in all major oceans except the Indian Ocean and 70% of the ocean's fish stocks were either being fully exploited, were overfished, or were recovering from being overfished.

2 The results of intensive fishing include: declines in targeted fish stocks and consequent changes in relative abundance of species; changes in size structure of fish populations; declines in pelagic and benthic animals captured incidentally as by-catch; acceleration of nutrient transfer to deep water through dumping of dead by-catch; increased numbers of scavengers in marine food chains receiving large amounts of by-catch; and destruction of seabed habitats through benthic trawling.

3 Fish stocks are also affected by construction of dams that eliminate spawning grounds (e.g. salmonids), and by loss of coastal spawning and nursery grounds due to land reclamation or pollution.

4 Human activities result in the release of a variety of pollutants into the sea. These substances, which may cause deleterious changes, include petroleum hydrocarbons, plastics, pesticides and related chlorinated hydrocarbons, metals, fertilizers, and radioactive wastes. Sewage outfalls deliver many of these pollutants as well as human wastes, detergents, and pathogenic bacteria and viruses. Power plants also release heated effluents that elevate ambient seawater temperatures.

5 Although oil spills are among the most visible types of marine pollution and ecological damage in the immediate site may be severe, populations generally recover within 5 to 10 years. In the open ocean, the accumulated concentration of petroleum hydrocarbons is too low to cause measurable effects.

6　Nonbiodegradable plastic materials are now found throughout the oceans. These include lost fishing nets, which may continue to entangle animals for years, and plastic materials that are mistaken for prey and ingested by turtles and seabirds.

7　Toxic synthetic pesticides like DDT and related compounds (dioxins, PCBs) that enter the marine environment are not readily degraded; they persist for long periods and enter marine food chains. Because they are stored in fat tissues, these compounds show biomagnification, with higher trophic level animals accumulating concentrations that may be lethal. Past incidences of pesticide and PCB poisoning in marine organisms have led to bans on usage and production in some countries.

8　Accumulations of heavy metals (e.g. mercury, copper) resulting from industrial outfalls may cause serious human health problems. Historically, the damaging effects caused by humans eating mercury-contaminated seafood were shown in the 1950s in Minamata, Japan, where more than 2000 people were directly affected. Now regulations and monitoring programmes exist to limit and detect unacceptable concentrations of these metals in marine products.

9　Sewage disposal is a major form of coastal pollution throughout the world. Nutrients in human wastes and those in detergents and fertilizers enrich local waters. This eutrophication may be beneficial in some cases, but often the amount of nutrients delivered leads to excessive plankton blooms that eventually decay and cause oxygen depletion. Pathogenic organisms in human wastes, like the cholera virus, can be filtered out of water near sewage outfalls by mussels and clams, and then be transmitted to humans who consume this seafood.

10　Radioactive wastes do not presently occur in concentrations that threaten marine life, although it is known that some organisms, particularly seaweeds and bivalves, can accumulate radionuclides from waters around nuclear plants.

11　Power plants release heated water that elevates ambient seawater temperature and thereby affects marine communities within the immediate area. In some cases the heated effluent is used to enhance growth rates of organisms grown in culture, but often the community changes are detrimental or unwanted.

12　Some marine communities and ecosystems have been changed through the deliberate or accidental transplantation of species. Increased commercial shipping has accelerated the rate of introduction of species into new environments, with an estimated 3000 species of marine plants and animals being carried daily across oceans in the ballast tanks of ships. Many do not survive but some, like the ctenophore *Mnemiopsis* or the zebra mussel, have major impacts on their new environments.

13　Estuaries and mangrove swamps are productive coastal ecosystems that constitute important spawning and nursery grounds for many fish, harbour shellfish populations, and provide rich feeding grounds for birds. As well, mangrove swamps buffer coastlines from erosion and inundation during tropical storms. However, these ecosystems are often heavily affected by human activities such as land reclamation, disposal of sewage and industrial wastes, and eutrophication.

14   Coral reefs are declining throughout the world. Expanding human populations near reefs and increasing tourism have accelerated development and have brought growing pressure to exploit reef resources. Coral reefs are detrimentally affected by increased sedimentation resulting from land development and subsequent erosion, and from eutrophication stemming from sewage disposal and agricultural runoff. In some locales, the coral is mined as building material. Destructive fishing techniques remove large numbers of fish, change species composition on the reef, and damage corals directly. Corals world-wide have been affected by elevations in seawater temperature, and global warming is a potential danger to reef communities.

Now try the following questions to consolidate your understanding of this Chapter.

QUESTION 9.5   The subject of human impacts on marine biota is much too large to cover fully in this chapter. What other impacts can you think of that might have been included?

QUESTION 9.6   Is it possible to harvest fish from the sea without causing changes in marine ecosystems?

QUESTION 9.7   Does sewage disposal at sea have any beneficial impact?

QUESTION 9.8   Estuary A receives freshwater with a high sediment load, whereas Estuary B receives relatively clear river water. Both receive approximately equal amounts of urban pollutants. Which estuary would have lower concentrations of these pollutants in the water, and which would have lower pollutant concentrations in the sediments?

Approx. time unit began (millions of years ago)	Eras	Periods	Epochs	Major biotic events
0.01	Cenzoic	Quaternary	Holocene	Modern man appears
2			Pleistocene	Early humans appear (2–3 mya)
5		Tertiary	Pliocene	
25			Miocene	
40			Oligocene	Early anthropoids appear (40 mya)
				Heteropods appear
55			Eocene	Marine mammals, shelled pteropods appear (55 mya)
				Seabirds appear (60 mya)
65			Paleocene	Dinosaurs become extinct (65 mya)
	Mesozoic	Cretaceous		Diatoms and silicoflagellates appear (100 mya)
				Ammonoid cephalopods become extinct
140				Coccolithophorids and forams appear; number of dinoflagellate species increases
200		Jurassic		Marine reptiles and primitive birds appear (200 mya)
				Most nautiloid cephalopods go extinct
240		Triassic		Dinosaurs and mammals appear; most nautiloid cephalopods become extinct
290	Paleozoic	Permian		Trilobites become extinct
				Teleost fishes appear (300 mya)
360		Carboniferous		
410		Devonian		Land plants appear (420 mya)
435		Silurian		Dinoflagellates, barnacles appear; bivalves become abundant
				Elasmobranchs (sharks), ammonoid cephalopods appear (400–450 mya)
500		Ordovician		Tintinnids evolve, coral reefs form
				Primitive fish appear (550 mya)
570		Cambrian		Trilobites dominate; ostracods, echinoderms, nautiloid cephalopods appear
				Marine algae become highly diversified
	Precambrian			Radiolaria appear (600 mya)
				Abundant fossils of marine invertebrates (e.g. jellyfish, sponges, molluscs)
				Oldest fossils of shelled marine animals (650 mya)
				First multicellular seaweeds (800 mya)
1000				Calcareous algae (1500 mya)
2000				Multicellular life appears (2100–1900 mya)
				Earliest photosynthetic organisms (2800–2500 mya)
3000				Oldest fossils of cyanobacteria (3500 mya)
				Life originates (3900–3500 mya)
4000				Atmosphere and ocean form (4400 mya)
5000				Earth's crust forms (4600 mya)

APPENDIX 2   CONVERSIONS

UNITS OF AREA

1 square centimetre (cm^2) $= 100$ mm^2 $= 0.155$ square inch
1 square metre (m^2) $= 10^4$ cm^2 $= 10.8$ square feet
1 square kilometre (km^2) $= 10^6$ m^2 $= 247.1$ acres
1 hectare (ha) $= 10\,000$ m^2

UNITS OF CONCENTRATION

molar concentration (M) $\equiv$ gramme molecular weight per litre
μg litre^{-1} $\equiv$ mg m^{-3}
parts per million (ppm) $\equiv$ mg litre^{-1}
parts per billion (ppb) $\equiv \mu g$ litre^{-1}
parts per trillion (ppt) $= 10^{-3} \mu g$ litre$^{-1} = 1$ nanogramme litre^{-1}
μg litre$^{-1} \div$ molecular weight $= \mu M = \mu$mol litre^{-1}

UNITS OF LENGTH

1 ångstrom (Å) $= 0.0001$ micron
1 nanometre (nm) $= 10^{-9}$ metres
1 micron (μ) $= 0.001$ millimetre (or 10^{-3} mm) $= 10^{-6}$ m
1 millimetre (mm) $= 1000$ microns $= 0.001$ metre
1 centimetre (cm) $= 10$ millimetres $= 0.394$ inch
1 decimetre (dm) $= 0.1$ metre
1 metre (m) $= 100$ centimetres $= 3.28$ feet
1 kilometre (km) $= 1000$ metres $= 3280$ feet
1 kilometre (km) $= 0.62$ statute mile $= 0.54$ nautical mile
1 inch (in) $= 2.54$ centimetres
1 foot (ft) $= 0.3048$ metre
1 yard (yd) $= 3$ feet $= 0.91$ metre
1 fathom $= 6$ feet $= 1.83$ metres
1 statute mile $= 1.6$ kilometres $= 0.87$ nautical mile
1 nautical mile $= 1.85$ kilometres $= 1.15$ statute miles

UNITS OF MASS

1 milligram (mg) $= 0.001$ gramme
1 kilogram (kg) $= 1000$ grammes $= 2.2$ pounds
1 tonne (t) $= 1$ metric ton $= 10^6$ grammes
1 pound (lb) $= 453.6$ grammes

UNITS OF TIME

1 minute (min) $= 60$ seconds (s)
1 hour (h) $= 3600$ s
1 day (d) $= 86\,400$ s
1 year (yr) $= 365$ d

UNITS OF VELOCITY

1 kilometre per hour = 27.8 centimetres per second

1 knot (kn) = 1 nautical mile per hour = 51.5 centimetres per second

UNITS OF VOLUME

1 millilitre (ml)   = 0.001 litre = 1 cm^3 (or 1 cc)

1 litre (l)   = 1000 cm^3 = 10^{-3} m

1 cubic metre (m^3) = 1000 litres

UNITS USED TO MEASURE SOLAR RADIATION ENERGY

1 einstein (E) = 6.02×10^{23} photons = 1 mole of photons

1 watt $m^{-2} \approx 4.16 \pm 0.42$ μ einsteins m^{-2} s^{-1}

  (The above relationship applies only to PAR)

1 joule m^{-2} s^{-1} = 1 watt m^{-2}

1 calorie cm^{-2} $min^{-1} \approx 700$ watts m^{-2}

1 langley = 1 calorie cm^{-2}

UNITS USED IN PRODUCTION STUDIES

1 calorie (cal) = 4.184 joules (J)

1 kilocalorie (kcal) = 1000 cal

1 gramme carbon (g C) $\approx$ 10 kcal

1 g C $\approx$ 2 g ash-free dry weight (where ash-free dry weight is dry weight
  less the weight of inorganic components such as shells)

1 g ash-free dry weight $\approx$ 21 kJ

1 g organic C $\approx$ 42 kJ

1 g ash-free dry weight $\approx$ 5 g wet weight

1 litre O_2 = 4.825 kcal

1 g carbohydrate $\approx$ 4.1 kcal

1 g protein $\approx$ 5.65 kcal

1 g fat $\approx$ 9.45 kcal

CUSHING, D. H. (1975) *Marine Ecology and Fisheries*, Cambridge University. This book begins with a review of marine production cycles, which are then related to the biology and population dynamics of commercial fish stocks; it concludes with a discussion of fluctuations in fish stocks caused by natural events and by human exploitation.

DUXBURY, A. C. and DUXBURY, A. (1994) *An Introduction to the World's Oceans*, (4th edition), Wm. C. Brown. A general, easy to read overview of the oceans, including introductory material on physical, geological, chemical and biological oceanography.

FRASER, J. (1962) *Nature Adrift, the Story of Marine Plankton*, G. T. Foulis. A well illustrated and informative account of planktonic organisms written in an easily understandable manner.

GAGE, J. D. and TYLER, P. A. (1991) *Deep-sea Biology: A Natural History of Organisms at the Deep-sea Floor*, Cambridge University. A recent review of deep-sea biology, including information on hydrothermal vent communities.

HARDY, A. (1970) *The Open Sea: Its Natural History. Part I: The World of Plankton*, (2nd edition), Collins. A classic account of plankton, delightfully written and illustrated with watercolour drawings done by the author while at sea.

LAWS, E. A. (1993) *Aquatic Pollution: An Introductory Text*, John Wiley. A thorough introduction to the sources and consequences of anthropogenic pollution in the sea and in freshwater.

MANN, K. H. and LAZIER, J. R. N. (1991) *Dynamics of Marine Ecosystems, Biological-Physical Interactions in the Oceans*, Blackwell. A comprehensive treatment of the links between water circulation patterns and biological processes; although the physical oceanography is at a fairly elemental level, some mathematical knowledge is necessary.

MARSHALL, N. B. (1980) *Deep Sea Biology: Developments and Perspectives*, Garland STPM Press. A descriptive account of deep-sea invertebrates and fish including morphological, behavioural and physiological adaptations.

MCCLUSKY, D. S. (1989) *The Estuarine Ecosystem*, (2nd edition), Blackie. An introduction to estuaries, emphasizing biological aspects and including a discussion on pollution and management.

NYBAKKEN, J. W. (1988) *Marine Biology: An Ecological Approach*, (2nd edition), Harper & Row. A well written, well organized treatment of marine biology, particularly recommended for its discussions of benthic communities.

PARSONS, T. R., TAKAHASHI, M., and HARGRAVE, B. (1984) *Biological Oceanographic Processes,* (3rd edition), Pergamon. A more advanced treatment of biological oceanography that emphasizes production processes; minimal mathematics.

RAYMONT, J. E. G. (1980) *Plankton and Productivity in the Oceans*, (2nd edition). Vol. 1, *Phytoplankton*; Vol. 2, *Zooplankton*, Macmillan. Classic, comprehensive reviews of plankton including descriptions, biology, distribution patterns, and abundance.

CHAPTER 1

QUESTION 1.1   The overlap occurs because plankton and nekton are separated strictly according to swimming ability, whereas the classification scheme in Figure 1.2 is based on size as well. Therefore, the figure shows that very large zooplankton with feeble swimming ability (e.g. some jellyfish) may be the same size, or even larger, than some nekton (e.g. fish).

QUESTION 1.2   Biological oceanography was slow to develop largely because of the inaccessibility of many areas of the oceans. Global systematic observations and collections at sea in all depths require large ships operating for relatively long time periods, as well as specially designed gear. These requirements make sea-going operations very expensive compared with land-based research.

QUESTION 1.3   Primitive fish first appeared in the seas about 550 million years ago.

QUESTION 1.4   At 3000 m depth, there is no light and consequently no plant life. The water temperature is low (2–4°C) and constant. Hydrostatic pressure is relatively high. Relative to conditions near the sea surface, nutrients like nitrate and phosphate are present in higher concentrations, but food supplies for animals are less abundant.

QUESTION 1.5   In 1960 the bathyscaphe *Trieste* carried two men to a depth of 10916 m in the Mariana Trench in the Pacific Ocean.

CHAPTER 2

QUESTION 2.1   The solar radiation received at the surface of the Arabian Sea is: (a) about 3700 μ E m^{-2} s^{-1} in September, and (b) about 2900 μ E m^{-2} s^{-1} in January.

QUESTION 2.2   The extinction coefficient, k, can be determined from:

$$k = \frac{\log_e 100 - \log_e 50}{10 \text{ m}} = \frac{4.6 - 3.9}{10 \text{ m}} = 0.07 \text{ m}^{-1}$$

QUESTION 2.3   This is due to the different penetration of various wavelengths in water. Red and yellow quickly disappear with depth, so that these colours can no longer be seen at diving depths. Green and blue wavelengths penetrate deeper and are still visible at the depths of reefs. Using a flash on underwater cameras restores the entire colour spectrum and the true colours of the reef are seen on the film.

QUESTION 2.4   Moonlight is obviously too weak to cause photosynthesis. However, notice that moonlight is sufficiently intense (even down to 600 m in clear oceanic water) to be seen by deep-sea fishes. Since they often respond to light by carrying out vertical migrations, the intensity of moonlight could have an effect on their movements in the water column. Moonlight could also facilitate detection of prey by predators that use vision.

QUESTION 2.5   The higher the salt content of the water, the lower the submerged weight of the organism. Consequently, organisms expend less energy to avoid sinking in water of higher salinity.

QUESTION 2.6   Both types of water (a and b) lie on the density contour described by a sigma-*t* value of 26.0. Using the equation given in the figure caption, $26.0 = (d - 1) \times 1000$, or $d = 1.026$ g cm^{-3}. The density of seawater described in (a) and in (b) is the same.

QUESTION 2.7   The initial freezing point will be slightly higher than that of average seawater with a salinity of 35, and freezing will start sooner. However, sea-ice formation itself results in increasing the salt content of the surrounding water, and this depresses the freezing point, thus inhibiting further ice formation.

QUESTION 2.8   This is known as the Antarctic Circumpolar Current, or West Wind Drift, that flows around the continent of Antarctica.

QUESTION 2.9   From Appendix 2, approximately 4.16 μ einsteins m^{-2} s^{-1} equals 1 watt m^{-2}. Therefore, 1 μ E m^{-2} s^{-1} is about 1/4 or 0.25 W m^{-2}, and 10 μ E m^{-2} s^{-1} is about 10/4 or 2.5 W m^{-2}.

QUESTION 2.10   This relates to the small temperature range of the sea compared with land. Many marine organisms, and residents of deep water in particular, experience only relatively small fluctuations in ambient temperature. Homoiothermic terrestrial animals, which are able to regulate internal body temperature, are better adapted to the wider environmental temperature range encountered on land.

QUESTION 2.11   (a)   This is caused by an excess of precipitation over evaporation in the rainy belt around the Equator.

(b)   The Arctic basin receives large amounts of freshwater from major rivers in Canada and Siberia.

QUESTION 2.12   A combination of low temperature and high salinity results in very dense water.

QUESTION 2.13   This is in the aphotic zone where there is no light. The water temperature will be 4°C or below, and the salinity will be about 35. This cold, high salinity water has a high density (from Figure 2.14, $\sigma_t \approx 27.75$). Hydrostatic pressures will exceed 200 atm.

CHAPTER 3

QUESTION 3.1   The volume of a sphere is calculated from: $4/3\pi r^3$. Cancelling out the expression $4/3\pi$, which is the same for both species, and converting diameter to radius by dividing by 0.5, gives r^3 values of $(0.5)^3$ for *Synechococcus* cell volume and $(25)^3$ for the dinoflagellate. Dividing $(25)^3$ by $(0.5)^3$ gives 125 000 *Synechococcus* cells, the number needed to produce the equivalent volume of one 50 μm-diameter dinoflagellate. So, although the concentrations of *Synechococcus* in the sea may be very high in terms of numbers per millilitre, their actual biomass (numbers $\times$ volume) may be quite low relative to other phytoplankton.

QUESTION 3.2   Since different wavelengths of light have different extinction coefficients and penetrate to different depths, algae with different accessory pigments can take advantage of trapping wavelengths that are not captured by chlorophyll *a* and may thus be able to extend their vertical range, or live at depths not inhabited by other photosynthetic species.

QUESTION 3.3   Calculating P_g for each species,

$$\text{Species 1}: P_g = \frac{2 \times 50}{10 + 50} = 1.6 \text{ mg C mg}^{-1} \text{ Chl } a \text{ h}^{-1}$$

$$\text{Species 2}: P_g = \frac{6 \times 50}{20 + 50} = 4.3 \text{ mg C mg}^{-1} \text{ Chl } a \text{ h}^{-1}$$

Therefore species 2 would be growing faster at the specified light intensity.

QUESTION 3.4   Calculating the critical depth from equation 3.6,

$$D_{cr} = \frac{500 \times 0.5}{0.07 \times 10} = 357 \text{ m}$$

Because the critical depth (357 m) is greater than the depth of mixing (100 m), there is net photosynthesis in the water column.

QUESTION 3.5   (a)   Obviously more carbon is being fixed photosynthetically in area B (50 mg C m^{-3} h^{-1}) than in area A (20 mg C m^{-3} h^{-1}). However, this is not a comparative measure of photosynthetic activity. Instead, the assimilation index can be used to compare areas A and B. This index is expressed as the amount of carbon fixed per quantity of chlorophyll a per hour, so for area A this would be 20 mg C 2 mg^{-1} Chl a h^{-1}, or 10 mg C mg^{-1} Chl a h^{-1}. The value of the assimilation index for area B is only 2 mg C mg^{-1} Chl a h^{-1}, indicating lower photosynthetic activity of the phytoplankton.

(b)   The difference in photosynthetic activity could be due, for example, to the phytoplankton in area B being at the end of a bloom, and those in area A growing in conditions of high nutrient concentrations at the beginning of a bloom.

QUESTION 3.6   The K_N values for nitrogen uptake are considerably higher in eutrophic waters compared with oligotrophic regions. Phytoplankton species living in oligotrophic waters can take up nitrate (or ammonium) at ambient concentrations of $< 0.1 \ \mu$ M. In contrast, phytoplankton in eutrophic waters generally need nitrate concentrations of $> 1.0 \ \mu$ M.

QUESTION 3.7   The stratification index is calculated from equation 3.13 as:

$$\log_{10} \frac{5000 \text{ cm}}{0.003 \times (3.3)^3} = 5.5$$

Frontal zones are usually formed when $S \approx 1.5$. The calculation indicates that no frontal zone will occur as a result of tidal flow over this bank.

QUESTION 3.8   Although much of the Indian Ocean lies between latitudes that receive high amounts of solar radiation, the basic circulation pattern is anticyclonic (see Figure 2.19). This results in convergence of surface water toward the central area of the gyre and deepening of the thermocline. Consequently, nutrient levels in the euphotic zone are comparatively low and primary production is nutrient-limited.

QUESTION 3.9   The many species of marine phytoplankton differ in their requirements for light and essential nutrients. They also contain different photosynthetically active pigments in different relative amounts and therefore can absorb different wavelengths. Certain species are shade-adapted and so live at deeper depths or under ice. Others are better able to carry out photosynthesis at lower (or higher) temperatures. Some can survive in environments where nutrient concentrations are relatively low. All of these differences may separate species spatially (in depth or geographically), or temporally as conditions change and favour one species over another. Thus what appears superficially to be a homogeneous environment is actually one that contains microenvironments of differing and variable light intensities and nutrient concentrations.

QUESTION 3.10   In equation 3.7,

$$(X_0 + \Delta X) = X_0 e^{\mu t}$$

and therefore,

$$\log_e(X_0 + \Delta X) - \log_e(X_0) = \mu t$$

and

$$\mu = \frac{\log_e(X_0 + \Delta X) - \log_e(X_0)}{t}$$

Substituting a value of 2.5 mg C m^{-3} for X_0, and ΔX being equal to 0.2 mg C m^{-3} in 1 hour gives:

$$\mu = \frac{\log_e(2.5 + 0.2) - \log_e 2.5}{1}$$

$$\mu = 0.9933 - 0.9163 \text{ h}^{-1} = 0.077 \text{ h}^{-1}$$

Using equation 3.10,

$$d = 0.69/0.077 \text{ h}^{-1} = 8.9 \text{ h}$$

QUESTION 3.11   (a)   A doubling time of about 9 hours translates to approximately 2.7 generations per day (from 24 h/9 h), and this is a rapid rate of growth.

(b)   This growth rate is typically found in phytoplankton living in tropical upwelling regions.

QUESTION 3.12   With the aid of graph paper, you can show that the growth rate of species A will already have reached 1 doubling per day at 0.2 μ M nitrate. The growth rate of species B will still be only about 80% that of species A at 0.4 μ M nitrate, and therefore species A will dominate at this concentration.

QUESTION 3.13   Different species of phytoplankton generally will be growing below their μ_{max} growth rates. This is because each will be limited by different K_N values for different nutrients. Therefore, the K_N values are more important than the different μ_{max} values in determining species diversity.

QUESTION 3.14   Theoretically yes, and this can be demonstrated experimentally. However, the logistics of applying and maintaining a large concentration of nutrient media over wide areas is generally economically prohibitive. In addition, this would not be realistic in the sea because there are so many different phytoplankton species with different requirements that it would probably be impossible to selectively enhance one particular type. As well, there are so many other variables (e.g. predation) controlling phytoplankton species composition in the sea that the results of adding a few nutrients would probably not favour a selected species. On the other hand,

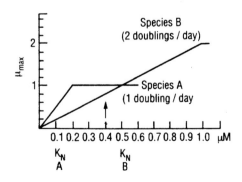

Question 3.12

nutrient addition in relatively small restricted areas could increase total production.

QUESTION 3.15 At this high latitude, there will be phytoplankton production only during those months when there is sufficient solar radiation to support photosynthesis. Ice is relatively transparent to light, but a thick ice cover will cause a reduction in the amount of PAR light available to phytoplankton, and the dominant species will be those that are adapted to live in low light levels (i.e. shade-adapted). Water temperatures will be very low ($< -1°C$), and this reduces the activities of plants (see Section 2.2). There are changes in salinity when sea-ice forms (see Section 2.4), and the gradient in salinity between the ice and surrounding water may affect species composition of the associated phytoplankton. In spite of the rigorous environmental conditions, certain phytoplankton are able to grow immediately under the ice and also within the ice fractures. They form an important part of the food chain in polar regions.

QUESTION 3.16 There are at least four ways in which this can happen, one of them general, the others more specific.

1. All types of phytoplankton may produce very large blooms that eventually decay, causing lowered oxygen concentrations and death of animals that cannot avoid these anoxic areas.

2. A few dozen species of dinoflagellates produce saxitoxin which is transferred via the food chain to vertebrate animals that are sensitive to this neurotoxin. Fish, birds, and marine mammals may suffer poisoning or death from the accumulation of saxitoxin, and humans develop paralytic shellfish poisoning from consuming shellfish that have fed on toxic dinoflagellates.

3. One diatom (*Pseudonitzschia*) produces a neurotoxin called domoic acid that has similar effects on marine animals and on humans who eat contaminated shellfish.

4. Ciguatera fish poisoning, which originates with a toxic dinoflagellate, is a common health problem in many tropical and subtropical countries.

CHAPTER 4

QUESTION 4.1 The pelagic environment lacks hiding places for animals seeking refuge from predators. Transparency permits background colours to be transmitted through an animal's tissues, and thus provides an ideal way of escaping detection by visual predators.

QUESTION 4.2 Asexual budding allows salps to respond to favourable environmental conditions by rapidly producing large numbers of new individuals, and it is therefore common to find salps in large swarms. However, all individuals produced by budding of one solitary form are genetically identical. The establishment of swarms of salp chains, in which many non-identical sexual individuals are in close proximity, favours cross-fertilization, and this process restores genetic variability in the population.

QUESTION 4.3 Red wavelengths of light are quickly absorbed and scattered in near-surface water. The only light present in deeper water is in the blue-green part of the spectrum. Red coloured bodies are difficult or impossible to see in blue light as the red is not reflected.

QUESTION 4.4   Most individuals migrate toward the surface at night, but the adults have a deeper daytime distribution than their young.

QUESTION 4.5   These copepods move into shallower depths in spring and early summer to take advantage of abundant food during the period of high primary productivity. In winter, when food is no longer abundant at the surface, these copepods move deeper where they subsist on stored fat or, in the case of *C. helgolandicus*, they may also feed carnivorously on other zooplankton. By living in deep water during winter, copepods avoid surface turbulence caused by storms, and their mortality due to predation may be reduced by remaining in dark waters.

QUESTION 4.6   (a)   The density of phytoplankton (as measured by chlorophyll *a*) is often inversely correlated with zooplankton numbers due to grazing by herbivores. However, the change in chlorophyll could also result from differences in nutrient concentration in the surface water over the 80-km distance that was sampled.

(b)   Zooplankton samples were taken at night-time from a shallow depth (3 m). Because of diel vertical migration, it would be expected that the numbers of zooplankton would decrease significantly at this depth during daylight hours.

QUESTION 4.7   Crustaceans have rigid exoskeletons. In order to increase in size and add more appendages, they must first shed the exoskeleton, expand the body, and then make a larger exoskeleton.

QUESTION 4.8   Most benthic invertebrates are slow-moving with limited mobility or are permanently attached to a substrate, and they remain in restricted regions throughout their adult lives. The production of meroplanktonic larvae, which are carried in ocean currents, ensures wider distribution for a species than could be attained by the benthic adults.

QUESTION 4.9   Among the phytoplankton, diatoms and silicoflagellates produce siliceous skeletons and coccolithophorids form calcareous plates, all of which can be found in sediments. Among the zooplankton, calcareous tests or shells are formed by foraminifera, heteropods, thecosomes, larval gymnosomes, and veliger larvae of benthic molluscs; and the radiolarians form siliceous skeletons. The skeletal remains of these animals are present in sediments, with those of forams, radiolarians and thecosomes being most common.

QUESTION 4.10   Carnivorous groups include the Cnidaria, Annelida, Chaetognatha, Amphipoda, and *Clione limacina* (a gymnosome). The zooplankton that are predominantly herbivorous include the euphausiids, salps, most of the copepods, and both species of *Limacina* (thecosomatous pteropods).

QUESTION 4.11   If plankton were evenly dispersed, it would be in such dilute concentrations that it would be difficult for grazers and predators to obtain enough food. When food occurs in patches, animals can expend less energy to obtain much more food, and this enables them to grow faster and thus enhances their own survival. This explanation only applies, however, to those organisms that are capable of actively locating high concentrations of food. Some planktonic animals do not necessarily rely on patches of prey and instead utilize passive foraging methods. For example, many medusae, siphonophores, and ctenophores have tentacles that can be extended for considerable distances into the surrounding water to capture more dispersed prey.

QUESTION 4.12   *Neocalanus cristatus* undergoes a seasonal migration that is associated with reproduction and development. Figure 4.22 shows the epipelagic summertime distribution of the copepodite V stage, which will begin to migrate deeper at the end of the summer before maturing into stage VI (the adult). The adults overwinter and lay eggs in deep water, at depths between 500 and 2000 m; young stages of copepodites migrate upward as they develop.

CHAPTER 5

QUESTION 5.1   (a)   Total numbers of individual organisms will decrease in succeeding trophic levels; for example, there are many more herbivorous zooplankton than there are fish and top-level predators.

(b)   There is considerably more total primary production than secondary production.

QUESTION 5.2 E_T would equal 0.166 (from 25 g C m^{-2} yr^{-1} divided by 150 g C m^2 yr^{-1}). Expressed as a percentage, about 17% of the net primary production is being transferred to the production of herbivorous copepods.

QUESTION 5.3   In general, as the numbers of prey diminish in higher levels, the predators become more active and spend more energy in seeking food. Respiration losses become relatively higher in more active animals. Consequently, the production of higher trophic level species is lower relative to production in preceding levels, and E_T values are lower.

QUESTION 5.4   As less of the primary production is transferred to the next trophic level in the pelagic food chain, the value of E_T will decrease.

QUESTION 5.5   (a)   The biomass in all trophic levels is approximately one order of magnitude higher in the Antarctic Ocean.

(b)   The surface waters of the Antarctic Divergence are upwelled and therefore high in nutrients, and this leads to high primary productivity; the dominant producers are chain-forming diatoms that are consumed directly by large zooplankton (euphausiids), and these animals are the major food of baleen whales; and, because the food chain is short, comparatively less energy is lost between primary producers and the highest trophic level.

QUESTION 5.6   Employing equation 5.2 and setting $n = 2$, the maximum amount of herring which could be produced by this food chain would be:

$$P_{(n+1)} = 300 \text{ g C m}^{-2} \text{ yr}^{-1} \times (0.1)^2 = 3 \text{ g C m}^{-2}\text{yr}^{-1}.$$

QUESTION 5.7   A number of other planktonic protozoans are bactivorous; these include foraminiferans, radiolarians, and ciliates (especially tintinnids). In addition, salps and a number of different types of meroplanktonic invertebrate larvae can capture and consume bacteria.

QUESTION 5.8   The North Pacific gyre is a region of low primary and secondary productivity because of low nutrient concentrations (see Section 3.5.1); consequently, there are few nutritive substrates available for bacterial growth. Freshwater localities receive both nutrients and organic materials from runoff. Freshwater primary production is high (as shown in Figure 5.8), and bacteria can utilize organic materials derived from the phytoplankton as well as from runoff.

QUESTION 5.9   From equation 5.4,

$$P_t = (80 - 30)\frac{0.15 + 0.6}{2} + (30 \times 0.6 - 80 \times 0.15),$$

$$= 50(0.375) + (18 - 12),$$

$$= 18.75 + 6,$$

$$= 24.75 \text{ mg m}^{-2}.$$

Therefore, average production per day was 0.56 mg m^{-2} day^{-1}, or about 2% day^{-1}.

QUESTION 5.10   (a)   Using equation 5.10 to solve for K_1,

$$K_1 = \frac{5.0 \text{ mg}}{7.5 \text{ mg}} \times 100\% = 67\%$$

(b)   Using equation 5.11 to solve for K_2,

$$K_2 = \frac{5.0 \text{ mg}}{(7.5 \text{ mg})(0.9)} \times 100\% = 74\%$$

QUESTION 5.11   In general, large diatom chains are consumed by relatively large zooplankton such as euphausiids. On the other hand, small flagellates are generally consumed by protozoans. Thus a shift in the size of the primary producer would probably cause a marked change in the type of dominant herbivores, with subsequent changes at all higher trophic levels. The length of the food chain would be expected to increase, with more trophic levels being added to the system in which flagellates are the major primary producers.

QUESTION 5.12   (a)   The forcing functions are the amount of light available and the nutrient concentrations.

(b)   Physiological functions include the reaction of light with phytoplankton; grazing by zooplankton and zooflagellates; predation by ctenophores, salmon, and microzooplankton; bacterial decomposition; and the growth rates of all organisms. Note that phasing functions, like temperature and light extinction coefficients, are not shown in this figure, but they are included in the actual model.

QUESTION 5.13   A light extinction coefficient of 0.7 m^{-1} indicates low light levels that severely slow the growth of phytoplankton. Thus zooplankton production is limited by the small amount of primary production.

QUESTION 5.14   (a)   If 30% of total production is regenerated, then we have:

300 g C m^{-2} yr^{-1} $\times$ 0.3 = 90 g C m^{-2} yr^{-1} of regenerated production, and

300 g C m^{-2} yr^{-1} $\times$ 0.7 = 210 g C m^{-2} yr^{-1} of new production. The f-ratio is calculated from new production divided by total production. Therefore:

$$f\text{-ratio} = 210/300 = 0.7.$$

(b)   This would be a region of moderately high production (see Table 3.5) where a relatively large amount of new nitrogen is entering the euphotic zone. A coastal area, such as a continental shelf-break, could have such values.

QUESTION 5.15   The exponential increase in sustainable fish yield is due to two factors, one being the increase in total primary production in eutrophic compared with oligotrophic conditions (as shown in Figure 5.21a). The second reason is because of the increased relative amount of new production to total production, as more new nitrate enters eutrophic waters from below the nutricline; this is shown in Figure 5.21b.

QUESTION 5.16   Among the phytoplankton, the coccolithophorids build plates composed of calcite. Among the zooplankton, the foraminifera have calcitic tests and planktonic molluscs (heteropods, thecosomatous pteropods, and veliger larvae) usually produce shells of aragonite. The endoskeletons of fish and marine mammals also contain some $CaCO_3$. In addition, very large amounts of calcium carbonate are present in corals and the shells of benthic molluscs.

QUESTION 5.17   No. Carbon dioxide concentration is never limiting to photosynthesis because of the reservoir of CO_2 that can be released from bicarbonate and carbonate ions. However, nitrate concentrations may often be low enough to limit protein manufacture by phytoplankton and thus limit total primary production.

QUESTION 5.18   Open ocean regions typically have six trophic levels, as shown in Figure 5.3. The answer can be derived from multiplying the transfer efficiencies in the successive trophic levels, as:

$$1000 \times 20/100 \times (10/100 \times 10/100 \times 10/100 \times 10/100)$$
$$= 0.02 \text{ g wet wt m}^{-2} \text{ yr}^{-1}$$

QUESTION 5.19   In evolutionary terms, the short food chain (III in Figure 5.3) that culminates in baleen whales is the most recent; diatoms first appeared about 100 million years ago and whales about 55 mya. This food chain is also the most efficient in that it is the one that delivers the most energy from primary producers to terminal consumers. Parts of the open ocean food chain (I in Figure 5.3) are among the oldest in an evolutionary sense; green algae were present long before dinoflagellates and diatoms, and marine protozoans are known from 600 mya. However, 400 million years ago, the top predators of open ocean food chains would have been jellyfish, pelagic cephalopods (e.g. ammonites) or primitive fish (e.g. sharks), all of which were present before the appearance of bony fish (teleosts).

QUESTION 5.20   The food chain of the region would have been shortened by one trophic level. If the average ecological efficiency were 10% between trophic levels, then theoretically the abundance of planktivorous fish available for harvesting should increase by an order of magnitude, if all other factors remain the same. However, in such circumstances, new predatory species often move into an area and replace those that were removed.

QUESTION 5.21   Converting annual fish catch to grammes m^{-2} gives:

$$0.5 \text{ tonnes} \times 10^6 \text{ g tonne}^{-1} = 0.5 \times 10^6 \text{ g wet wt/10 000 m}^2.$$

Converting wet weight of fish to dry weight gives:

$$(0.5 \times 10^6 \text{ g}) \times 0.2 = 10^5 \text{ g}/10\,000 \text{ m}^2, \text{ or } 10 \text{ g dry wt m}^{-2}.$$

Converting dry weight to carbon gives:

$$10 \text{ g} \times 0.5 = 5 \text{ g C m}^{-2}\text{yr}^{-1}.$$

Employing equation 5.2, and setting $n = 2$ because there are two trophic transfers from phytoplankton to zooplankton to sand eels, the average ecological efficiency can be calculated from:

$$5 \text{ g C m}^{-2}\text{yr}^{-1} = (200 \text{ g C m}^{-2}\text{yr}^{-1})E^2,$$

and

$$E = \sqrt{5/200} = 0.16, \text{ or } 16\%.$$

QUESTION 5.22   Case (a) might best be studied in an enclosed experimental ecosystem, where pesticides in varying concentrations could be added to enclosures containing large volumes of water and several trophic levels. No experimental container would be large enough to consider impacts of damming an estuary, so a computer simulation model would be the best approach for Case (b). In Case (c), physiological properties of plants and animals can be studied in the laboratory.

QUESTION 5.23   Using equation 5.9 gives:

$$A = \frac{5 \text{ mg} - 0.75 \text{ mg}}{5 \text{ mg}} \times 100\% = \frac{4.25}{5} \times 100\% = 85\% \text{ efficiency}$$

QUESTION 5.24   Theoretically yes, but the danger lies in adding so much nutrient that it could cause an excessively large phytoplankton bloom. Most of this production could not be eaten by the resident oysters. Large amounts of phytoplankton would probably die and, as a result of the decomposition processes, a large oxygen demand could create anoxic conditions and the death of the oysters. Nutrient additions in such cases would have to be made with considerable caution and understanding of the system.

QUESTION 5.25   The productivity of the area would eventually decrease because it would be robbed of the nitrogen that would have been regenerated from the fish community.

CHAPTER 6

QUESTION 6.1   Approximately $40\,000$ km (800 nets $\times$ 50 km length), extending from the surface to a depth of 8 or 10 m. Tied end to end, the 800 driftnets would stretch around the world about $1\frac{1}{2}$ times.

QUESTION 6.2   New-born young can develop faster in warm water; in cold water, more energy would have to be used to maintain body temperature. Food concentrations, however, are generally much higher in summer months in cold-water environments than in tropical waters throughout the year, and both adults and juveniles benefit from rich feeding grounds.

QUESTION 6.3   The major predators of Antarctic penguins are marine species, especially the leopard seal and toothed whales. There are no mammalian ground predators in the Antarctic, but one bird, the Antarctic skua, preys on the eggs and chicks of penguins.

QUESTION 6.4   Their low fecundity makes them especially vulnerable to over-harvesting as they cannot rapidly replace their numbers through reproduction.

QUESTION 6.5   Both countries are situated along an intense continental upwelling zone that produces an almost continual supply of new nutrients to the euphotic zone. These nutrients are transferred through the food chain to produce larger numbers of fish.

QUESTION 6.6   (a)   Yes, any growth rate that is slower than that of the predators would theoretically result in 100% mortality due to predation.

(b)   In actuality, this would not happen for several reasons. As more and more prey are eaten and become scarcely distributed, it becomes energetically costly to seek them out, and predators generally switch to more abundant food. If the predators grow much more rapidly than the prey, they may turn to eating larger food sizes, thus obtaining more energy per item ingested.

QUESTION 6.7   In general, fish grow most efficiently when they are small. The growth curve in Figure 6.11 indicates that older animals (age 3+) do not increase greatly in size, but they do continue to eat prey items, some of which are smaller fish. Therefore, the biomass yield will be highest and the fishery more efficient when large numbers of small fish are harvested rather than smaller numbers of large fish. For specific fisheries, however, the answer to this question may vary with the size at maturity, temperature, specific fishing methods, and the value of the catch (e.g. salmon are larger, but more valuable than herring). Note that five out of eight of the world's marine fisheries in Table 6.2 are based on species of small fish (anchovies, pilchards, capelin, and herring).

QUESTION 6.8   The total marine catch is about 84×10^6 tonnes per year (see Section 6.7.1). Thus mariculture produces about 6% of this figure.

QUESTION 6.9   Productive areas such as the leeward side of islands (Section 3.5.6), the mouth of estuaries (Section 3.5.5), and cold core rings (Section 3.5.1) should also support good fisheries because these are regions in which nutrient-rich subsurface water is mixed up into the euphotic zone.

QUESTION 6.10   Theoretically, the lower down on the food chain that marine organisms are harvested, the much greater the potential harvest. There are, in fact, some zooplankton fisheries in the world (e.g. for Antarctic krill). However, it becomes more and more economically costly to harvest very small-sized organisms in the oceans. Therefore, unless special techniques are developed, it is probable that fish (including shellfish and squid) will remain the most easily harvested protein of the sea.

QUESTION 6.11   The adult population size would change by 100% (99.90% of 10^3 indicates that one fish will survive per spawning female; a mortality of 99.95% indicates that only one fish per two spawning females will survive.) Notice that the change in mortality is statistically almost insignificant, but that the change in adult population numbers is 100%. These differences become even more dramatic when dealing with species, such as cod, that may lay more than 10^6 eggs per female per year.

QUESTION 6.12   The removal of piscivorous fish from top trophic levels may increase the numbers of planktivorous fish and larger planktivorous invertebrates, including commercially undesirable species such as jellyfish. The targeting and removal of large numbers of one species may confer an ecological advantage to competing species which may then increase in number (review the consequences of removing large numbers of baleen whales from the Antarctic in Section 5.2). Harvesting much of the biomass from intermediate trophic levels may decrease the numbers of top-level predators such as tunas, toothed whales, and sharks. The incidental capture and dumping of large numbers of undesirable 'trash' fish may increase the number of pelagic and benthic scavengers. At present, none of these potential ecological consequences have been included in fisheries management theories but, in fact, commercial harvests do change the ecology of heavily fished regions.

QUESTION 6.13   Because mariculture is carried out in coastal regions, species should be tolerant to relatively wide temperature and salinity ranges. Because larvae require different conditions from the adults and are more sensitive to environmental change, it is easier to culture those species that have few life stages in their life cycle. Those species with tolerance to living and growing in crowded conditions are preferred by culturists, as are those with fast growth rates and ready marketability.

QUESTION 6.14   From Table 5.1 (footnote), the total surface area of the ocean is about 362×10^6 km^2. The whale sanctuary around Antarctica is 28×10^6 km^2, which represents about 8% of the world's ocean.

CHAPTER 7

QUESTION 7.1   These categories have been established according to the type of photosynthetic pigments contained in different algal species and, to a lesser degree, on their external colour. Red algae, for example, contain large amounts of phycoerythrin and phycocyanin in addition to green chlorophylls. These red pigments absorb blue-green wavelengths, but reflect red. It is the reflected wavelengths that give algae characteristic colours.

QUESTION 7.2   Worms are more or less elongated, relatively slender, soft-bodied animals that are limbless, or nearly so. These anatomical features are ideally suited for living in or on soft sediments, which are so prevalent in the sea. Some groups (e.g. polychaetes and pogonophorans) have moved beyond the typical worm-like form to become more specialized species that are sessile tube-dwellers.

QUESTION 7.3   There are many forms of defence against predation. Some examples include: the hard calcareous shells of clams and snails; the production of unpleasant compounds by sponges; the stinging nematocysts of cnidarians; the large spines of sea urchins; the burrowing behaviour of infaunal species; the cryptic, or inconspicuous, growth forms of hydroids and bryozoans.

QUESTION 7.4   Using $B = X \times \overline{w}$, the biomass of clams sampled at 225 days is 378 m^{-2} × 9.910 mg, or 3746 mg m^{-2}. Biomass values calculated for all intervals are given below in the completed table.

The first part of equation 5.4 considers the biomass produced and then lost through predation or other mortality; when this is combined with the change in biomass between intervals, the net production can be obtained. Thus, to calculate biomass produced, but then lost, during the interval between 50 and 225 days, use:

$$\text{Biomass loss from the population} = (X_1 - X_2)\left(\frac{\overline{w}_1 + \overline{w}_2}{2}\right),$$

$$= (990 - 378)\left(\frac{5.364 + 9.910}{2}\right) = 4674 \text{ mg m}^{-2} \text{ per 175 days,}$$

or, divided by 175 days, biomass loss $=27$ mg m^{-2} day^{-1}. Combining this expression with the change in biomass gives the equation for net production,

$$P_1 = (X_1 - X_2)\left(\frac{\overline{w}_1 + \overline{w}_2}{2}\right) + (B_2 - B_1)$$

Thus the net production in the interval between 50 and 225 days is:

$$P_{175 \text{ days}} = (990 - 378)\left(\frac{5.364 + 9.910}{2}\right) + (3746 - 5310)$$

$$= 612(7.637) - 1564 = 3109.84 \text{ per 175 days,}$$

$$\text{or } P = 17.77 \text{ mg m}^{-2} \text{ day}^{-1}$$

Biomass losses and net production values for all intervals are given in the completed table.

Table 7.2 Production data from a population of the clam *Mactra* in the North Sea.

t (days)	X (no. m^{-2})	Mean weight (mg)	Biomass (mg m^{-2})	Biomass loss (mg m^{-2} day^{-1})	Net production (mg m^{-2} day^{-1})
0	7045	1.416	9976		
				411	317
50	990	5.364	5310		
				27	18
225	378	9.910	3746		
				14	66
398	289	44.286	12 799		
				12	36
616	246	73.542	18 091		

QUESTION 7.5 Slow-moving or sessile invertebrates without planktonic larvae can be carried great distances if the eggs, young or adults are attached to ships or to floating objects like wood, seaweed, or bottles. Eggs of shallow-living species may also be transported on the feet of birds. In addition, the juveniles or small adults of a few snail and bivalve species can drift in the water column by producing long mucous threads for suspension.

QUESTION 7.6 (a) Table 4.1 lists 8 phyla with holoplanktonic members; Table 7.1 lists 16 phyla of marine benthos. Although the Tables ignore some exceptional species (e.g. there is one planktonic echinoderm and one genus of benthonic chaetognaths), there are many more types of benthic animals than planktonic ones.

(b) The difference can be attributed largely to the much greater physical variety in benthic habitats compared to the more homogeneous water column.

QUESTION 7.7 Benthic species with lecithotrophic planktonic larvae produce (a) relatively few young that remain for only short times in the plankton and (b) therefore do not disperse far from the adult population. (c) Larval mortality is low relative to planktotrophic larvae. (d) Adult population size tends to be relatively constant over long intervals. (e) These factors suggest that the P/B ratio will be relatively low. All of the traits listed here are characteristic of K-selection.

QUESTION 7.8 Upward growth of stromatolites takes place at about 0.5 mm per year. Therefore,

$$1.5 \text{ m} = 1500 \text{ mm}$$

and,

$$\frac{1500 \text{ mm}}{0.5 \text{ mm yr}^{-1}} = 3000 \text{ years.}$$

Solving the equation gives an age of 3000 years.

CHAPTER 8

QUESTION 8.1 There is a greater water exchange bringing in more nutrients and plankton where the tidal range is higher. Increased nutrients lead to higher benthic primary production, and consequently more food for grazing animals. Food also increases for benthic suspension feeders when there are increased amounts of phyto- and zooplankton. Therefore the biomass per unit area of benthic plants and animals should be greater in areas with a high tidal range.

QUESTION 8.2 Interstitial species are very small, and body size limits the number of young that can be produced per individual. If fecundity is very low, mortality rates must also be low to ensure survival of populations. Direct development or brood protection ensures that the progeny remain in a favourable environment, and that they are not vulnerable to pelagic predators or filter-feeding benthos.

QUESTION 8.3 Not necessarily. H_2S is a by-product of sulphate reduction by bacteria and its occurrence, although objectionable to humans, is a natural process. However, H_2S production could be indicative of pollution where there are large quantities of decomposing organic matter that use up oxygen.

QUESTION 8.4 The smaller the size of the sediment particles, the greater the surface area per unit volume of sediment. Consequently, there is more space for attachment of surface-growing algae.

QUESTION 8.5 Many freshwater and marine species are physiologically excluded from living in estuaries because of the salinity stress. However, for those that can tolerate the salinity regime, productive estuaries offer a rich food supply. They may also provide a more protected environment, and one in which there is less competition than in areas with many species.

QUESTION 8.6 (a) The west coasts of continents are characterized by upwelling, and the average water temperatures are too low to permit coral reef growth.

(b) These rivers discharge huge amounts of freshwater carrying very high loads of sediment into the sea. The resulting mixed water of reduced salinity and high turbidity is carried northward by prevailing currents. Both low salinity and high turbidity prevent or inhibit the growth of coral reefs.

QUESTION 8.7 Yes, but these are not reef-building corals. Ahermatypic corals do not support a symbiotic relationship with zooxanthellae, and consequently they do not require light for nutrition and growth.

QUESTION 8.8 Yes. The nutrient concentration in water close to land should be higher than in the middle of a tropical ocean, and consequently the barrier reef should have a higher productivity.

QUESTION 8.9 The net primary productivity of mangrove swamps $(350-500 \text{ g C m}^{-2} \text{ yr}^{-1})$ is much higher than that of phytoplankton productivity in open tropical waters, the latter being about $75 \text{ g C m}^{-2} \text{ yr}^{-1}$.

QUESTION 8.10 Large sponges are epifaunal, attaching to firm substrates, and all are suspension feeders that create water currents to filter small plankton and bacteria from the water. The combination of hard substrates and abundant suspended organic particles most frequently occurs in shallower areas, so it is not surprising that such sponges are most successful above 2000 m. Sea cucumbers are either epifaunal or infaunal, depending on the species, but most are deposit feeders or detritivores; thus this group can successfully live on, or in, deep-sea soft substrates, where they obtain sufficient nourishment from detritus or organic material in the sediments.

QUESTION 8.11 The biomass of both benthos and zooplankton diminishes rapidly with increasing depth. In both groups, biomass declines by one to two orders of magnitude from the surface to 1000 m, then decreases by another order of magnitude from 1000 m to 4000 m.

QUESTION 8.12 Assuming that a dead diatom cell sinks at twice the rate of a living cell and thus at a maximum rate of about 60 m per day, it would take 95 days to reach 5700 m depth. In actuality, sinking cells are often aggregated in mucoid masses, and the sinking rate would be somewhat faster because of the larger size of the aggregate.

QUESTION 8.13 Except for size, *Tindaria* has many of the characteristics of a *K*-selected species. It lives in a predictable environment of constant temperature and salinity, and it is slow-growing, reaches sexual maturity very late, and has a long life span. The small size of *Tindaria* indicates that it produces few young. The P/B ratio for *Tindaria* would be low. In comparison, wood-boring clams are good examples of *r*-selected species based on the characteristics described in the text.

QUESTION 8.14 The temperature at 2500 m depth would usually be between 2 and 4°C.

QUESTION 8.15 Biomass is typically $<20 \text{ g m}^{-2}$ at 2500 m. The biomass of the Galapagos Vent community is $20-30 \text{ kg m}^{-2}$, or 2000–3000 times higher. At all hydrothermal vent sites so far studied, the biomass is at least 500–1000 times greater than in normal deep-sea communities.

QUESTION 8.16 Both offer a great deal of cover for small animals. Many animals specifically adapted to these communities live in their shelter and derive a protective refuge from predators, such as birds and migrating fish.

QUESTION 8.17 Although some individuals do lose the competition for space with faster growing species, the system is usually kept in a state of flux by various physical and biological factors. For example, intense grazing by herbivores reduces algal standing stock, and predators remove animals from the reef. Faster-growing organisms may face heavier predation because it is more efficient for animals to consume more abundant, rather than less abundant food. Grazing animals also may remove encrusting organisms, thus continually creating new space for settling larvae or algal spores. Success of larval recruitment for each species on the reef may also change over time, thus keeping the populations in a state of dynamic flux. Fluctuations in the physical environment too, such as temperature or salinity changes or storm-related variability, will favour certain species at one time and other species at other times.

QUESTION 8.18   A sea level rise of 2 m per 100 years is equivalent to 2 cm per year, and at least some corals and coral reefs are capable of such growth rates. Increased water depth and concomitant elevations in seawater temperature could affect growth and survival of some coral species. It might be well to keep in mind, however, that reef corals have survived for over 250 million years in spite of much larger sea level changes during this geologic period.

QUESTION 8.19   Both communities are found in upper intertidal zones with fluctuating salinity, and both are dominated by salt-tolerant, erect, flowering plants (angiosperms) with aerial storage of biomass. The roots of mangroves and marshgrasses facilitate sedimentation by slowing water velocities and by retaining sediment, and thus they stabilize coastlines. These plants contain large quantities of refractory structural elements that are difficult for herbivores to digest, and consequently both of these communities have detritus-based food chains. In addition, both communities support some terrestrial animals as well as marine species.

QUESTION 8.20   Crustaceans are among the most abundant of zooplankton, and all of them moult their exoskeletons between each growth increment. Copepods, for example, have twelve distinct life stages that are separated by moulting, so each copepod that reaches maturity has contributed 11 moults to the downward flux of organic material. Overall, crustacean moults are extremely abundant materials in the water column. However, their nutritive value for animals is considered to be low; most are probably converted to bacterial biomass through the actions of chitinoclastic bacteria.

QUESTION 8.21   Although many organisms have a limited food supply and live under very high pressures, there are some other properties of the deep sea which are advantageous to life. These include constant darkness, which helps in predator avoidance; constant temperature, which eliminates metabolic adjustments to change; and constant salinity, which eliminates the need for osmotic adjustments to varying salt concentration.

QUESTION 8.22   Chemosynthetic bacteria appeared before photosynthetic algae, and the earliest marine food chains must have been based on chemosynthesis.

QUESTION 8.23   (a)   Hadal trenches, hydrothermal vents, and cold sulphide seeps all have high proportions of endemic species that are not found outside the particular community.

(b)   All of these communities are spatially isolated from similar habitats; for example, hydrothermal vents may be separated by hundreds of kilometres. Thus dispersal success may be very limited and the fauna of these communities tends to be isolated. As well, all of these communities have special environmental conditions that require a high degree of adaptation, and many animals may be excluded from the sites by these constraints.

CHAPTER 9

QUESTION 9.1   Cold environments with little wave action would have relatively slow rates of recovery from large oil spills. This is because organisms in such habitats typically have slow growth rates and long life cycles, and because the oil will not be dispersed as quickly where there is little wave action and little turbulence. As well, biological degradation of oil proceeds more slowly at low temperatures.

QUESTION 9.2   One obvious way is to recycle plastic products, other ways are to restrict dumping of trash at sea from commercial and fishing vessels and to prohibit the spillage of pellets from plastics factories and during shipment. The development and use of new types of biodegradable plastic should also be encouraged.

QUESTION 9.3   The greatest impact would probably occur in summer. This is because the power demand (for air conditioning) is highest, while the temperature of ambient receiving waters is maximal. For many organisms, the sea temperature at that time is already near their upper thermal limit, and the additional heat may exceed this limit.

QUESTION 9.4   There are only a few, one being the Mackenzie River in Canada that empties into the Arctic Ocean. There are also several large rivers in Siberia that enter the Arctic Ocean, but at least some of these carry pollutants from the hinterlands.

QUESTION 9.5   Some other anthropogenic impacts on marine organisms include changes in coastal salinity due to dams; the impacts of offshore constructions such as oil rigs; the effects of tourists on beach ecology. There are also beneficial changes brought about by humans including the construction of artificial reefs and mariculture.

QUESTION 9.6   Probably not, but it is possible to minimize impacts by reducing fish harvests to some sustainable level.

QUESTION 9.7   Sewage disposal may have beneficial effects. In spite of the unwanted impacts, sewage contains nutrients that can potentially enrich oligotrophic waters and enhance food chains. However, such enrichment should not be made in confined areas such as coral atolls, where enhanced growth of phytoplankton, benthic algae, and seagrasses may cause coral mortality.

QUESTION 9.8   Because some pollutants are adsorbed onto sediment particles which settle on the seabed, Estuary A would have lower concentrations of pollutants in the water, but higher levels of pollutants in the sediments.

Abiotic factors Nonbiological; referring to chemical, physical and geological features of the environment.

Abyssal zone The benthic zone between about 2000 m and 6000 m depth.

Abyssopelagic zone The water column between 4000 m and 6000 m depth.

Accessory pigments Plant pigments other than chlorophyll that capture photons of light used in photosynthesis; e.g. carotenes, xanthophylls, and phycobilins.

Adsorption The adherence of ions or molecules in a fluid to the surfaces of particles suspended in the fluid.

Advection Horizontal or vertical movement of water.

Aerobic Living in oxygenated conditions.

Agnatha The class of primitive fish that includes hagfish and lampreys.

Ahermatypic corals Non-reef-building corals that lack symbiotic zooxanthellae.

Algae A diverse group of marine plants ranging from unicellular planktonic species to large benthic seaweeds; all lack true roots, stems, and leaves, and none produce flowers or seeds.

Amphipods Laterally compressed, planktonic or benthic crustaceans.

Anadromous Referring to fish that breed in freshwater but spend most of their adult life in the sea.

Anaerobic Living in the absence of oxygen.

Angiosperms Flowering plants, including species of mangroves, marshgrasses, and seagrasses.

Anoxic Without oxygen.

Anthozoa A class of the Phylum Cnidaria that includes sea anemones and corals.

Anticyclonic Moving in a clockwise direction in the Northern Hemisphere and in an anticlockwise direction in the Southern Hemisphere.

Aphotic zone That part of the ocean in which sunlight is absent.

Appendicularia See **Larvacea**.

Aragonite A form of calcium carbonate present in shells of pteropods and heteropods and in coral skeletons.

Ascidiacea See **Tunicates**.

Assimilated food That portion of ingested food that is absorbed and utilized by an animal, the remainder being discarded as faeces.

Assimilation efficiency The percentage of ingested food that is assimilated by an animal.

Assimilation index A measure of primary productivity in which plant growth is expressed in terms of amount of carbon fixed per unit of chlorophyll *a* per unit time.

Atoll A type of coral reef that grows around a subsided island and encloses a shallow lagoon.

Attenuation A decrease in the energy of light due to absorption and scattering in the water column.

Autotroph(ic) Referring to organisms that synthesize their own organic material from inorganic compounds; also known as primary producers.

Auxospore A reproductive cell of diatoms that re-establishes the initial size of the species after a period of asexual division.

Auxotroph(ic) Phytoplankton that require certain organic compounds, such as vitamins, for growth.

Bacillariophyceae The algal class of diatoms.

Bacterioplankton Planktonic bacteria.

Bactivores Animals that feed primarily on bacteria.

Baleen whales Those species that use specialized plates of horny material (baleen) to filter-feed.

Barnacle A type of benthic filter-feeding crustacean having calcareous plates and living permanently attached to a substrate.

Barrier reef A type of coral reef that lies some distance offshore, with water between the reef and land.

Bathyal zone The benthic zone between 200 m and about 2000 m depth.

Bathypelagic zone The water column between 1000 m and 4000 m depth.

Benthic Pertaining to the seafloor environment.

Benthic boundary layer The layer of water immediately above the seafloor and extending upward from ten to several hundred metres above the bottom.

Benthos Plants or animals that inhabit the benthic environment.

Bioaccumulation The build-up over time of substances (e.g. metals, chlorinated hydrocarbons) that cannot be excreted by an organism.

Bioerosion The breakdown of substrates, such as the calcium carbonate of coral reefs, by a variety of living organisms.

Biological indicators Pelagic organisms that live within relatively narrow temperature–salinity ranges and whose presence is indicative of a specific water mass with those environmental characteristics.

Bioluminescence The production of light by living organisms.

Biomagnification The increased tissue concentration of a bioaccumulated pollutant that occurs at successive trophic levels, resulting in top-level predators having the highest concentrations of substances like chlorinated hydrocarbons.

Biomass The number of individual organisms (in some area or volume or region) multiplied by the average weight of the individuals.

Biotic factors Biological; referring to environmental influences that arise from the activities of living organisms, such as competing species, predators, etc.

Bioturbation Disturbance of soft sediments by the movements and feeding activities of infauna.

Bipolar species Those species that live in both Antarctic and Arctic waters, but are not present in mid-latitudes.

Bivalves Molluscs with the shell divided into two valves; e.g. clams, mussels.

Bloom The sudden appearance of a high concentration of phytoplankton resulting from increased reproduction as a response to favourable conditions.

Blue-green algae (or **bacteria**) Photosynthetic organisms belonging to the Cyanobacteria.

Brachiopoda A phylum of benthic, sessile, filter-feeding animals with a bivalved calcareous shell.

Brackish water Water of reduced salinity resulting from a mixture of freshwater and seawater.

Bryozoa A group of sessile colonial animals belonging to the Phylum Ectoprocta.

By-catch Unwanted marine animals that are caught incidentally by commercial fisheries operations.

Calcification The process whereby calcium and carbonate ions are combined to form calcareous skeletal materials.

Calcite A form of calcium carbonate present in the shells of Foraminifera and most benthic molluscs.

Calorie The quantity of heat required to raise the temperature of 1 g of water through one Centigrade degree at 15°C.

Carnivore An animal that feeds exclusively or primarily on other animals.

Catadromous Referring to fish that breed in the sea but spend most of their adult life in freshwater (e.g. eels).

Catch-per-unit-effort (CPUE) The amount of fish caught for a given amount of fishing effort.

Cetacea The order of marine mammals that includes whales, porpoises, and dolphins.

Chaetognaths An animal phylum of holoplanktonic, unsegmented 'arrow worms'.

Chemoautotrophs Bacteria that utilize the energy contained in such compounds as methane and hydrogen sulphide to reduce carbon dioxide and make organic material.

Chemosynthesis The fixation of carbon from CO_2 into organic compounds by using energy derived from the oxidation of inorganic compounds such as ammonia, methane, and sulphur.

Chitin A horny substance forming the hard part of crustacean exoskeletons; biochemically, a polymer of the carbohydrate glucosamine.

Chitinoclastic bacteria Those that decompose chitin.

Chlorophyll A group of green plant pigments that capture photons of light to be used in photosynthesis.

Chlorophyll maximum The depth at which the concentration of chlorophyll is highest per unit volume of water.

Chondrichthyes The class of fish that includes skates, rays, and sharks.

Chordata A phylum of animals that includes planktonic salps and larvaceans, benthic tunicates and vertebrates.

Ciguatera fish poisoning (CFP) An illness common in tropical and subtropical countries, acquired by eating fish that have accumulated toxins from dinoflagellates attached to seaweeds.

Ciliates Planktonic or benthic protozoans that have hair-like structures called cilia which are used for locomotion and, in some species, for feeding.

Cladocera Planktonic crustacea with a bivalved exoskeleton.

Cnidaria A phylum that includes jellyfish, sea anemones, and corals; formerly Coelenterata.

Coccolithophorids Small, flagellate, unicellular phytoplankton having calcareous plates (coccoliths) in their cell walls.

Coelenterates Animals of the Phylum Cnidaria.

Cohort A group of organisms produced at the same time; one generation.

Cold core ring A rotating body of water with a relatively cool temperature and high productivity.

Community An ecological unit composed of the various populations of micro-organisms, plants, and animals that inhabit a particular area.

Compensation depth The depth at which the amount of carbon fixed in organic material by photosynthesis is equal to that which is consumed by the plants during respiration over a 24-hour period; also, the lower boundary of the euphotic zone.

Compensation light intensity The amount of light at which photosynthetic production just balances respiratory losses in the plants.

Competition The interaction among organisms that results when a necessary resource is in limited supply.

Continental shelf The zone bordering a continent, extending from the line of permanent immersion to the depth (usually 200 m) at which there is a marked increase in the slope.

Continental slope The relatively steep downward slope from the outer edge of the continental shelf to the flat ocean floor.

Convergence The situation in which different water masses come together, usually resulting in the sinking of surface water.

Copepodites Life stages of copepods following the naupliar larvae; copepodite VI is the adult stage.

Copepods A group of small planktonic, benthic or parasitic crustaceans; holoplanktonic species are usually the numerically dominant group of zooplankton captured by nets in most marine areas.

Coral A benthic animal (often colonial) belonging to the Phylum Cnidaria, Class Anthozoa, that forms a calcareous exoskeleton.

Cosmopolitan species Those species with a very broad geographical distribution; present in extensive areas of the Atlantic, Pacific, and Indian oceans.

Countershading Colour difference in the dorsal and ventral surfaces of an animal; a protective mechanism against visual predators.

Crinoids A class of echinoderms that includes feather stars and sea lilies.

Critical depth The depth at which the total photosynthetic production taking place in the water column (from the sea surface to the critical depth) is just balanced by the total respiratory losses of the phytoplankton within the same depth layer.

Critical phase That period in the life of a fish between hatching and the absorption of the yolk sac.

Crustacea A large class of primarily aquatic arthropods characterized by having a segmented body, paired appendages, and a chitinous exoskeleton.

Ctenophores Gelatinous zooplankton having eight longitudinal rows of fused cilia ('ctenes') used in swimming.

Cyanobacteria A class of photosynthetic organisms, some of which are capable of nitrogen-fixation.

Cyclonic Moving in a counterclockwise direction in the Northern Hemisphere and in a clockwise direction in the Southern Hemisphere.

Cypris A larval stage of barnacles that succeeds the naupliar stages.

Cysts Dormant stages of dinoflagellates that germinate under favourable conditions to produce swimming cells.

DDE Dichloro-diphenyl-ethane, a breakdown derivative of the pesticide DDT.

DDT The chemical pesticide dichloro-diphenyl-trichloro-ethane.

Decapods A group of large crustaceans that includes crabs, lobsters, shrimp, and prawns.

Decomposer An organism that breaks down dead organic material to inorganic forms.

Decomposition The breakdown of organic materials into inorganic elements by the mediation of bacteria.

Deep scattering layer A sound-reflecting layer caused by aggregations of animals.

Demersal Pelagic species that live near the seafloor; see also epibenthic.

Denitrification The formation of reduced nitrogen compounds from nitrate.

Density In physical terms, mass per unit volume; in ecological terms, numbers of individuals per unit volume or area.

Deposit feeding Feeding on organic particles located on or in the sediments of the seafloor.

Detritivores Animals that feed primarily on detritus.

Detritus Organic debris.

Diapause A period of suspended development or growth, accompanied by greatly decreased metabolism.

Diatomaceous ooze A sediment in which at least 30% of the particles are the skeletal remains of diatoms.

Diatoms Unicellular phytoplankton with an external skeleton of silica.

Diel Referring to events that occur with a 24-hour periodicity.

Diel vertical migration The vertical migration of pelagic species that occurs with a 24-hour periodicity.

Dinoflagellates Unicellular plankton having two flagella and, in some species, a cellulose test.

Disphotic zone The area of low light lying between the euphotic and aphotic zones.

Dissolution The breakdown of calcareous skeletal material to dissolved calcium and carbonate ions.

Diurnal Referring to events that occur during daytime.

Diurnal tide A tide with one high water and one low water each tidal day.

Divergence The horizontal flow of water away from a coast or away from a common centre, usually resulting in upwelling.

Diversity See **Species diversity**.

DOM Abbreviation for dissolved organic matter.

Domoic acid A neurotoxin produced by the diatom *Pseudonitzschia*.

Doubling time The time required for a population to double in size.

Downwelling The sinking of water.

Echinodermata A phylum of marine animals that includes starfish, sea urchins, and sand dollars among others.

Echinoids A class of echinoderms that includes sea urchins and sand dollars.

Echiura A phylum of benthic marine worms (echiurids).

Ecological efficiency The amount of energy extracted from a given trophic level divided by the energy supplied to that trophic level.

Ecosystem An ecological unit composed of the abiotic environment, together with one or more communities of organisms living in a large geographic area.

Eddy A circular movement of water.

Egestion The voiding of unutilized food as faeces.

El Niño Episodic climatic changes that include warming of the equatorial Pacific Ocean, and suppression of upwelling into the euphotic zone off the coast of Peru by intrusions of this warm, nutrient-poor, surface water.

Elasmobranchs Fish with cartilaginous skeletons; sharks, skates, and rays.

Endemic Organisms restricted to specific habitats.

Enteropneusts Benthic marine worms belonging to the Phylum Hemichordata.

Entrainment Mixing of salt water into fresh water, as in an estuary.

Environment A collective term for the conditions in which an organism lives, including abiotic features (e.g. light, temperature) and biotic features (e.g. predators, competitors).

Epibenthic Referring to pelagic species that live in association with the seafloor.

Epifauna Animals that live on, or attach to, a substrate surface.

Epipelagic zone The upper region of the sea from the surface to about 200–300 m depth.

Epiphytes Plants that grow on the surfaces of other plants.

Epipsammic algae Those species that live on sand grains.

Epontic algae Algae that grow within sea ice.

Equilibrium species Those species that are usually of relatively large size, have slow growth rates and long life spans, produce few young, and have relatively constant population sizes that are at or near the carrying capacity of the environment in which they live; K-selected species.

Estuary A semi-isolated coastal area that is diluted by freshwater discharge.

Euphausiids Shrimplike, holoplanktonic crustaceans; 'krill'.

Euphotic zone The surface waters of the oceans that receive sufficient light to support photosynthesis.

Eurybathic Able to tolerate a wide range in depth (pressure).

Euryhaline Able to tolerate a wide range in salinity.

Eurythermic Able to tolerate a wide temperature range.

Eutrophic Referring to areas that contain high nutrient concentrations and support high biological productivity.

Excretion The elimination of wastes produced from metabolic processes, usually in the form of urea or ammonia.

Extinction coefficient The ratio between the intensity of light at a given depth and the intensity at the sea surface.

Exudation The release of dissolved metabolites by phytoplankton.

f-**ratio** The ratio of new production to total (new + regenerated) production.

Fecundity The rate of production of eggs or young.

Femtoplankton Planktonic organisms (viruses) of 0.02–0.2 μm.

Filter feeding See **Suspension feeding**.

Food chain A linear sequence of organisms in which each is food for the next member in the sequence.

Food web A schematic depiction of the feeding interactions in a community.

Foraminifera Planktonic or benthic protozoans with a calcareous exoskeleton and pseudopodia.

Foraminiferan ooze A sediment containing 30% or more foraminiferan skeletons.

Fringing reef A type of coral reef attached directly to a land mass and not separated from it by a lagoon.

Frustule The external skeleton of a diatom.

Gelatinous zooplankton Fragile planktonic animals without rigid exoskeletons and with high water concentrations in their gelatinous tissues; e.g., 'jelly'fish, siphonophores, ctenophores, salps.

Generation time The number of generations produced per unit time.

Gramme carbon (g C) An expression of biomass in terms of the weight of carbon in a sample.

Grazing The consumption of plants by herbivores.

Gross photosynthesis The total amount of photosynthetic production before subtracting losses due to respiration.

Gross primary production The total quantity of organic tissue (or of carbon) fixed by photosynthesis.

Growth efficiency The amount of growth attained per unit of ingested (gross growth efficiency) or assimilated (net growth efficiency) food.

Gymnosomes Holoplanktonic, shell-less, carnivorous snails.

Gyre A circular motion of water, larger than an eddy.

Habitat The place inhabited by a plant or animal species.

Hadal zone The benthic zone from 6000 m to the deepest areas of the ocean.

Hadalpelagic The pelagic zone from 6000 m to the deepest areas of the ocean.

Half-life The length of time required for the radioactivity of a substance to decline by one-half.

Halocline The zone showing the greatest change in salinity with depth.

Herbivore An animal that feeds exclusively or primarily on plants.

Hermaphrodite An animal that produces both male and female gametes.

Hermatypic coral A reef-building coral that contains zooxanthellae in its tissues.

Heteropods Holoplanktonic snails characterized by having a single swimming fin.

Heterotroph(ic) Referring to organisms that require organic materials for food.

High water The maximum height reached by a rising tide.

HNLC areas Ocean regions with high nitrate concentration but low chlorophyll concentration.

Holoplankton Planktonic organisms that spend their entire lives in the water column; permanent residents of the plankton community.

Holothuroidea A class of echinoderms that includes sea cucumbers.

Homoiothermic Warm-blooded; able to regulate internal body temperature.

Hydroids Benthic colonial cnidarians, some of which produce free-swimming medusae.

Hydrostatic pressure The pressure exerted at a given depth by the weight of the overlying column of water.

Hypersaline Referring to water with a salinity of 40 or more.

Ichthyoplankton Planktonic fish eggs and fish larvae.

Infauna Animals that live within the sediments of the seafloor.

Infrared Invisible wavelengths of light, longer than about 780 nm, which are responsible for heating the ocean.

Ingestion The act of swallowing food.

Interstitial fauna Animals that live in the spaces between adjacent particles in a soft-bottom substrate.

Intertidal zone The zone between high and low tide marks that is periodically exposed to air.

Isopods An order of crustaceans generally having a flattened body and with both benthic and planktonic species.

Isothermal Of equal temperature.

Jellyfish See **Medusae**.

K-**selection** A life history pattern in which the species survive by being well-adapted and efficient, rather than by producing large numbers of young; see **Equilibrium species**.

Kelp A group of very large brown algae that grow subtidally in mid- and high latitudes.

Keystone species A species that maintains community structure through its feeding activities, and without which large changes would occur in the community.

Krill Euphausiids; often specifically *Euphausia superba* of the Antarctic Ocean. The word comes from an old Norwegian term *kril*, once applied to such diverse animals as vermin and larval fish.

Lagoon A shallow body of water encircled by an atoll, or lying between a land mass and a barrier reef.

Larvacea Zooplankton of the Phylum Urochordata that build houses of mucus and filter-feed on nanoplankton.

Lecithotrophic larvae Meroplanktonic larvae that do not feed on planktonic food.

Littoral zone The intertidal zone.

Low water The minimum height reached by a falling tide.

Macrobenthos See **Macrofauna**.

Macrofauna Benthic animals larger than 1.0 mm.

Macrophytes Large, visible plants (e.g. mangroves, rock algae, seagrasses).

Macroplankton Zooplankton of between 2 cm and 20 cm in size.

Mangal Mangrove swamp.

Mangroves Referring to a variety of salt-tolerant trees and shrubs that dominate many intertidal regions in tropical and subtropical latitudes.

Mariculture The artificial cultivation of marine species.

Marine snow Detrital aggregates >0.5 mm consisting of faecal pellets, empty larvacean houses, pteropod feeding webs, and other materials derived from living organisms, as well as associated bacteria.

Medusae Bell-shaped zooplankton of the Phylum Cnidaria; 'jellyfish'.

Megalopa A larval stage of crabs that follows the zoea stages.

Megaplankton Zooplankton of between 20 cm and 200 cm in size.

Meiobenthos See **Meiofauna**.

Meiofauna Animals between 0.1 mm and 1.0 mm in size that live in sediments; interstitial fauna.

Meroplankton Plankton that spend only part of their life cycle in the water column, usually the eggs and larvae of benthic or nektonic adults.

Mesopelagic zone The water column from the bottom of the epipelagic zone (200–300 m) to about 1000 m depth.

Mesoplankton Plankton of between 0.2 mm and 20 mm in size.

Mesotrophic Referring to a region with moderate concentrations of nutrients and moderate biological productivity.

Microbenthos See **Microfauna**.

Microbial loop Referring to the regeneration of nutrients, and their return to the food chain, that is mediated by bacteria and protozoans.

Microfauna Benthic animals smaller than 0.1 mm; mostly protozoans.

Microphytes Microscopic benthic plants.

Microplankton Plankton of between 20 µm and 200 µm in size.

Mineralization The breakdown of organic compounds into inorganic materials.

Mixed layer A layer of water that is mixed by wind and is therefore isothermal.

Mixotrophy Employing more than one type of feeding strategy in order to exploit different food resources.

Mollusca A phylum of animals that includes snails, clams, and squid among others.

Mucus A sticky exudate composed mainly of proteins and polysaccharides.

Mysids An order of shrimplike crustaceans, mostly epibenthic.

Mysticetes Baleen whales.

Nanoplankton Plankton in the size range of 2–20 µm, including some phytoplankton and some protozoans.

Nauplius A free-swimming larval stage of crustaceans.

Neap tides Tides occurring near the first and last quarters of the Moon, when the tidal range is least.

Nekton Pelagic animals capable of swimming against a current; adult squid, fish, and marine mammals.

Nematocysts Stinging cells on the tentacles of cnidarians.

Nemertea A phylum of unsegmented marine worms, all with a proboscis.

Neritic Referring to inshore waters shallower than 200 m in depth that overlie continental shelves.

Net photosynthesis The amount of photosynthetic production in excess of respiration losses.

Net primary production That portion of the gross primary production which is incorporated within the body of the primary producer and thus appears as growth.

Neuston Organisms that inhabit the uppermost few millimetres of the surface water.

New nitrogen Nitrogen that enters the euphotic zone from outside regions, especially nitrate entering in upwelled water.

New production Photosynthetic production based on new nitrogen.

Nitrification The oxidation of ammonia to nitrite and nitrate.

Nitrogen fixation The conversion of dissolved nitrogen gas to organic nitrogen compounds, usually mediated by cyanobacteria.

Nutricline The depth zone where nutrient concentrations increase rapidly with depth.

Nutrient Any of a number of inorganic or organic compounds or ions used primarily in the nutrition of primary producers; e.g., nitrogen and phosphorus compounds.

Oceanic Referring to offshore waters in areas deeper than 200 m.

Odonticetes Predatory toothed whales, dolphins and porpoises.

Oligohaline Referring to water of low salinity, usually less than about 5.

Oligotrophic Referring to a region with low nutrients and low biological productivity.

Omnivore An animal that eats both plant and animal foods.

Ophiuroids A class of echinoderms that includes brittle stars.

Opportunistic species Those species that are usually of relatively small size, have rapid growth rates and short life spans, produce many young, and have variable population sizes that are below the carrying capacity of the environment in which they live; *r*-selected species.

Osmoregulation Referring to physiological mechanisms that maintain the internal salt and fluid balance of an organism within an acceptable range.

Osmosis The movement of water through a semipermeable membrane separating two solutions of differing solute concentrations, making the concentration of water equal on both sides of the membrane.

Osteichthyes The class of fish that includes all those species with bony skeletons; the teleosts.

Ostracods A class of Crustacea characterized by having a bivalved exoskeleton.

Overfishing When the quantity of fish harvested exceeds the amount that can be resupplied by growth and reproduction.

Oxygen minimum layer A zone of low oxygen concentration, usually between 400 m and 800 m depth.

PAR Abbreviation for photosynthetically active radiation; wavelengths of between about 400 nm and 700 nm that are used by plants in photosynthesis.

Paralytic shellfish poisoning (PSP) A sometimes fatal paralysis resulting from the ingestion of shellfish containing saxitoxin, a neurotoxin produced by certain dinoflagellates and acquired by the shellfish feeding on these toxic algae.

Parthenogenesis Reproduction without fertilization, resulting in cloned offspring.

Patchiness A spatial pattern in which individuals are not distributed either uniformly or randomly, but are clustered in 'patches' of variable size.

***P/B* ratio** The ratio of annual production to average annual biomass for a particular species; high ratios are indicative of highly productive short-lived organisms (e.g. phytoplankton), and low ratios indicate large, slow-growing organisms (e.g. fish).

Pelagic Referring to the ocean water column and the organisms living therein.

Photocyte A cell in which bioluminescent light is produced.

Photoinhibition The inhibition of photosynthesis by high light intensities.

Photophore A complex organ in which bioluminescent light is produced.

Photosynthesis The process whereby plants utilize carbon dioxide, water, and solar energy to manufacture energy-rich organic compounds.

Photosynthetic quotient The amount (in moles) of O_2 produced by photosynthesis divided by the amount (in moles) of CO_2 taken up in the process.

Phytodetritus Nonliving particulate matter derived from phytoplankton or benthic plants.

Phytoplankton Microscopic planktonic plants; e.g. diatoms, dinoflagellates.

Picoplankton Plankton measuring 0.2–2.0 μm in size, mostly bacteria.

Pinnipedia The order of marine mammals that includes the seals, sea lions, and walruses, all having four swimming flippers.

Plankton Plants or animals that live in the water column and are incapable of swimming against a current.

Planktotrophic larvae Meroplanktonic larvae that depend on feeding on planktonic food (e.g. bacteria, phytoplankton) for their growth.

Pleuston Organisms that float passively at the sea surface and whose bodies project partly into the air.

POC Abbreviation for particulate organic carbon.

Pogonophora A phylum of benthic marine worms, all of which lack a mouth and gut.

Poikilothermic Cold-blooded; unable to regulate body temperature.

Polar The Arctic or Antarctic regions.

Polychaetes Marine segmented worms belonging to the Phylum Annelida; some are planktonic, but most are benthic.

Polyp An individual organism of a colonial cnidarian such as a hydroid or coral colony.

Population All the individuals of one species that inhabit the same geographic area.

Population density The number of individuals per unit area (or per unit volume of water) in a population.

Predation The act of an animal feeding upon another animal.

Primary consumers Herbivorous animals.

Primary producers Plants.

Primary production The amount of organic material synthesized from inorganic substances per unit volume of water or unit area.

Prochlorophytes Photosynthetic picoplankton lacking a nucleus and closely related to Cyanobacteria.

Production That part of assimilated energy which is retained and incorporated in the biomass of the organism.

Productivity The rate at which a given quantity of organic material is produced by organisms.

Protists A collective term for unicellular organisms that have cells with a true nucleus, including diatoms, dinoflagellates and protozoans.

Protozoa Unicellular animals.

Pteropod A holoplanktonic snail having two swimming wings.

Pteropod ooze Seafloor sediments composed of more than 30% $CaCO_3$ from pteropod or other pelagic mollusc shells.

Pycnocline The water layer in which density changes most rapidly with depth.

***r*-selection** A life history pattern in which the species survive by producing very large numbers of offspring, and opportunistically move into new suitable habitats whenever they become available; see **Opportunistic species**.

Radiolaria Planktonic protozoans with a siliceous skeleton and pseudopodia.

Radiolarian ooze A sediment formed from the remains of radiolarian skeletons.

Recruitment The addition of new (juvenile) individuals to a population.

Red tide A red coloration of seawater caused by high concentrations of certain species of micro-organisms, usually dinoflagellates, some of which release toxins.

Refractory materials Those that are resistant to decomposition.

Regenerated production Photosynthetic production based on nitrogen that is recycled within the euphotic zone.

Regime shift A long-term change in marine ecosystems and/or in biological production resulting from a change in the physical environment.

Respiration A metabolic process carried out by all organisms in which organic substances are broken down to yield energy; it is the opposite reaction to photosynthesis, and results in a release of carbon dioxide.

Resting spore A spore formed by diatoms or dinoflagellates which remains dormant for some period of time before reforming an active planktonic cell.

Salinity The total amount of dissolved material (salts) in seawater measured in g kg^{-1} of seawater (formerly denoted as parts per thousand, ppt, or ‰, but it is a dimensionless number that is now reported without units).

Salps Barrel-shaped gelatinous zooplankton of the Phylum Urochordata.

Saltmarsh An intertidal community dominated by emergent vegetation rooted in soils.

Saxitoxin A collective term for various neurotoxins produced by certain species of dinoflagellates.

Seagrass A collective name for certain marine flowering plants that grow in intertidal soft-substrate communities.

Seamount A submerged, isolated, mountain that rises from the seafloor.

Secondary consumers Carnivorous animals.

Secondary production The amount of organic material produced by animals from ingested food.

Self-shading The reduction in light caused by increasing numbers of phytoplankton.

Semidiurnal tide A tide with two high waters and two low waters each tidal day.

Sessile Referring to animals that are permanently attached to a substrate.

Siliceous Containing silica.

Silicoflagellates Small, flagellate, unicellular phytoplankton with a siliceous skeleton.

Siphonophores Pelagic colonial cnidarians.

Sipuncula A phylum of unsegmented marine worms, mostly benthic.

Sirenia The order of herbivorous marine mammals that includes manatees and dugongs.

Species A distinctive group of interbreeding individuals.

Species diversity The number of species in a particular area; or a measure derived from combining the number of species with their relative abundance in an area.

Species succession Successive changes in the relative abundance of species in a community which result from environmental change.

Spreading centre A region along which new seafloor is being produced.

Spring tides Tides occurring near the times of the new and full moon, when the tidal range is greatest.

Standing crop See **Standing stock**.

Standing stock The biomass of organisms present per unit volume or per unit area at a given time.

Stenobathic Referring to organisms that can tolerate only a narrow depth (pressure) range.

Stenohaline Referring to organisms that can tolerate only a narrow salinity range.

Stenothermic Referring to organisms that can tolerate only a narrow temperature range.

Stock/recruitment theory Fisheries management theories based on the relation between the numbers of adult fish and the predicted numbers of juvenile fish that enter the adult population.

Sublittoral zone The benthic zone extending from the low tide mark to the outer edge of the continental shelf.

Substrate A solid surface on which an organism lives or to which it is attached (also called substratum); or, a chemical that forms the basis of a biochemical reaction or acts as a nutrient for microorganisms.

Subtidal zone See **Sublittoral zone**.

Supralittoral zone The narrow benthic zone just above high water mark, immersed only during storms.

Supratidal zone See **Supralittoral zone**.

Suspension feeding Obtaining food by filtering particles out of the surrounding water.

Symbiosis A close physiological association between two species, often for mutual benefit.

Tanaids Small benthic marine crustaceans.

Teleosts Fish with bony skeletons.

Thecosomes Holoplanktonic snails having paired swimming wings and usually a calcareous shell.

Thermocline The water layer in which temperature changes most rapidly with increasing depth.

Tidal range The difference in height between consecutive high and low waters.

Tide The periodic rise and fall of the sea surface due to gravitational attractions of the Sun and Moon acting on the rotating Earth.

Tintinnids Planktonic, ciliate protozoans having a proteinaceous outer shell.

Top-level predators Animals that have no natural predators, other than humans.

Transfer efficiency The annual production in one trophic level divided by the annual production in the preceding (lower) trophic level; a measure of the efficiency with which energy is transferred between trophic levels.

Trench A narrow, relatively steep-sided depression in the seafloor that lies below 6000 m in depth.

Trochophore A free-swimming larval stage of polychaete worms and some molluscs, characterized by having bands of cilia around the body.

Trophic level The nutritional position occupied by an organism in a food chain or food web; e.g. primary producers (plants); primary consumers (herbivores); secondary consumers (carnivores), etc.

Tunicates Sessile benthic animals belonging to the Phylum Chordata.

Turbidity Reduced visibility in water due to the presence of suspended particles.

Turbulence Physical mixing of water.

Ultraviolet Invisible radiation with wavelengths of less than 380 nm.

Upwelling A rising of nutrient-rich water toward the sea surface.

Veliger A free-swimming larval stage of molluscs.

Vestimentifera A group of benthic marine worms related to the Pogonophora and characteristically found at hydrothermal vents and cold seeps.

Visible spectrum Visible radiation with wavelengths of approximately 400–700 nm.

Warm core ring A rotating body of water with relatively warm temperature and low productivity.

Water mass A large volume of seawater having a common origin and a distinctive combination of temperature, salinity, and density characteristics.

Xenophyophoria Large, unicellular, benthic protists.

Zoea A planktonic larval stage of crabs with characteristic spines on the exoskeleton.

Zonation Parallel bands of distinctive plant and animal associations that develop in intertidal regions.

Zoobenthos Benthic animals.

Zooflagellates Colourless, heterotrophic, flagellated protists.

Zooplankton Planktonic animals.

Zooxanthellae Photosynthetic micro-organisms, usually dinoflagellates, that live symbiotically in the tissues of organisms such as corals and molluscs.

ACKNOWLEDGEMENTS

We wish to gratefully acknowledge the assistance of The Open University oceanography course team in preparing this volume: Angela Colling, John Phillips, Dave Park, Dave Rothery, John Wright. They generously passed on their experience in writing previous volumes in this series, and their advice and critiques of our early drafts were invaluable.

Colour figures were generously provided by Dr. F. J. R. Taylor, University of British Columbia (*Colour Plates 1–3, 5, 6, 9–11, 38*); Suisan Aviation Co., Tokyo (*Plate 4*); NSF/NASA (*Plates 7, 8*); Dr. L. P. Madin, Woods Hole Oceanographic Institution (*Plates 12, 14, 15, 25–27*); Dr. G. R. Harbison, Woods Hole Oceanographic Institution (*Plates 13, 22*); R. W. Gilmer (*Plates 16, 17, 20, 21, 28*); Dr. M. Omori, Tokyo University of Fisheries (*Plate 23*); Dr. A. Alldredge, University of California, Santa Barbara (*Plate 24* taken by J. M. King); Dr. T. Carefoot, University of British Columbia (*Plates 31, 36, 37*); Department of Energy, Mines and Resources, Canada (*Plate 33*); Dr. J. B. Lewis, McGill University (*Plates 34, 35*); Dr. R. R. Hessler, Scripps Institution of Oceanography (*Plates 39, 40*); and P. Lasserre, Station Biologique de Roscoff (*Plate 41*).

In-text photographs were kindly provided by Dr. F. J. R. Taylor, University of British Columbia (*Figure 3.3d*); R. Gilmer (*Figure 4.3*); Dr. O. Roger Anderson, Lamont-Doherty Geological Observatory of Columbia University (*Figures 4.5, 4.6*); R. Brown, Department of Fisheries and Oceans, Canada (*Figure 4.17*); Fisheries and Oceans (Canada) (*Figure 6.17*); C. E. Mills, Friday Harbor Laboratories (*Figure 9.3*); B. E. Brown, University of Newcastle upon Tyne (*Figure 9.4*).

We are extremely grateful to Mrs. Barbara Rokeby who patiently produced the many drafts of figures and who contributed several original drawings. The following line figures were reprinted or modified and redrawn from previously published material, and grateful acknowledgement is made to the following sources:

Figure 1.2 J. McN. Sieburth *et al.* (1978) in *Limnology and Oceanography*, **23**, American Society of Limnology and Oceanography; *Figures 1.5, 1.6* C. W. Thomson and J. Murray (1885) *Report on the Scientific Results of the Voyage of HMS Challenger during the years 1873–76, Narrative*, Vol. I, First Part; *Figure 1.7* E. Haeckel (1887) *Report on the Scientific Results of the Voyage of HMS Challenger during the years 1873–76, Zoology*, Vol. XVIII; *Figure 2.5* G. L. Clarke and E. J. Denton (1962) in *The Sea, Ideas and Observations on Progress in the Study of the Seas*, Interscience; *Figure 2.6*, The Open University (1989) *Ocean Circulation*, Pergamon; *Figures 2.7, 2.9, 2.11, 2.14, 2.16, 2.18, 6.6* The Open University (1989) *Seawater: Its Composition, Properties and Behaviour*, Pergamon; *Figure 2.10* R. V. Tait (1968) *Elements of Marine Ecology*, Butterworths Scientific Ltd; *Figure 2.19* A. N. Strahler (1963) *Earth Sciences*, Harper & Row Pubs; *Figures 2.13, 2.15* W. J. Emery & J. Meinke (1986) *Oceanographica Acta*, **9**, Gauthier Villars; *Figure 2.12* H. V. Sverdrup *et al.* (1942) *The Oceans*, Prentice Hall Inc.; *Figures 3.1a–d,i* E. E. Cupp (1943) *Marine Plankton Diatoms of the West Coast of North America*, University of California; *Figures 3.1e–h* D. L. Smith (1977) *A Guide to Marine Coastal Plankton and Marine*

Invertebrate Larvae, Kendall/Hunt; *Figure 3.3* M. V. Lebour (1925) *The Dinoflagellates of Northern Seas*, Marine Biological Association of the U.K.; *Figures 3.5, 3.6, 3.18, 5.9* T. R. Parsons *et al.* (1984) *Biological Oceanographic Processes*, Pergamon; *Figure 3.8* U. Sommer (1989) in *Plankton Ecology*, Springer-Verlag; *Figure 3.9* P. Tett; *Figure 3.10* T. R. Parsons (1979) in *South African Journal of Science*, **75**, South African Research Council; *Figures 3.13, 3.14* R. D. Pingree (1978) in *Spatial Patterns in Plankton Communities*, Plenum; *Figures 3.16, 7.1, 7.2, 8.13* (1984) *Oceanography, Biological Environments*, The Open University; *Figure 3.17* A. K. Heinrich (1962) in *Journal Conseil International pour l'Exploration de la Mer*, **27**, Conseil International pour l'Exploration de la Mer; *Figure 4.7* E. N. Kozloff (1987) *Marine Invertebrates of the Pacific Northwest*, University of Washington; *Figure 4.8* Zhuang Shi-de and Chen Xiaolin (1978) in *Marine Science and Technology* (in Chinese), **9**, State Oceanographic Administration of the People's Republic of China; *Figure 4.10* J. Fraser (1962) *Nature Adrift*, Foulis; *Figure 4.15* M. Omori (1974) in *Advances in Marine Biology*, **12**, Academic Press; *Figure 4.16* E. Brinton (1967) in *Limnology and Oceanography*, **12**, American Society of Limnology and Oceanography; *Figure 4.18* J. Fulton (1973) in *Journal of the Fisheries Research Board of Canada*, **30**, Fisheries Research Board of Canada; *Figure 4.19* R. Williams (1985) in *Marine Biology*, **86**, Springer International; *Figure 4.20* D. L. Mackas *et al.* (1985) in *Bulletin of Marine Science*, **37**, University of Miami; *Figure 4.21* P. H. Wiebe (1970) in *Limnology and Oceanography*, **15**, American Society of Limnology and Oceanography; *Figure 4.22* S. Nishizawa (1979) in *Scientific Report to the Japanese Ministry of Education*, No. 236017; *Figure 4.23* Sir Alister Hardy Foundation for Ocean Science, Annual Report 1991; *Figure 5.2* R. W. Sheldon *et al.* (1972) in *Limnology and Oceanography*, **17**, American Society of Limnology and Oceanography; *Figure 5.4* A. Clarke (1988) in *Comparative Biochemistry and Physiology*, **90B**, Pergamon; *Figure 5.6* J. H. Steele (1974) *The Structure of Marine Ecosystems*, Harvard University; *Figure 5.8* B. C. Cho and F. Azam (1990) in *Marine Ecology Progress Series*, **63**, Inter-Research; *Figure 5.14* W. H. Thomas and D. L. R. Seibert (1977) in *Bulletin of Marine Science*, **27**, University of Miami; *Figure 5.17* T. R. Parsons and T. A. Kessler (1986) in *The Role of Freshwater Outflow in Coastal Marine Ecosystems*, Springer-Verlag; *Figure 5.18* Yu. I. Sorokin (1969) in *Primary Productivity in Aquatic Environments*, University of California; *Figure 5.19* T. R. Parsons and P. J. Harrison (1983) in *Encyclopedia of Plant Physiology*, **12D**, Springer-Verlag; *Figure 5.21* R. L. Iverson (1990) in *Limnology and Oceanography*, **35**, American Society of Limnology and Oceanography; *Figure 6.1* R. Payne; *Figure 6.2* N. P. Ashmole (1971) in *Avian Biology*, **I**, Academic Press; *Figures 6.7a, 8.5* Friedrich (1969) *Marine Biology*, Sidgwick & Jackson; *Figure 6.7b, 6.8b* N. B. Marshall (1954) *Aspects of Deep Sea Biology*, Hutchinson; *Figure 6.8a* C. P. Idyll (1964) *Abyss*, Thomas Crowell; *Figure 6.9* F. S. Russell *et al.* (1971) in *Nature*, Macmillan; *Figure 6.14* R. S. K. Barnes and R. N. Hughes (1988) *An Introduction to Marine Ecology*, Blackwell; *Figure 6.15* Department of Fisheries and Oceans, Canada; *Figure 7.3* G. Thorson (1971) *Life in the Sea*, Weidenfeld & Nicolson, by permission of the Estate of Gunnar Thorson; *Figure 7.4* T. Fenchel (1969) in *Ophelia*, **6**, Marine Biological Laboratory, Helsingoer, Denmark; *Figure 8.1* J. Connell (1961) in *Ecology*, **42**, Ecological Society of America; *Figure 8.2* R. Paine (1966) in *American Naturalist*, **100**, American Society of Naturalists; *Figure 8.7* A. Remane (1934) in *Zoologischer Anzeiger*, Suppl. **7**,; *Figure 8.9*

D. E. Ingmanson and W. J. Wallace (1973) *Oceanology: An Introduction*, Wadsworth; *Figure 8.15* J. W. Nybakken (1988) *Marine Biology, An Ecological Approach*, Harper & Row; *Figure 8.16* R. D. Turner and R. A. Lutz (1984) in *Oceanus*, **27**, Woods Hole Oceanographic Institution.

Data in Table 5.2 were taken from Laws (1985) in *American Scientist*, **73**, Sigma Xi, The Scientific Research Society; data on seabird numbers in Figure 6.10 were kindly provided by F. Chavez, Monterey Bay Aquarium Research Institute; data in Table 7.2 were taken from Birkett (1959) in *Conseil L. International Exploration de la Mer* (Unpublished Report C. M., No. 42); Table 8.1 is adapted from R. S. K. Barnes and R. N. Hughes (1988) *An Introduction to Marine Ecology*, Blackwell.

Index

Note: page numbers in italics refer to illustrations; in bold to tables